IET ENERGY ENGINEERING 159

Transforming the Grid Towards Fully Renewable Energy

Other volumes in this series

Volume 1 **Power Circuit Breaker Theory and Design** C.H. Flurscheim (Editor)
Volume 4 **Industrial Microwave Heating** A.C. Metaxas and R.J. Meredith
Volume 7 **Insulators for High Voltages** J.S.T. Looms
Volume 8 **Variable Frequency AC Motor Drive Systems** D. Finney
Volume 10 **SF_6 Switchgear** H.M. Ryan and G.R. Jones
Volume 11 **Conduction and Induction Heating** E.J. Davies
Volume 13 **Statistical Techniques for High Voltage Engineering** W. Hauschild and W. Mosch
Volume 14 **Uninterruptible Power Supplies** J. Platts and J.D. St Aubyn (Editors)
Volume 15 **Digital Protection for Power Systems** A.T. Johns and S.K. Salman
Volume 16 **Electricity Economics and Planning** T.W. Berrie
Volume 18 **Vacuum Switchgear** A. Greenwood
Volume 19 **Electrical Safety: A guide to causes and prevention of hazards** J. Maxwell Adams
Volume 21 **Electricity Distribution Network Design, 2nd Edition** E. Lakervi and E.J. Holmes
Volume 22 **Artificial Intelligence Techniques in Power Systems** K. Warwick, A.O. Ekwue and R. Aggarwal (Editors)
Volume 24 **Power System Commissioning and Maintenance Practice** K. Harker
Volume 25 **Engineers' Handbook of Industrial Microwave Heating** R.J. Meredith
Volume 26 **Small Electric Motors** H. Moczala *et al.*
Volume 27 **AC–DC Power System Analysis** J. Arrillaga and B.C. Smith
Volume 29 **High Voltage Direct Current Transmission, 2nd Edition** J. Arrillaga
Volume 30 **Flexible AC Transmission Systems (FACTS)** Y.-H. Song (Editor)
Volume 31 **Embedded Generation** N. Jenkins *et al.*
Volume 32 **High Voltage Engineering and Testing, 2nd Edition** H.M. Ryan (Editor)
Volume 33 **Overvoltage Protection of Low-Voltage Systems, Revised Edition** P. Hasse
Volume 36 **Voltage Quality in Electrical Power Systems** J. Schlabbach *et al.*
Volume 37 **Electrical Steels for Rotating Machines** P. Beckley
Volume 38 **The Electric Car: Development and future of battery, hybrid and fuel-cell cars** M. Westbrook
Volume 39 **Power Systems Electromagnetic Transients Simulation** J. Arrillaga and N. Watson
Volume 40 **Advances in High Voltage Engineering** M. Haddad and D. Warne
Volume 41 **Electrical Operation of Electrostatic Precipitators** K. Parker
Volume 43 **Thermal Power Plant Simulation and Control** D. Flynn
Volume 44 **Economic Evaluation of Projects in the Electricity Supply Industry** H. Khatib
Volume 45 **Propulsion Systems for Hybrid Vehicles** J. Miller
Volume 46 **Distribution Switchgear** S. Stewart
Volume 47 **Protection of Electricity Distribution Networks, 2nd Edition** J. Gers and E. Holmes
Volume 48 **Wood Pole Overhead Lines** B. Wareing
Volume 49 **Electric Fuses, 3rd Edition** A. Wright and G. Newbery
Volume 50 **Wind Power Integration: Connection and system operational aspects** B. Fox *et al.*
Volume 51 **Short Circuit Currents** J. Schlabbach
Volume 52 **Nuclear Power** J. Wood
Volume 53 **Condition Assessment of High Voltage Insulation in Power System Equipment** R.E. James and Q. Su
Volume 55 **Local Energy: Distributed generation of heat and power** J. Wood
Volume 56 **Condition Monitoring of Rotating Electrical Machines** P. Tavner, L. Ran, J. Penman and H. Sedding
Volume 57 **The Control Techniques Drives and Controls Handbook, 2nd Edition** B. Drury
Volume 58 **Lightning Protection** V. Cooray (Editor)
Volume 59 **Ultracapacitor Applications** J.M. Miller

Volume 62 **Lightning Electromagnetics** V. Cooray
Volume 63 **Energy Storage for Power Systems, 2nd Edition** A. Ter-Gazarian
Volume 65 **Protection of Electricity Distribution Networks, 3rd Edition** J. Gers
Volume 66 **High Voltage Engineering Testing, 3rd Edition** H. Ryan (Editor)
Volume 67 **Multicore Simulation of Power System Transients** F.M. Uriate
Volume 68 **Distribution System Analysis and Automation** J. Gers
Volume 69 **The Lightening Flash, 2nd Edition** V. Cooray (Editor)
Volume 70 **Economic Evaluation of Projects in the Electricity Supply Industry, 3rd Edition** H. Khatib
Volume 72 **Control Circuits in Power Electronics: Practical issues in design and implementation** M. Castilla (Editor)
Volume 73 **Wide Area Monitoring, Protection and Control Systems: The enabler for smarter grids** A. Vaccaro and A. Zobaa (Editors)
Volume 74 **Power Electronic Converters and Systems: Frontiers and applications** A.M. Trzynadlowski (Editor)
Volume 75 **Power Distribution Automation** B. Das (Editor)
Volume 76 **Power System Stability: Modelling, analysis and control** A.A. Sallam and Om P. Malik
Volume 78 **Numerical Analysis of Power System Transients and Dynamics** A. Ametani (Editor)
Volume 79 **Vehicle-to-Grid: Linking electric vehicles to the smart grid** J. Lu and J. Hossain (Editors)
Volume 81 **Cyber-Physical-Social Systems and Constructs in Electric Power Engineering** S. Suryanarayanan, R. Roche and T.M. Hansen (Editors)
Volume 82 **Periodic Control of Power Electronic Converters** F. Blaabjerg, K. Zhou, D. Wang and Y. Yang
Volume 86 **Advances in Power System Modelling, Control and Stability Analysis** F. Milano (Editor)
Volume 87 **Cogeneration: Technologies, optimisation and implementation** C.A. Frangopoulos (Editor)
Volume 88 **Smarter Energy: From smart metering to the smart grid** H. Sun, N. Hatziargyriou, H.V. Poor, L. Carpanini and M.A. Sánchez Fornié (Editors)
Volume 89 **Hydrogen Production, Separation and Purification for Energy** A. Basile, F. Dalena, J. Tong and T.N. Veziroğlu (Editors)
Volume 90 **Clean Energy Microgrids** S. Obara and J. Morel (Editors)
Volume 91 **Fuzzy Logic Control in Energy Systems with Design Applications in MATLAB®/Simulink®** İ.H. Altaş
Volume 92 **Power Quality in Future Electrical Power Systems** A.F. Zobaa and S.H.E.A. Aleem (Editors)
Volume 93 **Cogeneration and District Energy Systems: Modelling, analysis and optimization** M.A. Rosen and S. Koohi-Fayegh
Volume 94 **Introduction to the Smart Grid: Concepts, technologies and evolution** S.K. Salman
Volume 95 **Communication, Control and Security Challenges for the Smart Grid** S.M. Muyeen and S. Rahman (Editors)
Volume 96 **Industrial Power Systems with Distributed and Embedded Generation** R. Belu
Volume 97 **Synchronized Phasor Measurements for Smart Grids** M.J.B. Reddy and D.K. Mohanta (Editors)
Volume 98 **Large Scale Grid Integration of Renewable Energy Sources** A. Moreno-Munoz (Editor)
Volume 100 **Modeling and Dynamic Behaviour of Hydropower Plants** N. Kishor and J. Fraile-Ardanuy (Editors)
Volume 101 **Methane and Hydrogen for Energy Storage** R. Carriveau and D.S.-K. Ting
Volume 104 **Power Transformer Condition Monitoring and Diagnosis** A. Abu-Siada (Editor)
Volume 106 **Surface Passivation of Industrial Crystalline Silicon Solar Cells** J. John (Editor)

Volume 107 **Bifacial Photovoltaics: Technology, applications and economics** J. Libal and R. Kopecek (Editors)
Volume 108 **Fault Diagnosis of Induction Motors** J. Faiz, V. Ghorbanian and G. Joksimović
Volume 110 **High Voltage Power Network Construction** K. Harker
Volume 111 **Energy Storage at Different Voltage Levels: Technology, integration, and market aspects** A.F. Zobaa, P.F. Ribeiro, S.H.A. Aleem and S.N. Afifi (Editors)
Volume 112 **Wireless Power Transfer: Theory, technology and application** N. Shinohara
Volume 115 **DC Distribution Systems and Microgrids** T. Dragičević, F. Blaabjerg and P. Wheeler
Volume 117 **Structural Control and Fault Detection of Wind Turbine Systems** H.R. Karimi
Volume 119 **Thermal Power Plant Control and Instrumentation: The control of boilers and HRSGs, 2nd Edition** D. Lindsley, J. Grist and D. Parker
Volume 120 **Fault Diagnosis for Robust Inverter Power Drives** A. Ginart (Editor)
Volume 121 **Monitoring and Control Using Synchrophasors in Power Systems with Renewables** I. Kamwa and C. Lu (Editors)
Volume 123 **Power Systems Electromagnetic Transients Simulation, 2nd Edition** N. Watson and J. Arrillaga
Volume 124 **Power Market Transformation** B. Murray
Volume 125 **Wind Energy Modeling and Simulation, Volume 1: Atmosphere and plant** P. Veers (Editor)
Volume 126 **Diagnosis and Fault Tolerance of Electrical Machines, Power Electronics and Drives** A.J.M. Cardoso
Volume 128 **Characterization of Wide Bandgap Power Semiconductor Devices** F. Wang, Z. Zhang and E.A. Jones
Volume 129 **Renewable Energy from the Oceans: From wave, tidal and gradient systems to offshore wind and solar** D. Coiro and T. Sant (Editors)
Volume 130 **Wind and Solar Based Energy Systems for Communities** R. Carriveau and D.S.-K. Ting (Editors)
Volume 131 **Metaheuristic Optimization in Power Engineering** J. Radosavljević
Volume 132 **Power Line Communication Systems for Smart Grids** I.R.S Casella and A. Anpalagan
Volume 139 **Variability, Scalability and Stability of Microgrids** S.M. Muyeen, S.M. Islam and F. Blaabjerg (Editors)
Volume 145 **Condition Monitoring of Rotating Electrical Machines** P. Tavner, L. Ran and C. Crabtree
Volume 146 **Energy Storage for Power Systems, 3rd Edition** A.G. Ter-Gazarian
Volume 147 **Distribution Systems Analysis and Automation, 2nd Edition** J. Gers
Volume 155 **Energy Generation and Efficiency Technologies for Green Residential Buildings** D. Ting and R. Carriveau (Editors)
Volume 157 **Electrical Steels, 2 Volumes** A. Moses, K. Jenkins, Philip Anderson and H. Stanbury
Volume 158 **Advanced Dielectric Materials for Electrostatic Capacitors** Q. Li (Editor)
Volume 160 **Microgrids for Rural Areas: Research and case studies** R.K. Chauhan, K. Chauhan and S.N. Singh (Editors)
Volume 166 **Advanced Characterization of Thin Film Solar Cells** N Haegel and M. Al-Jassim (Editors)
Volume 172 **Lighting Interaction with Power Systems, 2 Volumes** A. Piantini (Editor)
Volume 905 **Power System Protection, 4 Volumes**

Transforming the Grid Towards Fully Renewable Energy

Edited by
Oliver Probst, Sergio Castellanos and Rodrigo Palacios

The Institution of Engineering and Technology

Published by The Institution of Engineering and Technology, London, United Kingdom

The Institution of Engineering and Technology is registered as a Charity in England & Wales (no. 211014) and Scotland (no. SC038698).

First published 2020

The Institution of Engineering and Technology
Michael Faraday House
Six Hills Way, Stevenage
Herts, SG1 2AY, United Kingdom

www.theiet.org

British Library Cataloguing in Publication Data
A catalogue record for this product is available from the British Library

ISBN 978-1-83953-021-0 (hardback)
ISBN 978-1-83953-022-7 (PDF)

Typeset in India by MPS Limited

To my wife Silvia, my love and companion, my rock and my pillar, whose unconditional support for this project is greatly appreciated. To Lucy, who never stopped believing in me and who is going to be a great engineer. To Adrian, whose creative strokes are making a valuable contribution to architecture in Germany and Mexico. To Alex, whose critical mind holds great promises. To all my students and graduates of many years, who believe humankind is capable of sustainable development and whose enthusiastic commitment is making a difference in the world today.

– Oliver Probst

This book is the product of selfless contributions from many academics and thought leaders eager to act on climate, to whom I owe an enormous debt of gratitude for their time, passion, and commitment. I wholeheartedly hope this book – that spans technical, economic, and social dimensions – can inspire our current leaders to be bold in their decision-making and the future generations to continue fighting for a clean, sustainable, and equitable future.

– Sergio Castellanos

To my wife Diana, for being an endless source of love, support, and encouragement. To my kittens Sol and Luna and their purrs for being the soundtrack that accompanied me throughout this project.

– Rodrigo Palacios

Contents

About the editors

Oliver Probst is a professor of physics and renewable energy at Instituto Tecnológico y de Estudios Superiores de Monterrey, Mexico, where he has been training renewable energy professionals and entrepreneurs for over 20 years. As a wind energy consultant, he has contributed to the development of nearly 2,000 MW of operating wind farms in Mexico and the United States. His research focuses on grid integration of renewable energy, wind resource assessment and wind turbine technology. Oliver holds a Dr. rer. nat. in Physics from the University of Heidelberg, Germany.

Sergio Castellanos is an incoming assistant professor in the Civil, Architectural and Environmental Engineering department at The University of Texas at Austin. He previously held a professional researcher position at the University of California, Berkeley, where he was part of the California Institute for Energy and Environment (CIEE), Energy and Resources Group (ERG), and Berkeley's Social Science Data Lab (D-Lab). His current research focuses on energy systems planning, sustainable transportation, and cleantech manufacturing – all with a strong emphasis on equity. Sergio holds a PhD from the Massachusetts Institute of Technology in Mechanical Engineering.

Rodrigo Palacios is an energy systems modelling expert currently with Iniciativa Climática de México, where he works on the design of pathways for a decarbonized power sector in Mexico based on quantitative modelling and outreach to government and private industry. Rodrigo holds a PhD in Engineering with a focus on optimization from Universidad de Burgos, Spain, and MSc degrees in labour safety and renewable energy from Universidad Camilo José Cela and CEU San Pablo, Spain, respectively.

Preface

The call for renewable energy emerged in the early seventies, partly motivated by the first and second oil crises, but only in the recent past have renewable energy technologies gained enough momentum to create credible hope for a long-lasting transformation of the electricity sector (and eventually, the whole energy sector) into one guided fundamentally by sustainability. To properly design the pathways into eventual sustainability, it is instructive to understand the mechanisms that have allowed renewables to become disruptive, but also to become critically aware of the flaws and limitations of approaches used so far. Re-developing a sector as large and enabling as the electricity system is potentially rewarding but is not without perils, as the largely unregulated growth of information technologies painfully reminds us. What has worked in the early days may be poisonous in the long run. Much as a growing biological organism, with its delicate balance between promoters and inhibitors, a renascent electricity system has to be guided carefully in its transformation process, making sure the net worth gained is clearly positive. While this may sound to some as an exhortation to leave things as they are, the message carried is a completely different one: let us make a decided move towards a clean economy (of which a sustainable electricity system will be an important part), but let us also make sure the process is as educated, inclusive and data-driven as possible. What, then, are the guidelines for this admittedly challenging process?

First of all, there must be no doubt about what is meant by sustainability. Following the early definition of the Brundtland Commission, sustainable development strives to satisfy the necessities of the present without compromising the possibilities of future generations of satisfying theirs. This precludes a never-ending growth of material goods and energy use from being considered sustainable. Burying toxic waste with extremely long-term horizons such as spent nuclear fuel (including very long-lived isotopes of artificial elements such as Plutonium) to fix a short-term problem (supposing it does) also compromises the manoeuvre margins of future generations. Second, generating and using electricity and energy in general is only one facet (albeit an important one, given the enabling nature of energy technologies) of the human endeavour, so space and resources have to be dedicated to other uses, most notably food production. The energy sector transformation should enable the food sector to be more efficient and sustainable, rather than enter a competition with it for arable lands and other resources. Third, a living planet needs more than one species (i.e. humanity, including the few species held and cultivated in massive numbers for food production) to thrive and create a worthy environment, so a generous portion of the remaining land and resources has to be allotted for the natural world, fostering biodiversity, natural services such as clean air and water, and providing room for reflection and inspiration for those humans seeking sense in a crowed busy world.

Sustainability, however, is not embraced by everybody, and mainstream thinking is still largely dominated by the cornucopian idea that human ingenuity is an unlimited resource which can somehow make up for the finite amounts of energy, matter and land available on our planet. Combined with a strong faith in markets, which have undoubtedly demonstrated their ability to leverage seemingly unlimited resources (albeit often for limited groups of people only), cornucopian philosophy sees the world as an ever-growing process, like a child with a broken growth hormone regulation system. While a robust growth is critically important for a newborn of most species in order to reach food independence and the ability to safeguard its life as fast as possible, regulation is needed all the way. Once a pigeon can fly, growing larger makes only limited sense, since the lift/weight ratio will become unfavourable at larger bodies, unless the original design is abandoned, which is generally not possible with living species. An eagle can of course grow much larger and carry a larger total weight, but its design and food supply scheme are completely different. An eagle is not a pigeon grown larger. Human-made systems are of course not limited in principle by one fixed design, but most human infrastructure is generally designed and built for decades, if not generations, and changes are mostly incremental. This is precisely the situation of the electricity grid.

Given the conservative nature of grid operators, the electricity system seems to be an unlikely candidate for taking the lead in the transformation of society into a sustainable one. However, the contrary may be true. The electricity sector has long been characterized by a long-term planning perspective, with financial and operational horizons often spanning several decades. It also sticks with depreciated assets, as long as their continuous operation makes sense, thereby saving valuable resources and keeping costs low. Investments in the electricity sector are long term and low return, prioritizing guaranteed income streams spanning long periods over speculative earnings. The electricity system is a place where continuity and transgenerational responsibility come naturally.

However, the conservative nature of the electricity sector also implies a considerable dose of scepticism regarding new technologies, even though changes are common in the electricity grid, most notably in operation and maintenance, but also in response to modernization requirements of generation, transformation and transmission assets. Integrating variable renewable energy generation into the grid no longer is a novelty in many parts of the world, and operators have learned to deal with certain degrees of variable generation without major problems, often using the same dispatch guidelines as before the introduction of renewables. Being technically minded, grid engineers are also often open to technical progress, and advanced dispatch models based on stochastic modelling and forecasts are likely to be met with interest and enthusiastic response, provided an adequate training programme is set up and maintained. Barriers to an increased penetration of renewable energy seem to exist more on a political level, particularly if a larger participation of renewables is accompanied by mechanisms fostering competition between an incumbent utility and new entrants.

The drivers for change are clear: not only is the transformation of the electricity sector an urgent mandate dictated by climate change and environmental degradation, but also is there now a clear business case for renewables everywhere. Electricity from wind and solar is now cheaper (often very significantly) than from new fossil-fuel plants, and dramatically so compared to new nuclear plants, and new generation capacity can be built much faster with wind and solar than with any other technology, providing the energy transition with the speed needed to cope with climate change.

As technologies and the societal context evolve, so must the planning and regulation process. How is this achieved? The key are adaptive regulation and policies. Much like the growth hormone supply of a growing child has to be continuously adjusted before being cut off completely, rules and policies regulating the continuous development of the electricity sector must evolve over time as the results obtained are compared against the desired outcomes and corrections are applied as necessary. Such is the case of renewable energy development in Germany, where the now famous feed-in law first led to a spectacular growth in onshore wind energy installations in the 1990s, followed by a second focus on residential photovoltaics, and the current emphasis being on bioenergy and offshore wind. While not being without criticism, the German approach can be considered a hallmark of sustainable development, and its role as a worldwide game changer can hardly be overestimated. Mexico, on the other side of the Atlantic and with a quite different cultural background, also has re-invented its electricity sector based on prudent and adaptive regulation, albeit based on a very different approach*. California, of course, deserves the credit not only for kicking off the renewable energy transition altogether but also for having remained on the forefront of innovative and adaptive regulation and policy since starting the process in the 1980s. Examples from around the world abound, and some of the experts of our cosmopolitan team of authors will share important key lessons learned in the process.

Retrospectively, of course, many things look easier than they were at the time, and the efforts undergone by the pioneers of renewable energy in California, Germany, Mexico and elsewhere can hardly be overstated. Intruding into the comfort zone of highly regulated monopolies did require a dose of disruption in order to create a vision for renewables and overcome the natural scepticism with which new ideas and competitors are generally met. Technical proficiency of new technologies and providers is also evidently an issue that should not be belittled or overlooked, but barriers to innovations are mostly political rather than technical. It

*Unfortunately, the electricity grid is not free from interference from policymakers, and Mexico is not an exception. In spite of hallmark achievements with the large-scale deployment of renewables through competitive and transparent processes, the current federal administration is now putting a renewed focus on generation from fossil fuels such as fuel oil and coal, as well as pushing back competition, with the stated intention of returning to a largely monopolistic role of the state utility.

has become almost colloquial to consider the opening of electricity systems dominated by vertically integrated utilities, i.e. electricity companies in charge of the full-life cycle of electricity generation, delivery and commercialization as a *de-regulation* process, generally accompanied by a call for energy markets. Though both de-regulation and energy markets have been chosen by a number of countries or regions to foster their energy transitions, it is easy to get the wrong idea, so a few words of caution may be in order. First of all, de-regulation does not mean the absence of rules, but rather a re-structuring of an electricity sector originally dominated by a vertically integrated 'regulated' utility into one where competition between different providers and technology can happen in principle, and where an independent board or commission, ideally with both democratic legitimacy and technical proficiency, oversees the development of the sector based on a host of objectives, generally defined by politics. The dramatic failure of the early California electricity market markedly demonstrates the poor results lax regulation or lax oversight can bring about. Fortunately, much has been learned on the subject since the California failure, and modern market designs heavily tap into the lessons learned over the past decade.

Electricity markets are generally not designed primarily to promote renewable energies but rather to foster competition, hopefully leading to consumer benefits. In many countries and regions nowadays, some kind of market is operational, each with its own approach to renewable energy promotion, and its specific problems. Negative electricity prices are making a big splash in Germany, and the *missing money* problem occurring in energy-only markets with increasing renewable energy penetration, notably in the US, has prompted a large body of publications, both scientific and divulgatory, sometimes creating the impression that certain mysterious forces of nature are impeding renewable energy penetration beyond certain critical levels. Though it has to be acknowledged that market design and operation may be as difficult as grid operation and control, it is important to realize that such problems can be solved with prudent intervention of regulators, fuelled and fertilized by a continuous public consultation process with competent and interested parties. Negative prices are possible when market rules allow the placement of negative bids, otherwise they do not exist[†]. They may make sense for conventional generators with limited flexibility, as the penalty implied by negative bids may be lower than the costs incurred by ramping or investing in increased flexibility. Once generators have adapted (and often they do within a few years, exemplified by many coal-fired plants in Germany), negative prices cease to be a problem for them. Similarly, placing negative bids makes sense for subsidized renewable energy generators whose income is still positive due to their market incentives (until a certain threshold is met which takes away the subsidy), but become

[†]There are a few pathologies, though, where negative prices can arise even in the presence of exclusively positive bids, driven by flow restrictions, which in turn give rise to *congestion*. In grids with large accumulated lags in transmission upgrades, this may become a problem. It is clear, however, that such problem arise as a consequence of lack of action by policymakers and regulators, rather than because of intrinsic problems with market design.

irrelevant in unsubsidized markets. This may then be a prudent moment to abolish negative bids altogether. Missing money, i.e. 'insufficient' payments to generators in competitive energy-only electricity markets like ERCOT in Texas and elsewhere in the US, may indeed disincentivize investments in new generating capacity and other necessary upgrades and be therefore detrimental to grid stability in the long run. To address this problem, often the creation of a capacity market has been proposed, where a generator gets paid for the amount of capacity it holds available during the most critical hours of the year, often accompanied by random tests in case of conventional generators, as well as hefty penalties for all generators non-complying with their committed capacity. Countries with considerable over-capacities such as the UK or Germany have so far refrained from establishing capacity markets, whereas others, such as the wholesale electricity market in Mexico, included a capacity component from its inception. In any case, market design will continue to be a subject in which heated debates are carried out (some caused by diverging interests, others because of fundamentally different priorities and world views), so it is important to conduct this discussion on a rational level, using scientific input as much as possible for decision-making and being very clear about the underlying assumptions and premises of each approach.

In the present book, an attempt has been made to bring all the different facets of the complex topic of the clean energy transition together into one consistent piece, taking the reader from a reflection on global sustainability and the future of humankind all the way to new approaches to the organization of the electricity sector. Some of the tools required for the modernization of the sector, both technical and regulatory, will be visited on the way. This book heavily taps into the combined vision of an internationally renowned team of researchers who have generously shared their knowledge with us.

In the first part of the book, comprising the first three chapters, the problem of transitioning to a sustainable electricity sector is laid out and dimensioned. In the first chapter, authored by one of the editors (OP), a reflection on the maturity of humankind in the face of the truly existential challenge of climate change is presented; the author then proposes to view the clean energy transition of the electricity system as an enabling tool for a deep decarbonization of the world economy and the transition to sustainability, an idea reinforced and further developed by Anders Arvesen in Chapter 3. Chapter 2, written by sustainability pioneer Mark Diesendorf, sets out to answer the fundamental question whether an electricity sector with very high penetrations of variable renewable generation is possible with current or near-future technology, a question decidedly answered affirmatively. Professor Diesendorf provides a detailed discussion of the technological factors constraining the grid integration of variable generation from renewables and shows that all the fundamental requirements can readily be met with available technologies. He further refutes some of the popular myths regarding reliability and security, showing that the key to designing electricity systems with high penetration of variable generation lies with creating flexible resources and that design thinking marked by traditional dispatch schemes is likely to be part of the problem and not part of the solution. Chapter 3, authored by Andres Arvesen, contributes a life-cycle vision of different common generation[‡] technologies (including renewables, fossil-fuel options and nuclear energy), organized in the framework of the United Nations' Sustainable Development Goals. The author shows that this methodology is uniquely suited to detect trade-offs between different technologies, thereby providing guidance to policymakers and stakeholder who choose to support their decisions with data-driven approaches.

Part II of this book, represented by Chapters 4–9, is intended as a presentation of the comprehensive tools available for the large-scale integration of (variable) clean energy technologies into the grid, particularly for scenarios with very high shares of renewable energy, the central topic of this book. Misconceptions still abound, by both policymakers and engineers trained in traditional design thinking, making a comprehensive description of the available design tools mandatory. Chapter 4, written by Pedro Crespo de Granados, Christian Skar and Raquel Alonso Pedrero, provide specific answers to the question of how to design a cost-effective European grid dominated by variable clean generation, derived from comprehensive modelling of both the electricity and the natural gas grids in Europe. They make a convincing case for an aggressive expansion of the transmission links between countries in order to tap into the complementarity of variable renewable generation in different parts of Europe. Large-scale grid expansion is shown to be by far the most cost-effective solution. Generation from natural gas, though becoming less significant in overall volumes, is still expected to play a decisive role for balancing and regulation services, making the design of proper remuneration of such services a key element in the energy transition. Chapter 5, written by Ricardo Cruz, Daniel Chacón and Rodrigo Palacios, discusses the role of Distributed Generation (DG), with a strong emphasis on the emerging Mexican market. DG has long been an integral part of electric grids worldwide, but typically its role has been limited to specific functions such as emergency generation or cogeneration (also known as Combined Heat and Power). With the emergence of cost-effective small-scale renewable power generation, particularly from photovoltaic systems, distributed generation is now reaching a completely new level. Given the appropriate legislative and regulatory framework, renewable DG may become a main driver of the clean energy transition. In case of Mexico, the regulatory changes made possible by the power sector reform have led to an enthusiastic response from both residential and commercial users, fuelled both by record-low prices of photovoltaic systems and high solar radiation levels. Evidently, a departure from the traditional grid design is not without challenges, and new approaches, collectively known as *smart grids*, are required, offering great prospects for supporting deep decarbonization and increasing reliability, while providing greater consumer choices and a more efficient use of electricity. These issues are dealt with in Chapter 6 by Alexandra von Meier, Mohini Bariya, Jesús Valdez and Jonathan Mayo, who provide a comprehensive account of the technical challenges posed by deep decarbonization scenarios and the technological toolkit already available for building a

[‡]In this preface, as well as in the rest of the book, "generation" will be used in the colloquial sense commonly employed in energy studies. In a physical sense, of course, energy can only be *converted*.

renewable electricity sector with high reliability. Chapter 7, authored by Therese Peffer, deals with a specific set of tools and strategies, collectively known as *demand response* (DR), which is likely to be part of any modern electric grid in the near future. Smart interventions on the load side can often be conducted without a measurable impact on user comfort or productivity, while providing a very cost-effective way of balancing load and generation, particularly during peak load periods. The author shows how the enabling role of modern Internet-based technologies is key to a widespread application of DR, while insisting on the necessity of properly understanding user behaviour. Chapter 8, written by David Fernandes, Ricardo de Azevedo and Oliver Probst, addresses another key topic of the renewable energy transition, storage technologies and regulations. Although there may be little need for storage in electricity grids with low renewable energy penetration levels, deep carbonization scenarios will generally require a fair amount of storage. However, even at moderate penetration levels storage systems can add significant value to generators, consumers and the electric grid by offering a host of services ranging from frequency regulation to load shifting to transmission upgrade deferrals. The last chapter of Part II, Chapter 9 authored by Michel Rivero, Alberto Reyes, Mauricio Escalante and Oliver Probst, deals with another critical element of deep carbonization scenarios, the accurate forecasting of the wind and solar resource. The authors first review the relevant time scales required by grid services as well as regulation and planning tasks, and then go on to contrast these scales with time scales occurring in atmospheric phenomena determining the wind and solar resource. Different approaches for forecasting and their applicability to the different grid task are discussed.

Part III is dedicated to practical solutions for the renewable energy transition. It starts with Chapter 10 by Guillermo Zúñiga, who shares first-hand experiences obtained with the implementation of the electricity sector reform in Mexico, after reviewing the general principles guiding the transition from vertically integrated to liberalized power sectors and summarizing some of the international experience. Commissioner Zúñiga stresses the necessity of re-regulating the sector after liberalization in order to achieve the required levels of competition and transparency needed to reduce the cost of electricity, decarbonize the sector and increase consumer benefit. The impressive reduction in the cost of wholesale electricity from wind and solar in Mexico (with a record low of less than 20US$/MWh achieved in the 2017 clean energy auction) and the similarly impressive growth in solar-photovoltaic distributed generation (discussed at length in Chapter 5) can be interpreted as a direct result of these liberalization and re-regulation efforts. Chapter 11, written by James Bushnell, deals with the interplay of electricity markets and (renewable) power technologies operating at or near zero marginal cost. Though some concerns exist that markets with high shares of very low marginal cost technologies may disincentivize investments in generation, as already commented upon above, the author makes a convincing case that markets with high shares of renewables expose existing shortcomings, rather than creating fundamentally new problems, and points to feasible solutions. Chapter 12, written by Paul van Baal and Matthias Finger, addresses a key topic of the renewable energy transition, mobility. Though the benefits of using a renewable electricity sector to leverage a deep decarbonization of transport activities seem convincing, there are considerable challenges involved in this *sector coupling*. The authors point to solutions through holistic approaches where technological trends are closely aligned with the development of institutions. Chapter 13, authored by Mary Collins, provides a comprehensive development perspective for a deeply decarbonized electricity sector, connecting the global perspective, marked by the urgency to engage in the restructuring of the world economy in the face of climate change, with specific actions promising to foster this transition. A special emphasis is laid on the practical organization of technological change through the creation of innovation clusters to develop products and services, create and integrate supply chains, and develop a strong and well-trained workforce, while stressing the importance of creating a consistent market pull for sustainable products and services through incentives and long-term investment security. Chapter 14, written by Sergio Castellanos and Daniel Kammen, makes the closing statement of the book, reviewing what has been achieved on the way to a renewable electricity grid and, ultimately, a sustainable human society on a well-functioning planet, and pointing to the agenda for the next decades.

We hope the readers of this book will find ideas and inspirations for their own efforts towards sustainability, some reinforcing their existing knowledge of the subjects, others educating themselves for the first time on some of the key topics. In either case, this book should be viewed as a roadmap, not a blueprint for specific detailed actions, and the inspired collaboration of all educated humans subscribing to the project of sustainability is indispensable. The editors welcome constructive criticism and suggestions from the reader, hoping that a continuing dialogue will help overcome the barriers keeping us from safeguarding our only home in the cosmos.

The editors, February of 2020

Part I

Problem statement and potentials

Chapter 1

A clean electricity sector as a major driver of a sustainable economy

Oliver Probst[1]

The transformation of electricity sectors has been occurring over the last two decades because of deregulation and liberalization of electricity markets on the one hand and a greater interest in renewable energy technologies on the other. Although being largely independent phenomena, renewables and markets are nowadays closely intertwined [1,2], calling for novel approaches to grid management and expansion planning. At present, with renewables reaching records lows in electricity pricing [3,4], the main concern has shifted from economics to resource variability and adequacy, predictability, and grid-friendliness. As conventional generators slowly retreat, often fighting a fierce rearguard battle, new roles have to be found for existing generating capacity, and their potentially grid-stabilizing capacities have to be appropriately remunerated in the context of a complex yet short-sighted electricity market.

Yet, challenges are more substantial than expected; with solar photovoltaics becoming the new Dorado of electricity, gold rushers indifferent to environmental and social concerns may cause a similar threat to sustainability as strip mines and myriads of hydro-fracking pods, unless grid expansion planning duly reconciles the different objectives, including a permanent and effective protection of wildlife. Opportunities, on the other hand, are large: not only can electrified transport, fueled by clean sources of electricity, trigger the long overdue transition from the century-old internal combustion engine, but also can the convergence of the electricity and the transport sector, enabled by ever more powerful information technologies, provide the synergies required for an energy system dominated by variability.

Little doubt exists that an energy sector based exclusively on renewable energy sources, possibly with the exception of a few niche applications, is both desirable and feasible, at least in theory. Opinions diverge [5,6], however, as to whether this goal is achievable in practice and whether the timescale of the transition can be made short enough to allow for an effective mitigation of climate change. It is also often questioned whether the transition to a clean electricity sector can make a contribution large enough to meet the greenhouse gas (GHG) emission reductions required not to overspend the remaining global GHG emission budget. Although it is undoubtedly

[1]Tecnológico de Monterrey, School for Engineering and Sciences, Monterrey, N.L., Mexico

true that total GHG emissions have increased in most parts of the world, and even countries such as Germany or Denmark with global leadership in the promotion of renewables have achieved little reductions in recent years, but the conclusions drawn from these observations differ wildly. While some consider this to be proof that "the clean energy transition can't be done," others believe that the large growth phase in clean generation is yet to come and that consumption can be curbed.

In the present chapter, an overview of the topics surrounding this debate is provided, and a case will be made that the transition to sustainability is possible if technological and social progress can be achieved simultaneously. Although social progress traditionally lags far behind technological advances, improved empirical insights into human behavior, coupled with opportunities provided by information technology, may in principle pave the way out of the current planetary crisis.

1.1 Sustainability and the eternal growth paradigm

A sustainable ecosystem [7], by definition, is a system that can sustain itself over indefinite periods of time without collapsing or transforming itself into something else. It does so by counteracting on external perturbations, tending to restore equilibrium. Large perturbations, conversely, will generally lead to the destruction of the system, followed by a transitional phase and possibly a new equilibrium. Undoubtedly, the planet Earth has undergone a number of large perturbations in its geological history [8], giving rise to considerable changes in climate, land cover, and species populating the planet. Equally clear is that the changes brought about by the human species *Homo sapiens* in the last roughly 12,000 years, and more importantly, in the last 200 years since the onset of the industrial revolution, are unprecedented in magnitude and speed [9]. Modern humans now populate most of the inhabitable land, harvest about half of the Earth's biological production, and mine the underground for fossil fuels and minerals at almost any conceivable location. Consequently, other species have been brought to extinction, have been relegated to a few relatively protected areas, or have been subjected to a large-scale industrial utilization in benefit of the *H. sapiens*. It requires little imagination to predict that, if unchecked, this trend will lead to a planetary state where literally all resources available will be used by one species.

A number of questions naturally arise: first, can the rate at which GHGs are injected into the atmosphere (and the oceans, causing oceanic acidification the consequences of which we have not even begun to fully grasp) be slowed fast enough to keep the planet from running into a hot-house state [8,9], while still striving to bring prosperity to the whole of the world population*? The traditionally claimed links between gross domestic product (GDP) and development on the one hand, and

*Nowadays, the goal of organizations promoting international development seems to have become somewhat more humble: the World Bank now advocates "to end extreme poverty and promote shared prosperity in a sustainable way," while the International Monetary Fund works to "foster global monetary cooperation, secure financial stability, facilitate international trade, promote high employment and sustainable economic growth, and reduce poverty around the world."

between GDP and resource consumption, particularly primary energy consumption, on the other do not seen to leave much room for policies savings us from a runaway greenhouse effect. This topic is explored in Section 1.2.1.

Although keeping the planet from a climate catastrophe seems to be a task formidable enough to worry about other issues, sustainability goes of course much deeper. While satisfying real (and made-up) needs of the human population is likely to remain the main focus of the human endeavor, an increasing number of people are nowadays interested in ethical approaches than include nonhuman forms of life, particularly animals [10]. While traditionally, and for good reasons, most of the attention regarding animal welfare in the context of sustainability (if any) has centered on *biodiversity*, it is about time to translate the currently, widely recognized fact that sentience is not unique to humans into specific actions. As stated by the *Cambridge Declaration on Consciousness*

> [t]he absence of a neocortex does not appear to preclude an organism from experiencing affective states. Convergent evidence indicates that nonhuman animals have the neuroanatomical, neurochemical, and neurophysiological substrates of conscious states along with the capacity to exhibit intentional behaviors. Consequently, the weight of evidence indicates that humans are not unique in possessing the neurological substrates that generate consciousness. Nonhuman animals, including all mammals and birds, and many other creatures, including octopuses, also possess these neurological substrates [11].

While the recognition of animal sentience may be trivial to some, providing legal credit to these findings is both challenging and far-reaching. Evidently, a legal recognition of consciousness would provide animals with a certain rights [12], entitling them to well-being and, ultimately, protection from untimely and forced death. Denying animal sentience is reminiscent of the "Southern Defense of Slavery" [13] which openly dismissed the idea of abolishing slavery on the grounds of its expected "damages to the economy," alongside with biblical references showing that the founding fathers of the Abrahamic religions had also held slaves. Protecting animals from exploitation and slaughter evidently would cause a dramatic disruption in the meat industry, as well as other industries heavily relying on animal testing, but the oft-quoted human ingenuity might be able to cope with it, with far-reaching consequences for sustainability. Replacing meat obtained from slaughtered animals by synthetic meats, both created from animal cells and with plant-based ingredients, will not only save billions of factory farm animals from unnecessary suffering, but also go a long way toward decarbonizing the world economy. The science on this is pretty clear [14]: producing 100 g of protein from beef emits 25 kg of CO_2 on average, whereas the same amount of protein produced from peas generates only 0.36 kg of CO_2, i.e., two orders of magnitude less. Although the variations among different farming systems and approaches are large, the general message is clear: making meat from plant products (or eating vegetables to start with) not only avoids an unimaginable scale of animal suffering but also gives us a better chance of meeting our GHG emission goals.

1.2 A road map toward sustainability

Transforming a sector as large and important as the energy sector is evidently a formidable endeavor, so it is convenient to break down the task into more tractable parts. Few people would question the assertion that a "clean" energy sector is desirable in principle, but considerable dissent exists as to whether such an undertaking is "feasible" or "viable," and if so, on what timescale such a transition should occur; Chapters 2 and 3 address these questions at considerable depth. So, what are the difficulties and barriers, and what are the approaches to overcome them?

1.2.1 Curbing natural resource consumption

At present, the vast majority of economists and decision-makers would assert that there is no alternative to economic development other than pursuing a continuing growth of economic throughput, generally measured in terms of the GDP of a nation.[†] The limitations of using GDP as the main indicator are increasingly acknowledged in the scientific literature, by stressing inequality [16] and the need for measuring human welfare in more encompassing terms [17], but the GDP approach is still by far the dominant one. GDP growth, on the other hand, is generally considered to be closely linked with energy consumption. Although a number of authors have concluded that no causal link exists between the growth of energy consumption and GDP, a recent meta-analysis by Handrich *et al.* [18] suggested that either energy consumption drives GDP growth or vice versa. The same authors argued, though, that *climate change emissions* may actually be decoupled from economic growth through an increased energy efficiency and renewable energy deployment, sustaining their findings with an econometric approach. This rigorous technical result can of course be readily interpreted in terms of the (1) substitutability between fossil and renewable energy and (2) the observation that actually delivered energy services, rather than primary energy consumption, should be the driver of economic growth. Although the evidence found by Handrich *et al.* [18] is somewhat encouraging to the extent that decoupling of economic growth and GHG emissions seems to be possible, it does confirm the old economic wisdom that consumption growth fuels prosperity. Evidently, perpetual resource consumption growth is a physical impossibility on a finite planet, so a clear dilemma seems to exist between the stated intention of most world governments and supranational agencies of promoting prosperity for an ever larger human population and the need to limit the negative impacts resulting from mining, extraction, and land appropriation. Interestingly, these completely data-driven econometric results are very much in line with the conclusions obtained by Garrett [15] from systems modeling, where both the Earth's atmosphere and the world economy are represented as well mixed systems. Garret concluded that efficiency

[†]Given the global nature of climate change, it may be more relevant to consider the Gross World Product (GWP) rather than GDPs of individual nations [15].

gains are unlikely to lead to significant reductions in CO_2 emissions, given a host of "rebound" effects (not modeled explicitly) that tend to eat up any reductions in energy consumption through increased levels of spending. Somewhat unexpectedly, but actually easy to understand, Garret was able to show that *only decarbonization* has the potential to reduce emissions, and that such an emission reduction can only be achieved if the logarithmic rate of change of decarbonization $|d\ \ln c/dt|$ is larger than the growth of wealth.

The way out[‡] of this dilemma may lie with the fact that we are measuring the wrong metric. As stated by Kubiszewski *et al.* [19], GDP was never designed to measure prosperity or development, and its shortcomings are well documented. As argued by these authors and others, other indicators may be more appropriated actual wealth creation. The main indicator put forward by the authors and widely used in the ecological economic sciences is the Genuine Progress Indicator (GPI), a metric based on the Index of Sustainable Economic Welfare (ISEW). As shown in Kubiszewski *et al.* [19], GDP and GPI show a very good ($R^2 = 0.98$) positive correlation in the annual per capita GDP range from US$2,000 to US$6,000/cap/year, whereas the correlation becomes negative ($R^2 = 0.61$) for larger annual per capita GDP values. On the timeline, it can be observed that global per capita GPI reached a maximum around 1975, with a slight decline later in 2004, in spite of a continuous near-linear increase of per capita GDP for the period of 1950–2004. This finding represents a strong statement: beyond a certain level of economic activity (as measured by GDP), no additional increase in net wealth is obtained.

Combining this finding with the strong correlation between energy use stated above, it can be summarized that increased energy consumption does fuel GDP growth, *but the bulk of it goes into activities that do not increase progress*. Conversely, this indicates a very significant potential for reducing energy services altogether, as opposed to "only" providing them more efficiently or through cleaner technologies. It is important to stress that this conclusion remains valid independently of the choice of the exact choice of the progress metric [20], as long as the trade-offs between resource consumption growth and its detrimental effects on the social and environmental spheres are appropriately measured. The considerable debate about the right metric for the measurement of human progress should not distract from the fact resource that consumption comes at a cost, and curbing it is an unnegotiable requirement of sustainable development. Whether GPI or other metrics are suitable for measuring the trade-offs between wealth and resource consumption may be debatable [20], but the conclusion is ultimately unavoidable: *there is no way around a deep decarbonization of the atmosphere*, both by

[‡]Somewhat ironically, the title of Garrett's paper is "No way out," insinuating that decoupling of GDP growth and primary energy use cannot be achieved. Although Garrett's modeling approach and conclusions seem rigorous, the silver lining may be that Garret's conclusion that rapid rates of deep decarbonization are not "realistic" may not be accurate. Of course, the current pace of progress on climate change mitigation may be much too slow to curb global emissions before a critical tipping point is reached, beyond which the Earth's climate will run away toward a "hot-house" state (see earlier).

decarbonizing the energy sector and by rebuilding large-scale living ecosystems, or a runaway greenhouse effect is likely to occur.

1.2.2 Electrification of energy services

The use of fire is at the heart of the human endeavor since hunters and gatherers roamed the plains of Africa. Fire has been and is still actively used to clear land from vegetation (nowadays an unduly impactful activity), drive animals out of their hideouts, cook food, smelt and treat materials, and certainly, drive thermal processes related to the provision of useful energy and energy services. Combustion is so intertwined with human culture and thinking that many people have difficulties imaging a human society without it. The expansion of human capabilities is, however, increasingly dependent on the availability of energy services related to the use of electricity. Electricity can (and should) be provided by clean renewable energy sources; therefore, massively migrating energy services to electricity-based technologies may be a suitable pathway to a deep decarbonization.[§] It is about time to leave our early ancestors' fascination with fire behind.

1.2.3 Deep decarbonization

While some still debate [6,21] whether deeply decarbonized power systems and industrial processes are possible or maybe even desirable, others have begun to work on practical implementations of the clean energy transition. Making steel with renewable energy, through direct reduction with hydrogen generated from wind power or other renewable energy sources [22,23], is already becoming a reality in European countries such as Germany and Sweden, driven both by high prices of natural gas and competitive wind technology. Direct-reduction steel-making technology has been around for decades, so completely overhauling the steel-making industry is a matter of political will on an international level, rather than a matter of disruptive innovations yet to come. Similarly, converting private vehicles to fuel cell drives, using either direct hydrogen or methanol generated from wind energy, is now a technological reality. Moving down this road is a matter of political will, and only to a limited degree a matter of cost, though opinions to the contrary abound. With individuals willing to spend US$50,000 on a new vehicle abounding in many developed and developing countries, fuel costs are more a matter of perception rather than an actual factor governing the utilization of private transport.[¶]

[§]This is not to say that other energy vectors for the transport and storage of renewable energy should not be pursued as well. A particularly noteworthy pathway is the use of clean synthetic fuels derived from renewables, such as hydrogen or derived fuels such as e-methane or e-methanol. A glimpse at these technologies is taken in Chapter 8 on storage.

[¶]As always, it is difficult to design a "one size fits it all" strategy. Even though the luxury car segment is thriving in most parts of the world, most people have more mundane things to worry about. However, subsidizing (fossil) fuels to keep private transport accessible to the low-income segment of the population is hardly an appropriate strategy for all. Luxury cars could be mandated to run on clean fuels or clean electricity. Luckily, the rise of electric mobility seems to occur largely in the luxury car segment, even without significant government intervention.

Deeply decarbonizing the electric grid, in spite of the fact that most renewable energy technologies naturally produce electricity, is still often contested on the grounds of the "intermittency" of clean energy technologies such as wind and solar. Obviously, the progress of some countries, e.g., Denmark, with high degrees of penetration of clean technologies seem to contradict these views. However, detractors are often quick to point to specific conditions that "made it easier" for the Danes or others to accomplish their achievements, in the case of Denmark because of the strong interconnections to its neighbors. However, these claims can be easily refuted. Uruguay, to name an unsuspected candidate, has moved from an electric grid largely dominated by thermal generation and hydro[||] to one using fossil-fueled thermal generation mainly for regulation.[**] More remarkably, this transition has occurred in less than a decade. Finally, Uruguay, though being interconnected with its neighbors (Argentina and Brazil), is only exporting and importing electricity according to market conveniences and is not relying on these interconnections to stabilize its system. Evidently, to achieve this remarkable feat, Uruguay has required some efforts by the grid operator, supported by universities and funding agencies, to redesign its dispatch operations originally based on a fundamentally deterministic approach to one based on probabilistic dispatch relying heavily on accurate advanced forecasting[††] techniques [25]. Although Uruguay heavily taps into its hydro plants for load balancing, its small geographic size precludes it from taking larger benefit from geographic dispersion of solar and wind resources, a tool available for larger countries, particularly those with a number of different climate zones (e.g., Mexico). A smaller hydro resource may then be readily replaced in those countries by the larger geographic dispersity of its solar and wind resource, given adequate transmission[‡‡] infrastructure.

1.3 The electric grid in the renewable energy era

1.3.1 Large-scale transmission revisited

Large-scale electric power transmission is at the heart of the electric grid, enabling transfer from centralized power plants to load centers and ultimately (through the distribution grid) to individual consumers. With new locations of generation emerging (out at sea where the wind blows steadily or in the deserts where clouds only occasionally screen the sun), often not previously covered by existing transmission, the expansion planning of the transmission grid [26,27] acquires a new dimension: not only must the energy be able to flow from the high-resource regions to the load centers without running into congestion, but also should the temporal complementarity of generation in different regions be taken advantage of. A robust

[||]With strong variations in the annual hydro resource, ranging from 100% of the required electricity consumption in wet years to only 25% in dry years.
[**]In 2018, thermal generation accounted for only 2.9% of all electricity generated in the country [24].
[††]Forecasting is dealt with at length in Chapter 9.
[‡‡]Transmission is discussed in Section 1.3.1 and discussed at length in Chapter 4.

grid allowing for large-scale smoothing effects among renewable energy power plants at different locations and with different temporal characteristics goes a long way toward creating a more predictable power generation timeline and reducing the need for storage. Evidently, a grid with nonexisting transmission constraints is impossible to build, and sizing line capacities is not a straightforward task either, given the complexity of power flow in a grid where both demand and generation are variable. Although advanced statistical modeling can narrow down some of these uncertainties, at least for a short and medium term, the planning of transmission infrastructure generally spans time horizons covering a number of technological revolutions and policy changes, making the process even more challenging. Last but not least, building new transmission remains a challenge in most countries because of the need of securing rights of way (ROW) along long corridors, motivating the search for solutions taking advantage of existing ROW. Such solutions include dynamic line rating (DLR) [28–30] technologies and reconductoring [31] with high-capacity conductors such as high-temperature low-sag (HTLS) conductors, with both approaches applying to thermally limited lines, flexible alternating current transmission systems (FACTS) [32], increasing stability-limited transmission lines by dynamically controlling line reactance, and conversion alternating-current to direct-current transmission lines. In Chapter 4, a roadmap is sketched on the different factors governing the efficient planning of transmission expansion for electric grids with a high penetration of renewables, providing the reader with significant insights into the technological and economic options available.

1.3.2 Distributed generation: moving out of the niche

At present, although the electric grid as we know it has been around for the better part of the twentieth century and until the present day, the unification of the different technologies, voltage levels, and standards has by no means been a straightforward process, spanning several decades until its completion. Distributed generation and physically separated microgrids have been around in the United States well into the twentieth century and still persist in many remote locations in the world. However, with the availability of small-scale solar, wind, or bioenergy plants, distributed generation [33] is making a strong reappearance in many parts of the world, including and especially in locations connected to the main grid. This potentially reduces the distances between generators and loads, avoiding transmission and distributions losses, reduces transmission congestion, and helps postpone investments in transmission upgrades. An equally important potential benefit is the utilization of the existing civilizatory footprint for power plant installation, reducing the impact on yet pristine or minimally disturbed natural areas. Other oft-quoted benefits include the possibility of combining distributed generation with local storage, improving load matching, reducing peak load, and diminishing traffic on the distribution network. Particular attention is also paid to the possibility of using a potentially large fleet of parked electric vehicles as a distributed large battery interacting with distributed generation [34]. In Chapter 5, experiences with distributed generation are described and critically examined, and some of the technical and regulatory challenges lying ahead are sketched.

1.3.3 Storage: the main driver or a nice-to-have?

Storage, often considered the panacea of renewable energy grid integration, is equally a technological, a regulatory, and an environmental challenge. Although scientifically sound options for storing energy in general and electricity in particular abound, few technologies have been used beyond a demonstration scale. Pumped hydro storage [35] is the main technology employed on a utility scale and has been used as such since the early twentieth century. Although well proven, pumped storage has been limited by the specific set of geographic conditions required for its installation, environmental concerns, and a large upfront investment, making it difficult for projects to materialize unless a long-term income stream can be secured. In electric grids with a competitive electricity market, where the specific conditions determining the income stream components are often highly volatile, there may be therefore little investor interest in pumped storage. Battery energy storage systems (BESS) [36], which have made a more recent appearance in electric grids, so far share the relatively high capital requirement with pumped storage; however, a number of features makes them potentially more attractive, such as relocatability, modular structure, and site independence. Storage systems can provide a number of services to the grid involving many different timescales, each covering a specific physical aspect of grid operation. Frequency regulation, to name one example, is critically important, and even small amounts of energy storage may make a large difference in weak grids if the power capacity is appropriately selected. Unfortunately, power plant participation in frequency regulation is often a requirement of the grid code rather than a remunerated service, so installing a storage facility next to a renewable energy plant will simply increase the cost but not create an income stream to pay for the initial investment. Dual, triple, or multiple uses of a storage facility can of course in principle improve the economics, and many services may actually be offered simultaneously, without too many trade-offs among them. However, a good understanding of these mechanisms by the regulator and correspondingly flexible regulations will often be required to make pervasive storage in the electric grid of the future a reality. Chapter 8 covers both basic technological aspects of storage systems and their prospects for the near and medium term, as well a discussion of grid services provided by storage and some recommendations on their regulations and remuneration.

1.3.4 Forecasting—beyond the crystal ball

At current levels of variable generation in most electric grids, the largest contribution to additional operating reserve requirements caused by variable generation such as wind and solar comes usually from forecast errors [37]. Although forecast modeling has greatly progressed over the last decade, its uncertainties are still typically larger than those associated with demand predictions. The larger the forecast error, the larger the operational reserves required to maintain the reliability standards, the costlier the operation of the grid. Electricity markets have reacted in some places by introducing shorter trading periods, also known as intraday trading, thereby tapping into the lower uncertainties conferred by updated forecasts and shorter forecast horizons. Demand response systems can provide additional reserves and therefore compensate for forecast errors, and storage can both provide increased reserve margins and flatten out variable generation. Progress, however, is needed on all fronts, and improvements in forecasting are potentially low-hanging fruit, given the relatively low cost of the modeling process compared with other options. To properly design novel strategies, it is first paramount to identify the state of the art of variable generation forecasting approaches on all timescales relevant to grid operations, ranging from the very short-term predictions required for real-time frequency regulation (typically in the order of seconds), secondary and tertiary regulation (minutes to hours), unit commitment (hours to days), and preventive maintenance planning (days to weeks). In addition to timescales, the spatial structures of the variable generation resource have to be understood properly and incorporated into markets and grid operation. Variations with both small time and length scales (say, subminute variations on a wind farm length scale) may be difficult to capture with physical modeling tools but may be subject to accurate modeling and prediction with multivariate stochastical techniques, enhanced by learning capabilities and possibly on-site measurements. The output of a solar photovoltaic power plant typically operating under clear-skype conditions, on the other hand, may be too strongly affected by unusual events such as the passage of a cloud for stochastic methods to be effective; similarly, physical methods are unlikely to predict cloud formation and characteristics as well as its dynamics with sufficient accuracy. In such cases, cloud cover observation methods (either ground- or satellite-based, or both) may provide the required accuracy. On large spatial scales, relevant to regional and interregional load balancing, the approaches may be quite different. Here, the preparation of separate forecasts by wind farm or solar plant operators may have be complemented by the grid operator by using a large-scale forecast model that adequately captures the spatiotemporal correlations, thereby anticipating the regional and interregional smoothing effects resulting from an incomplete correlation. Chapter 6 examines the complex topic of forecasting and its key role for fostering a high penetration level of renewables in the electric grid.

1.3.5 Smart grids, demand control, and energy efficiency

1.3.5.1 What can smart grids do for you?

Smart grids [38] are associated with a number of capabilities and functions, many of which are unrelated to renewable energy sources and technologies. Their potential for fostering and easing the transformation of the electric grid toward one based on true sustainability is however substantial. The services that can be provided by a smart grid range from taking advantage of spare capacities on transmission lines through DLR, transmission line capacity regulation through FACTS, fast reactive power and voltage control, demand response management, and grid storage, among others. While traditional grid upgrades and maintenance often rely on centralized planning and cost socialization, having led to substantial lags in grid modernization in many parts, monetizing the grid-stabilizing and -enhancing services may foster a more dynamic development than the one that would be possible under a central planning process. In Chapter 6, an overview of the technological options associated with smart grids and suitable for renewable energy promotion is provided, followed by an assessment of their present costs and their cost reduction potential. The focus is then laid on the services that can be provided by smart grid technologies and their relation to existing and potential new products traded on wholesale electricity markets.

1.3.5.2 Demand control and demand-side management

It is an oft-quoted fact that buildings account for a very significant fraction of the final electricity consumption, with a strongly increasing trend [39]. With real estate markets soaring in many parts of the world, this trend is likely to continue. Although the trend itself is an additional burden on an electricity system striving to become sustainable, there are also a number of opportunities for buildings to actually ease the transition, using a number of measures collectively known as demand response [40]. First of all, buildings have a considerable thermal inertia, even the fully glazed versions that prevail in many new constructions nowadays. Where architectural preferences do not favor a good thermal behavior, building energy efficiency codes [41,42] can be instated or enhanced to include thermal inertia requirements, though often probably not without a battle with special interest groups. Given the large fraction of electricity consumption in buildings accounted for by HVAC equipment, thermal inertia can be used favorably to shift load away from peak demand periods, or more importantly in an electric sector increasingly dominated by renewables, to low operating reserve periods. With traditional operating and control equipment for HVAC and lighting, adjusting to fixed peak demand periods could be achieved in a semimanual way, if the maintenance personnel is appropriately trained and incentivized; adjusting to a dynamically evolving operating reserve, however, requires a high degree of automation and a considerably more detailed understanding of the buildings behavior. Fortunately, smart buildings [43] are already becoming a reality in many parts of the world, both because of the potential of the associated technology for creating net savings and providing increased functionality, and the prestige value conferred to a building publicly recognized as innovative. Smart building technologies can tap into thermal inertia and remaining inefficiencies, possibly complemented by local storage technologies such as ice storage, to provide demand response on timescales relevant to grid operation. Interestingly, demand response can very effectively address one of the challenging aspects of the transition to a grid with a high fraction of renewables: frequency support, and regulation [44]. With a smaller number of conventional generators based on rotary machinery, directly coupled to the grid, accounting for electricity generation, frequency control in response to a loss of generation becomes more challenging, both because of the lesser amount of rotary inertia associated with the remaining generation and a reduced primary operating reserve. Smart buildings participating in a frequency control program could very effectively restore the missing inertia and primary operating reserve, at literally no discomfort to building occupants. Even a participation in secondary reserves and load-following, typically involving timescales of a few minutes, and other regulation tasks are in principle possible without a perceptible change in user comfort. In regulating tasks associated with longer timescales, other options such as increasing air velocity to compensate for slightly higher air temperatures are available. Even though user comfort is likely to be affected in only a minimal way by building demand response measures, the participation in grid control schemes can still be appropriately remunerated, creating a strong economic incentive. Where market response is slower than required, building codes can and should be implemented to secure the participation of one of the largest and fastest growing load sectors. A final consideration refers to data centers, a special segment of the building sector with a particularly large growth and considerably more stringent climate control requirements. Chapter 7 discusses opportunities for curbing and controlling demand in this sector and provides an account of the state of the art of the field.

1.4 Markets and regulations

1.4.1 Regulations: the importance of a long-term vision

Barriers to renewable energy integration are often not so much about renewables themselves—or their cost of generation—but about providing access to otherwise monopolistic markets. Even back in 1978, when renewables were a far cry from being considered competitive, the Public Utility Regulator Policy Act (PURPA) [45] opened the door to renewable energy participation in the United States by requiring utilities (generally enjoying regional monopolies and return-on-equity figures guaranteed by law) to purchase electricity generated from renewable energy sources at a price lower than the one charged by the utility to its consumers. Surprisingly to many, wind and solar power plants have been up to this challenge for 40 years. In Germany, the current famous feed-in law [46,47], instated in 1991, mandating regional utilities to buy electricity from renewables at a guaranteed fraction of the consumer tariff and providing a simplified administrative scheme for interconnection, has consistently been credited with being the most reliable and cost-effective measure for promoting renewable energy growth worldwide. Mexico opened its monopolistic electricity sector in late 1992 to allow for private sector participation under certain circumstances which allowed de facto electricity sales transactions between third parties, mostly through the self-generation scheme. A number of additional provisions in subsequent years, including an energy banking scheme and a capacity recognition of renewables, allowed renewable energy projects (mostly wind) to thrive. The 2013 constitutional energy reform, followed by the 2014 Power Industry Law that instated a market-based electricity sector, abolished all previous incentives to renewables, introducing the so-called clean energy certificates as their sole replacement [3]. To avoid a collapse of the renewable energy project pipeline, the Mexican government mandated the state utility to acquire long-term clean energy contracts through an annual competitive bidding process, leading to record-low (US$<20/MWh) prices for wind and solar electricity in 2017 [4].[§§] The lessons learned from the cases described are critically important for the financing of renewable energy projects worldwide. In all cases, the legal and regulatory provisions mentioned allowed for projects with a reliable long-term income stream, attractive to risk-adverse investors with a variety of backgrounds. Projects in grids with wholesale electricity markets, particularly in incipient markets, have often trouble in accessing financing because of the greater volatility of markets prices compared with regulated tariffs and the risk adversity of traditional project financiers. While mature markets may allow project financing of projects with a high merchant component and still keep the risk premium at an acceptable level, renewable energy projects may come to an end in relatively new markets if some kind of long-term stability mechanism cannot be provided by the government. An overview of the importance of regulations with a long-term vision is laid out in Chapter 10, and a detailed description of regulations at work is provided for the case of the Mexican electricity sector.

1.4.2 How can markets help renewables?

While electricity markets were not explicitly designed to foster renewables, currently both coexist in many parts of the world. Even without renewables, markets seldom work as envisaged in economists' textbooks, and a close oversight by an independent authority to secure reliability at all times is vital [2]. On the other hand, in many places renewables have received some kind of preferential treatment in early stages of development [46–48], but incentives are generally considered to be transient in nature, and in the steady state renewables are expected to compete with other generators on an equal footing, possibly adjusted for market-based incentives such as clean or explicitly renewable energy certificates. A prudent market design, however, remains vital even in countries and regions with a well-developed renewable sector if the transition to sustainability is not to become stalled. A potential problem arising in markets with high renewable energy penetration [49] is the fact that any technology with a very low operating or short-term marginal cost such as wind and solar will tend to push the fuel-based technologies up the merit-order ladder and lead to a net reduction of local marginal prices (LMPs) in the system, making it more difficult for all generators (including the renewables) to recuperate their investment and fixed cost. While this effect is inherent in energy-only markets and is not limited to markets with renewables, it becomes particularly prominent in markets with high renewable energy penetration. A solution allowing for the recovery of investment costs is the creation of a capacity market, already implemented in many systems. Capacity payments provide incentives for the construction of new power plants and also tend to increase

[§§]Unfortunately, this vision of an increasingly clean electricity sector is not shared by the current federal government who has canceled all competitive tenders for renewable generation and has publicly declared that increasing the generation market share of the incumbent state utility (rather than maximizing social welfare or decreasing emissions) should be the guiding principle of its electricity sector policy.

the regulation reserves required in electric grids with higher degrees of variability; they also tend to favor conventional generators over variable generation renewables, unless mechanisms for the reliable recognition of capacity contributions from renewables are instated. Other important points to be addressed are the role of storage and demand-side response (DSR) [43]. Both generator-associated and grid-based storage units can play a vital role in promoting a higher penetration of variable generation renewables but the different services provided by storage must be fully recognized to make a business case. DSR, on the other hand, has an enormous potential for participating in load balancing and frequency regulation, but its participation in most markets is incipient at best. Supposing a well-designed market, storage and DSR may represent the ultimate solution to the transition to an electric grid fully based on renewable energy generation. Chapter 11 provides a comprehensive account of the complex interplay between markets and renewables, showing that renewables hardly introduce additional difficulties into market design but rather expose existing weaknesses, and points to some solutions for the difficulties lying ahead.

1.5 Taming the beast: toward a concept of mobility

Energy consumption in transportation [50] continues to be one of the greatest threats to sustainability. Not only does transportation account for more than a fifth of the worldwide primary energy consumption, but also there is strong growth trend, particularly in developing nations. Personal mobility is still largely oriented toward the individual use of automobiles [50], with public transportation systems often being inefficient, costly, or even nonexisting. With an increasing fraction of the world population concentrating in large urban centers, there appears to be a demand for new and improved public or shared transport options. Overcoming the vices of a sedentary lifestyle also seems to point to the need for alternatives to personal automobility, including walking, biking, and electrobiking. New ways of professional collaboration through the systematic use of information technologies for sharing and collaborating, finally, avoid the need for commuting or at least substantially reduce it. Even with these clear opportunities lying ahead, and with both traditional car manufacturers and new technology-driven entrants exploring the new playing field, the barriers to sustainable mobility are manifold. Cities converted into a grid of highways are unlikely candidates for promoting pleasant walks to school or bike rides to work, sky-rocketing real estate prices in areas with sizeable sources of employment do not allow employees to live near their work, and traditional public transportation planning approaches are likely to fail in fast expanding cities where the target is constantly moving. Consequently, the problem of sustainable transportation is one of urban planning as much as it is one of regulation and engineering. Fortunately, there are a number of success stories to be drawn upon, from cities where biking has become the method of choice for transportation (such as Copenhagen and many Dutch cities) to walkable waterfronts and city centers.

Chapter 12 identifies some of the most pressing issues and sketch pathways that can bring about transformation in a relatively short period of time.

1.6 Where we go from here? The clean energy transition as the litmus test of human maturity

Although in many parts of the world the transition to clean electricity is well under way, fueled both by visionary regulations and falling prices for renewable energy technologies and project development, it is relatively clear that an accelerated pace is needed for reducing the overall carbon footprint of humanity. Although renewables have been growing strongly, overall carbon emissions have been growing as well, both because of increased electricity consumption and little inclination to pull out of fossil fuel utilization for electricity generation, and more importantly, for transportation fuels. Whereas the replacement of fossil transportation fuels by sustainable alternatives is still in its infancy, problems related to very high penetrations of renewables in the electric grid are comparatively minor and can be solved by a combination of increased transmission capacities, modest amounts of storage, demand response technologies, and a smarter management of the grid. Penetration levels of up to 80% with only minor grid upgrades have been claimed to be possible, with some even asserting that 100% penetration could be possible with current technologies. Others have concluded that even satisfying 100% of all energy needs with renewables is possible, based on detailed studies, e.g., for Germany [51] or the United States [5]. Evidently, such proof-of-concept studies have only limited value for guiding practical implementations, often because technology changes produce both winners and losers, but they provide a useful reference against which practical strategies can be compared. Deep decarbonization studies for the electric grid and even the whole energy system are no longer limited to the resource or technology aspects but they increasingly address policy implications and regulatory pathways [48,52]. In any case, a faster deployment of the energy transition as a whole is likely to depend on the speed with which electric grids around the world can become sustainable. At present, electrifying transportation altogether seems much more realistic than 26 years ago when the United Nations Framework Convention on Climate Change was signed in Rio de Janeiro, and it is likely to be a more sustainable option that feeding biofuels into the fossil fuel refinement and transportation network, or building a new infrastructure around hydrogen. The situation is somewhat different in those parts of the world where access to main electricity is currently difficult or nonexisting. Here the challenge, rather than to phase out fossil fuels from electricity generation, is to build an infrastructure based on clean energy without having to go through fossil-fueled solutions and provide access to the development opportunities only electricity-based services can provide. Affordable (generally solar-powered) generation kits and pay-as-you go schemes may make a difference where traditional financial approaches for electricity procurement do not work.

Not everything is about satisfying demand, however. Although most official planning documents consider consumption growth to be inevitable, ever-lasting growth is by no means a law of nature, and it is of course simply an impossibility in a finite world. Long-lasting (and often exponential) growth inevitably leads to an eventual collapse when the carrying capacity of the system is exceeded. Signs of exceeded carrying capacities can already been found around the world where salinated soils, depleted wells, and eroded landscapes force millions into migration. Limiting demand by requiring an efficient use of all resources will be important and imposing regulations onto those sectors that most contribute to consumption and consumption growth is a vital ingredient in the clean energy transition. Obvious candidates abound, such as car mileage and building energy consumption. With the digital revolution taking humanity by storm, however it is also important not to close one's eyes before the growing fraction of electricity consumption accounted for by data centers and the processing and transmission of often worthless (or even harmful) data. The grotesque growth of electricity use associated with the mining of cryptocurrencies is a particularly noteworthy example but it is unlikely to be the last instance of excessive resource consumption by a sector once believed to be immaterial.

References

[1] Mastropietro P, Rodilla P, Rangel LE, *et al.* Reforming the Colombian electricity market for an efficient integration of renewables: A proposal. *Energy Policy*. 2020;139:111346. Available from: http://www.sciencedirect.com/science/article/pii/S0301421520301038

[2] Peng D, and Poudineh R. Electricity market design under increasing renewable energy penetration: Misalignments observed in the European Union. *Utilities Policy*. 2019;61:100970. Available from: http://www.sciencedirect.com/science/article/pii/S0957178719303236

[3] Llamas A, and Probst O. On the role of efficient cogeneration for meeting Mexico's clean energy goals. *Energy Policy*. 2018;112:173–183. Available from: http://www.sciencedirect.com/science/article/pii/S0301421517306390

[4] Deign J. *Updated: Mexico's Energy Auction Just Logged the Lowest Solar Power Price on the Planet*. Greentech Media; 2017. Available from: https://www.greentechmedia.com/articles/read/mexico-auction-bids-lowest-solar-wind-price-on-the-planet

[5] Jacobson MZ, Delucchi MA, Cameron MA, *et al.* Low-cost solution to the grid reliability problem with 100% penetration of intermittent wind, water, and solar for all purposes. *Proceedings of the National Academy of Sciences*. 2015;112(49):15060–15065. Available from: https://www.pnas.org/content/112/49/15060

[6] Clack CTM, Qvist SA, Apt J, *et al.* Evaluation of a proposal for reliable low-cost grid power with 100% wind, water, and solar. *Proceedings of the National Academy of Sciences*. 2017;114(26):6722–6727. Available from: https://www.pnas.org/content/114/26/6722

[7] Chapin III FS, Torn M, and Tateno M. Principles of ecosystem sustainability. *American Naturalist*. 1996;148:1016–1037.

[8] Steffen W, Rockström J, Richardson K, *et al.* Trajectories of the earth system in the Anthropocene. *Proceedings of the National Academy of Sciences*. 2018;115(33):8252–8259. Available from: https://www.pnas.org/content/115/33/8252

[9] Kidder DL, and Worsley TR. A human-induced hothouse climate? *GSA Today*. 2012;22(2):4–11.

[10] Ethical theories and nonhuman animals. *Animal Ethics*. Available from: https://www.animal-ethics.org/ethics-animals-section/ethical-theories-nonhuman-animals/

[11] *The Cambridge Declaration on Consciousness*. Available from: http://fcmconference.org/img/CambridgeDeclarationOnConsciousness.pdf

[12] *Rights Theories: Different Positions*. Available from: https://www.animal-ethics.org/rights-theories-different-positions/

[13] *The Southern Defense of Slavery*. Available from: https://u-s-history.com/pages/h244.html

[14] Poore J, and Nemecek T. Reducing food's environmental impacts through producers and consumers. *Science*. 2018;360(6392):987–992. Available from: https://science.sciencemag.org/content/360/6392/987

[15] Garrett TJ. No way out? The double-bind in seeking global prosperity alongside mitigated climate change. *Earth System Dynamics*. 2012;3(1):1–17.

[16] Piketty T. Translated by Arthur Goldhammer. *Capital in the Twenty-First Century*. Cambridge, MA: The Belknap Press of Harvard University Press; 2014.

[17] Pasternak AD, and Livermore L. *Approved for Public Release; Further Dissemination Unlimited Global Energy Futures and Human Development: A Framework for Analysis*. October 2000. Available from: http://www.llnl.gov/tid/Library.htmzl

[18] Handrich L, Kemfert C, Mattes A, *et al. Turning Point: Decoupling Greenhouse Gas Emissions from Economic Growth*. Berlin, Germany: Heinrich-Böll-Stiftung; 2015.

[19] Kubiszewski I, Costanza R, Franco C, *et al.* Beyond GDP: Measuring and achieving global genuine progress. *Ecological Economics*. 2013;93:57–68. Available from: http://www.sciencedirect.com/science/article/pii/S0921800913001584

[20] Fox MJV, and Erickson JD. Design and meaning of the Genuine Progress Indicator: A statistical analysis of the U.S. fifty-state model. *Ecological Economics*. 2020;167:106441. Available from: http://www.sciencedirect.com/science/article/pii/S0921800919302411

[21] Heard BP, Brook BW, Wigley TML, *et al.* Burden of proof: A comprehensive review of the feasibility of 100 renewable-electricity systems. *Renewable and Sustainable Energy Reviews*. 2017;76:1122–1133. Available from: http://www.sciencedirect.com/science/article/pii/S1364032117304495

[22] Salzgitter Flachstahl, Linde in steel industry clean hydrogen project. *Fuel Cells Bulletin*. 2018;2018(12):12. Available from: http://www.sciencedirect.com/science/article/pii/S1464285918304681

[23] Bos MJ, Kersten SRA, and Brilman DWF. Wind power to methanol: Renewable methanol production using electricity, electrolysis of water and CO_2 air capture. *Applied Energy*. 2020;264:114672. Available from: http://www.sciencedirect.com/science/article/pii/S0306261920301847

[24] Informe Anual del Mercado Eléctrico 2018. (In Spanish). Available from: https://adme.com.uy/db-docs/Docs_secciones/nid_526/Informe_Anual_2018_V8.pdf

[25] Caporale X, Galusso FB, and Chaer R. Incorporation of forecasts to the optimal operation of dynamic systems using stochastic CEGH models. In: *2018 IEEE 9th Power, Instrumentation and Measurement Meeting (EPIM)*; 2018. pp. 1–4.

[26] Conlon T, Waite M, and Modi V. Assessing new transmission and energy storage in achieving increasing renewable generation targets in a regional grid. *Applied Energy*. 2019;250:1085–1098. Available from: http://www.sciencedirect.com/science/article/pii/S0306261919309213

[27] Ameur A, Berrada A, Loudiyi K, *et al.* Analysis of renewable energy integration into the transmission network. *The Electricity Journal*. 2019;32(10): 106676. Available from: http://www.sciencedirect.com/science/article/pii/S1040619019302817

[28] Morozovska K, Naim W, Viafora N, *et al.* A framework for application of dynamic line rating to aluminum conductor steel reinforced cables based on mechanical strength and durability. *International Journal of Electrical Power & Energy Systems*. 2020;116:105491. Available from: http://www.sciencedirect.com/science/article/pii/S0142061519309494

[29] Viafora N, Delikaraoglou S, Pinson P, *et al.* Chance-constrained optimal power flow with non-parametric probability distributions of dynamic line ratings. *International Journal of Electrical Power & Energy Systems*. 2020;114:105389. Available from: http://www.sciencedirect.com/science/article/pii/S0142061519309317

[30] Tarn-Santiso AV, Llamas A, and Probst O. Assessment of the potential for dynamic uprating of transmission lines in the Mexican National Electric Grid. *Electric Power Systems Research*. 2019;171:251–263. Available from: http://www.sciencedirect.com/science/article/pii/S0378779619300240

[31] Mbuli N, Xezile R, Motsoeneng L, *et al.* A literature review on capacity uprate of transmission lines: 2008 to 2018. *Electric Power Systems Research*. 2019;170:215–221. Available from: http://www.sciencedirect.com/science/article/pii/S0378779619300124

[32] Peng FZ. Flexible AC Transmission Systems (FACTS) and Resilient AC Distribution Systems (RACDS) in smart grid. *Proceedings of the IEEE*. 2017;105(11):2099–2115.

[33] Bayhan S, and Abu-Rub H. Smart energy management system for distributed generations in AC microgrid. In: *2019 IEEE 13th International Conference*

on Compatibility, Power Electronics and Power Engineering (CPE-POWERENG); 2019. pp. 1–5.
[34] Turan MT, Ates Y, Erdinc O, *et al.* Effect of distributed generation based campus model combined with electric vehicle charging stations on the distribution network. In: *2019 International Conference on Smart Energy Systems and Technologies (SEST)*; 2019. pp. 1–5.
[35] Menndez J, Fernndez-Oro JM, Galdo M, *et al.* Efficiency analysis of underground pumped storage hydropower plants. *Journal of Energy Storage*. 2020;28:101234. Available from: http://www.sciencedirect.com/science/article/pii/S2352152X20300323
[36] Kucevic D, Tepe B, Englberger S, *et al.* Standard battery energy storage system profiles: Analysis of various applications for stationary energy storage systems using a holistic simulation framework. *Journal of Energy Storage*. 2020;28:101077. Available from: http://www.sciencedirect.com/science/article/pii/S2352152X19309016
[37] Hodge BMS, Ela EG, and Milligan M. Characterizing and modeling wind power forecast errors from operational systems for use in wind integration planning studies. *Wind Engineering*. 2012;36(5):509–524. Available from: https://doi.org/10.1260/0309-524X.36.5.509
[38] Dileep G. A survey on smart grid technologies and applications. *Renewable Energy*. 2020;146:2589–2625. Available from: http://www.sciencedirect.com/science/article/pii/S0960148119312790
[39] Li X, Zhou Y, Yu S, *et al.* Urban heat island impacts on building energy consumption: A review of approaches and findings. *Energy*. 2019;174:407–419. Available from: http://www.sciencedirect.com/science/article/pii/S0360544219303895
[40] Huang P, Fan C, Zhang X, *et al.* A hierarchical coordinated demand response control for buildings with improved performances at building group. *Applied Energy*. 2019;242:684–694. Available from: http://www.sciencedirect.com/science/article/pii/S0306261919305574
[41] Baniassadi A, Heusinger J, and Sailor DJ. Energy efficiency vs resiliency to extreme heat and power outages: The role of evolving building energy codes. *Building and Environment*. 2018;139:86–94. Available from: http://www.sciencedirect.com/science/article/pii/S0360132318302841
[42] Probst O. Cooling load of buildings and code compliance. *Applied Energy*. 2004;77(2):171–186. Available from: http://www.sciencedirect.com/science/article/pii/S0306261903001090
[43] Cutsem OV, Dac DH, Boudou P, *et al.* Cooperative energy management of a community of smart-buildings: A blockchain approach. *International Journal of Electrical Power & Energy Systems*. 2020;117:105643. Available from: http://www.sciencedirect.com/science/article/pii/S0142061519323087
[44] Wang Y, Xu Y, and Tang Y. Distributed aggregation control of grid-interactive smart buildings for power system frequency support. *Applied Energy*. 2019;251:113371. Available from: http://www.sciencedirect.com/science/article/pii/S0306261919310451

[45] Yeazel SM. PURPAs midlife crisis: Will its 40th birthday be one of fundamental change? *The Electricity Journal*. 2018;31(5):9–21. Available from: http://www.sciencedirect.com/science/article/pii/S1040619018301234
[46] Hitaj C, and Lschel A. The impact of a feed-in tariff on wind power development in Germany. *Resource and Energy Economics*. 2019;57:18–35. Available from: http://www.sciencedirect.com/science/article/pii/S0928765517304165
[47] Leiren MD, and Reimer I. Historical institutionalist perspective on the shift from feed-in tariffs towards auctioning in German renewable energy policy. *Energy Research & Social Science*. 2018;43:33–40. Available from: https://www.sciencedirect.com/science/article/pii/S2214629618305152
[48] Kim JE, and Tang T. Preventing early lock-in with technology-specific policy designs: The Renewable Portfolio Standards and diversity in renewable energy technologies. *Renewable and Sustainable Energy Reviews*. 2020;123:109738. Available from: http://www.sciencedirect.com/science/article/pii/S1364032120300356
[49] Roos A, and Bolkesj TF. Value of demand flexibility on spot and reserve electricity markets in future power system with increased shares of variable renewable energy. *Energy*. 2018;144:207–217. Available from: http://www.sciencedirect.com/science/article/pii/S0360544217320042
[50] Kaza N. Urban form and transportation energy consumption. *Energy Policy*. 2020;136:111049. Available from: http://www.sciencedirect.com/science/article/pii/S0301421519306366
[51] Hansen K, Mathiesen BV, and Skov IR. Full energy system transition towards 100% renewable energy in Germany in 2050. *Renewable and Sustainable Energy Reviews*. 2019;102:1–13. Available from: http://www.sciencedirect.com/science/article/pii/S1364032118307913
[52] Miller G. Beyond 100% renewable: Policy and practical pathways to 24/7 renewable energy procurement. *The Electricity Journal*. 2020;33(2):106695. Available from: http://www.sciencedirect.com/science/article/pii/S1040619019303008

Chapter 2

Towards an electricity sector with 100% renewable energy generation

Mark Diesendorf[1]

Technical terms

Dispatchable technologies can supply power on demand, preferably with a short response time. All dispatchable sources have some kind of energy storage, e.g. dam, battery, thermal storage and fuel store.

Flexible technologies can be started and stopped rapidly at short notice and their power output or load can be varied rapidly to meet varying demand and to compensate for varying supply, e.g. hydro with dam, OCGT, battery, CST with thermal storage and demand response.

Reliability is a measure of the ability of the whole generation and transmission system to meet demand. Generation reliability is measured by, e.g. loss of load probability (the probability that supply fails to meet demand), or the proportion of unserved energy demand over a year.

Security is a measure of the ability of the power system to tolerate disturbances and hence maintain electricity supply to consumers.

A *baseload* power station is one that can operate 24/7 at close to its rated generating capacity, except when it breaks down or undergoes routine maintenance.

A *peak-load* power station is one that is flexible and dispatchable and is used to compensate for rapid variations in demand and supply by baseload and RElec power stations.

Penetration of RElec into the grid is defined in this chapter as the annual electricity generation from RElec divided by the total annual generation in the grid.

Capacity factor of a power station is its average power output (averaged over ≥ 1 years) divided by its generating capacity also known as rated power.

[1]School of Humanities and Languages, UNSW Sydney, Sydney, Australia

Once-through hydro denotes a hydro-electric system in which water flows one-way downhill, either from a dam or just down a river, to generate electricity.

Pumped hydro is a hydro-electric system in which water is circulated between an upper and lower reservoir, generating electricity on the downhill run.

Abbreviations

100RE	100% renewable energy (i.e. all energy, not limited to electricity)
100RElec	100% renewable electricity
100VRElec	100% variable renewable electricity
ACT	Australian Capital Territory
CST	Concentrated solar thermal
EV	Electric vehicle
FF	Fossil fuel
ICE	Internal combustion engine
LCOE	Levelised cost of energy
LOLP	Loss of load probability
NEM	National Electricity Market (of Australia)
OCGT	Open-cycle gas turbine
PV	Photovoltaics
RE	Renewable energy
RElec	Renewable electricity
VRElec	Variable renewable electricity

In large-scale electricity systems in many parts of the world, the levelised cost of energy generated by wind and solar photovoltaic (PV) farms is currently cheaper than that from new fossil fuelled and nuclear power stations and is still falling. Therefore, in many regions with low conventional hydro resources, wind and solar PV can economically provide the bulk of annual electricity generation, replacing baseload fossil fuelled and nuclear power stations. However, a grid powered entirely by renewable energy with high penetrations of wind and solar PV must contain a contribution to annual generation from flexible, dispatchable renewable power stations or other forms of storage, to maintain reliability of the whole generation system and to help control frequency. Flexible dispatchable renewable power stations all have some form of storage, whether mechanical, thermal, chemical, gravitational or electrical; they include hydro (both once-through and pumped), concentrated solar thermal with thermal storage, open-cycle gas turbines using renewable fuels and batteries. Both practical experience with grids operating for short periods (hours to days so far) on 100% renewable electricity and simulation models spanning a year or more of operations with real or synthetic hourly (or less) data on demand, wind, water and solar show that well-designed 100% renewable electricity systems, in which the majority of annual generation comes from variable renewables, can be designed to be reliable, to maintain frequency and to be economically competitive. Thus, a rapid transition is possible on technical and economic grounds, but strategic planning is needed and political barriers must be overcome.

2.1 Introduction

This chapter examines the technical and economic feasibility of transitioning large-scale electricity supply-demand systems to 100% renewable electricity (100RElec). By showing how 100RElec can satisfy the key criteria of reliability, security and affordability, and by arguing that a rapid transition is technically and economically possible this chapter refutes the principal myths propagated by critics of 100RElec.

Annual average generation of 80–100RElec is already well-established in countries and states/provinces with large hydro-electric resources, e.g. Iceland, Norway, New Zealand, Bhutan and Tasmania. However, providing a reliable 100RElec system is more challenging in regions that have little or no conventional (once-through) hydro potential and have large contributions from variable renewable electricity (VRElec), such as wind and solar photovoltaics (PV). Hence, this chapter focuses on 100RElec systems in which VRElec forms the major proportion of annual electricity generation.

Driven by economics and environmental concerns, the transition to 100RElec is occurring simultaneously on small, medium and large scales. For example, at the end of 2019, over 20% of Australian dwellings were powered, at least partially, by rooftop solar PV, with two Australian states, Queensland and South Australia, already at 35% [1]. At a medium scale, community RElec projects were until recently the principal contributors to the growth in RElec in Denmark (mostly wind, supplemented by bioenergy from agricultural residues) and Germany (wind supplemented by rooftop solar PV). In several states of the United States, large-scale wind and solar dominate the growing contributions of RElec to the grid. While recognising the importance of small- and medium-scale RElec in the transition, the focus of this book and this chapter is on large-scale systems, that is, transitioning the grid to 100RElec.

The focus on electricity, rather than renewable energy in general, deserves comment. A future low-carbon energy system is very likely to be one in which most energy is supplied in the form of electricity, the easiest form of energy to transform to renewable sources [2–4]. The latter two studies showed that most heat that is currently supplied to the industrial, commercial and residential sectors by direct combustion of fossil fuels (FF) can be provided more efficiently and with much less pollution by RElec, either via heat pumps or direct electrical heating. Furthermore, they showed that most transportation, whether public or private, can be provided by electric vehicles (EV). The principal exceptions are air and sea transport, together with long-distance rural road transport in remote areas, for which future energy is likely to be provided by synthetic fuels produced by using RElec to produce hydrogen by electrolysis or thermal decomposition and then combining it with nitrogen from the air to produce ammonia [5,6]. The combustion of hydrogen and ammonia yields no carbon emissions.

RElec is mainstream and it is no longer appropriate to call it 'alternative' energy. By the end of 2018, RElec supplied 26.2% of global electricity production, with 15.8% of global production derived from hydro (REN21 2019). In 2018, almost two-thirds (65.4%) of global investment in new power was in non-hydro RElec (US$272 billion). Together with 3.9% (US$16 billion) in hydro, total investment in RElec was approximately double total investment in FF electricity (22.8%) and nuclear electricity (7.9%) combined [7].

New VRElec is leading global investment in power capacity because it is already less expensive than new FF and much less expensive than nuclear electricity in many regions of the world [8–13]. Furthermore, the global weighted-average levelised cost of energy (LCOE) of on-shore wind and solar PV is expected to be less than the marginal *operating* costs of coal power in some regions by 2020. In other words, soon it will be economically viable to replace coal-fired power stations that have paid off their capital costs, with VRElec. The economic impact of adding sufficient storage/backup, to maintain reliability as the penetration into the grid of VRElec increases, is discussed in Section 2.3.4.

Complementing the economic benefits of RE are many environmental and social benefits. In addition to mitigating anthropogenic global climate change, RE reduces air and water pollution, respiratory and other diseases, water use and land degradation produced by FF use. It provides many new jobs. Once it has been constructed on a large scale, it can also stabilise energy costs because most RE technologies do not have a fuel cost.

The rapid progress of VRElec has disrupted existing energy technologies, both FF and nuclear, and triggered resistance by incumbents. Some incumbents and others have disseminated the following poorly substantiated criticisms of RElec via lobby groups, politicians and the media.

- Baseload power stations (code for coal and nuclear) are essential for the operation of electricity grids and VRElec cannot provide them.
- In the absence of baseload power stations, vast amounts of expensive storage are needed to maintain reliability of the system.
- Therefore, systems dominated by VRElec would be very expensive.
- Since historical energy transitions were very slow, the present transition must also be slow.

This chapter shows that these claims are either incorrect or, in the case of the last claim, unproven. To the contrary, well-designed 100RElec systems, in which the majority of electricity is generated by VRElec, can provide reliability, security and affordability, without baseload power stations and, in several regions, with relatively small amounts of storage. This is shown by both practical experience (Section 2.2) and computer simulation modelling (Section 2.3).

2.2 Progress in countries, states and cities

Table 2.1 shows the countries and states/provinces where VRElec supplied a large proportion of annual generation and the majority of RElec in 2017 or 2018.

Among countries, Denmark is the world leader in terms of per cent of VRElec in total generation and is on track towards its target of 100RElec and heat by 2035. Although it has negligible hydro potential within its borders, it has one of the highest levels of reliability and security of supply in Europe [19,20]. It balances the fluctuations in wind through its transmission interconnections to its neighbours, especially to hydro-electric storages in Norway, and bioenergy from its own agricultural residues. Attitudes to VRElec in the Danish electricity industry have evolved, as expressed by the following quotation from the Vice President Associated Activities, Energinet.dk:

Table 2.1 Countries and states with high proportions of VRElec generation, 2018

Region	VRElec as % of total generation or consumption	Comment
Country		
Denmark	49% of generation	Wind + small amount of PV Interconnected mainly to hydro storage in Norway
Scotland	40% of generation; 60% of consumption	VRElec mostly wind, supplemented solar PV. Dispatchable RElec: hydro 10% and biomass 5% of total generation; big electricity exports to England
Germany	25% of generation	Wind 17.3% and solar PV 7.2% of total generation. Dispatchable RElec: biomass 7%, hydro 2%; waste 1%
State		
Schleswig-Holstein, Germany	100 + % net	Wind; strong interconnection to neighbours
Mecklenburg-Vorpommen, Germany	100 + % net	Wind; strong interconnection to neighbours
South Australia, Australia	53%	Mostly wind, supplemented by solar PV. No hydro; weak interconnection to a neighbouring state
Iowa, USA	34%	Wind
Kansas, USA	36%	Wind
Oklahoma, USA	32%	Wind

Sources: Denmark [14]; Scotland [15]; Germany [16]; South Australia [17] and the USA [18].
Notes: 'Net' refers to annual averages, taking account of electricity trade with neighbouring jurisdictions. A 'dispatchable' technology can supply power on demand.

> We were once afraid of what would happen when wind energy generation reached 5% of the total consumption. We then worried about approaching 10% – would the system be able to cope? Some years later, we said that 20% had to be the absolute limit! However, in 2016, Danish wind turbines produced more than the total electricity consumption for 317 hours of the year, and we barely give this any thought [20].

Unfortunately, this understanding has not yet been reached in some other countries. Nevertheless, Scotland, Germany and several states or provinces have also made substantial progress in implementing VRElec.

In Scotland, VRElec supplied 40% of total electricity generation and the equivalent of 60% of electricity consumption in 2018. The difference results from Scotland's substantial exports of electricity to the rest of the UK. Wind generated 71% and solar 6% of Scotland's RElec generation. Already in 2017, it was reported that wind turbines in Scotland generated more electricity than the country needed for four consecutive days [21].

In Germany, the growth in VRElec is compensating for the phase-out of nuclear power. When the phase-out is completed in 2022, further growth in VRElec will begin to substitute for FF electricity. Two of Germany's northern states generate more than 100% VRElec (denoted hereinafter as 100VRElec) *net*, which means that they sell their excess electricity to their neighbours and buy electricity from their neighbours to fill gaps in local generation and so, averaged over the year, their generation is greater than 100VRElec.

South Australia, which has only low-capacity transmission connections to one neighbouring state, Victoria, has operated for several consecutive hours on 100VRElec. It has also operated reliably during three periods as an 'island'. However, as its annual average VRElec penetration approached 50%, the Australian Energy Market Operator (AEMO) imposed limits on the amount of wind generation permitted at any one time [22]. In addition, in times of high wind output, the state is required to ensure a certain capacity of gas generators remain on line. New synchronous condensers being installed may remove this need, leading to lower prices and less curtailment [23]. Meanwhile, with no potential for once-through hydro with dam, South Australia has installed several big batteries and is developing dispatchable renewables in the form of off-river pumped hydro.

On the scale of cities and towns, there are scores of projects involving local governments and community groups for either 100RElec or 100% RE (denoted as 100RE), the latter covering all energy [24]. The Australian Capital Territory (ACT), comprising the city of Canberra and a few farms, reached 100RElec *net* in 2020, based on its programme of reverse auctions with contracts-for-difference for wind farms in three states together with solar farms in their own territory and rooftop solar [25,26].

In addition, the Climate Group, in association with CDP, has a program called '100 RE' that is bringing together big businesses from around the world to commit to 100RElec [27]. Several companies – e.g. Google and Apple – have announced that their global facilities have already achieved that target with a combination of on-site generation and purchases of RElec from external wind and solar farms.

To sum up, several years ago, 100RElec, in which most electricity is VRElec, was widely regarded as a heroic aspirational target. However, in 2019, some medium-scale electricity grids with strong transmission interconnections to their neighbours, some towns and cities and a few large corporations had already achieved it in terms of annual average *net* generation. In addition, a few large-scale electricity grids have operated reliably on 100VRElec for periods of hours to days. Others are operating reliably, with VRElec penetrations of 30%–53% in terms of annual energy generation, and these penetrations are growing.

Next, to gain further insight into the feasibility of operating a large-scale electricity grid system continuously and reliably on 100RElec with majority of generation from VRElec, we consider scenarios and simulation modelling.

2.3 Scenario and simulation modelling: refuting the myths

2.3.1 Scenario and simulation modelling

The transition to 100RElec entails both building scenarios that span the period from the present or recent past to specified future years and simulating the operation of the electricity demand-supply system with various penetrations of RElec up to 100%. The simulations use hourly (or shorter) time steps spanning periods of one or more years. Since the vast majority of energy in a future 100RE system will be delivered in the form of RElec, some scenarios are needed which include the electrification of most of the heating and transport sectors (Section 2.1). The present section (Section 2.3) outlines the tool of simulation modelling and then discusses the contributions to global scenario studies and simulation modelling by two leading research groups.

Computer simulation modelling is a valuable tool for exploring three of the key challenges of 100RElec systems with high penetrations of VRElec: reliability, security and affordable system cost. In the simplest simulation models, real or synthetic time series data (usually hourly or half-hourly) is accessed on electricity demand and simultaneously on RE supply from wind, sunshine, water, etc. in the region of interest. Another necessary database comprises variable and dispatchable RElec power stations and other forms of storage that are appropriate for the region; this database includes their performance characteristics and, in some studies, their capital and operating costs. Dispatchable RElec technologies all have some form of storage, either mechanical, thermal, gravitational or chemical. Depending on weather in the region, they can be chosen, for example, from hydro with storage (both once-through and pumped), concentrated solar thermal (CST) with thermal storage, open-cycle gas turbines (OCGTs) with renewable fuels, batteries and compressed air. The model simulates the operation of the grid over one or more years and, at each time step, demand is balanced as far as possible with VRElec supply. Gaps in supply are filled with dispatchable renewables, other forms of storage and, in some models, demand response, to meet the reliability criterion of the system. Sensitivity analysis is conducted by varying the supply mix and strategies for using dispatchable renewables/storage and by modifying the configuration of transmission lines, e.g. adding key transmission links between areas of high wind and areas of high sunshine. Economic optimisation of the supply-demand system can be done either by performing thousands of runs with the above variations or by utilising an economic optimisation model at the outset. Some simulations are restricted to commercially available technologies, while others include technologies that look promising but are not yet economic.

Scores of simulation models have been published to date for states/provinces, countries, larger regions and the whole world. Selected lists are available in the studies by Ram and others [4,28–30].

One of the leading teams building energy scenarios and doing simulations of 100RElec systems is led by Mark Jacobson at Stanford University. In 2015, the team published a detailed scenario study showing how the whole US energy system (electricity, transportation, heating/cooling and industry) could operate on 100RE [3]. This group had previously performed detailed studies of all 50 US states. The 100RE technologies were based, directly, or indirectly via hydrogen, on RElec from wind, water and solar. The study addressed grid reliability with a simulation spanning 6 years. Wind and solar time series data with time steps of 30 s were obtained from a 3D global weather model that simulates real events. The study calculated both economic and social costs and found that 'the resulting 2050–2055 US electricity social cost for a full system is much less than for fossil fuels'.

A critique by Clack *et al.* claimed that 'this work used invalid modelling tools, contained modelling errors and made implausible and inadequately supported assumptions' [31]. The reply by Jacobson *et al.* [32] refuted the criticisms point by point and reaffirmed their original conclusions. In the view of the present author, Jacobson *et al.* [32] successfully refuted all but one of the Clack *et al.*'s [31] error claims. The exception is Jacobson's assumption of a huge and unrealistic increase in hydro capacity by installing additional turbines on existing dams, to assist in balancing variable RElec. However, this is a minor 'error', because the additional hydro could be replaced by alternatives such as CST with thermal storage, OCGTs fuelled by renewable hydrogen or ammonia and new off-river pumped hydro and batteries.

Jacobson *et al.* also published two papers on transitioning the energy sectors of 139 countries to 100RE based on wind, water, sunlight and storage by 2050. One paper aimed to provide transition roadmaps for all 139 countries [33], while the other matched demand and supply in 20 world regions and 139 countries within them [28]. Despite the emphasis on 139 roadmaps in [33], the primary value of this research was its top-down approach to a global transition. An important finding was that huge reductions in electricity demand (42% in 2050 compared with business-as-usual) can be achieved by:

1. the higher work to energy ratio of electricity over combustion;
2. eliminating energy industry self-use for the upstream mining, transporting and/or refining of coal, oil, gas, biofuels, bioenergy and uranium and

3. assumed policy-driven increases in end-use energy efficiency beyond those in the BAU case [32].

Subject to its assumptions, the paper also provides a valuable global view of the potential benefits of the transition in terms of mitigating carbon emissions, air pollution mortality and morbidity, the modest land area occupied by the RElec technologies and the reduced social cost of the transition. However, in the present author's view, the paper went too far in claiming that it provided roadmaps for the 139 individual countries. For many of these countries, other authors have published detailed scenario studies, often including simulation modelling of 100RElec, many of which were overlooked by Jacobson *et al.* [33] along with relevant geographical and demographic data. Inevitably top-down approaches can provide valuable macro views while failing to do justice to micro details.

Another leading team in building scenarios and performing simulations is based at Lappeenranta University of Technology. Recently, together with the Energy Watch Group, they published a major scenario study of the transition to a global 100RE system [4]. Like the global study by Jacobson *et al.* [33], they assumed the deep electrification of the heat and transport sectors. In addition, their study was the first to present a global 1.5°C scenario that is multiregional (9 major regions and 145 subregions) and cost optimal. By spanning the period 2015–2050 in 5-year time steps, the scenario allows the evolution of the transition to be tracked. An earlier study by the same group focused on the power sector used hourly time steps and ensured reliability via a constraint within the linear optimisation model used [34]. As a top-down model, the greatest value of this study for all energy consists of the calculations of huge savings in primary energy, substantial employment creation and economic aspects. Concerning the latter, this study found that the levelised cost of energy for the global 100RE system is approximately the same as for the current system; however, when the negative externalities of the FF system are included, the 100RE system is much less expensive, confirming the earlier result of Jacobson *et al.* [33].

Next, the issue of reliability is examined in more detail.

2.3.2 Reliability

Reliability is the ability of the whole electricity supply-demand system to meet demand. While systems in industrialised countries have generally a high level of reliability, no system is 100% reliable – which would require an infinite amount of backup.

Generation reliability is generally measured by either the loss of load probability (LOLP) – the expectation value of the probability that supply fails to meet demand – or by the proportion of unserved energy demand over a year or by the frequency and duration of failures to meet demand [35]. Some critics of RElec have ignored the standard electric power engineering definitions of reliability, invented their own definitions and then claimed incorrectly that 100RElec is 'unreliable'. In particular, Brook and Bradshaw [36, Table 3] selected, without justification, dispatchability of individual power stations as a qualitative indicator of performance of power stations, apparently assuming incorrectly that all power stations in a grid system have to be highly dispatchable to have a reliable system. Heard *et al.* [37, p. 1124] explicitly confused reliability with the presence of baseload power stations in the system:

> For this reason, generation that is constant (i.e., available at all times [baseload]) and/or fully dispatchable (able to be called-up or withdrawn at any time in response to demand changes) is deemed essential for system reliability.

This statement contains several additional egregious errors. Baseload power stations are inflexible in operation and so cannot be 'called up or withdrawn at any time'. System reliability is probabilistic concept, not deterministic. In addition, the authors ignored the fact that some of the simulation studies they purported to critique demonstrate reliable operation of 100% RElec systems that have high penetrations of VRElec without any baseload power stations, as discussed later.

The approaches in both studies [36,37], which have overlapping authors, are very similar: they set up arbitrary, inadequately justified criteria that disadvantage RElec compared with nuclear energy. This is a 'straw man' argument. Another example of this approach is Criterion 1 of Heard *et al.* [37], namely 'The electricity demand to which supply will be matched must be projected realistically over the future time interval of interest'. This criterion is true but redundant, because 100RElec simulations can be scaled up to any demand consistent with land-use constraints. There is ample land (and ocean area) available on Earth for many times the existing energy demand. It is true that some regions and countries lack sufficient land for many wind and solar farms (e.g. Japan, South Korea and the Netherlands) and that the world's renewable energy resource is not distributed uniformly. However, that is the same situation for FF and uranium. RElec is already traded by transmission line and, in future, liquid and gaseous fuels will be produced by using RElec and traded in tanker ships. Meanwhile, electricity demand is stabilising in the rich countries. The increase expected from the growth in EV may be offset to a large degree by improvements in energy efficiency (see Section 2.3.1) [4,33].

The erroneous approaches to reliability in the studies [36,37] and in other articles have been refuted in the peer-reviewed literature [29,30,38]. Computer simulations with hourly (or smaller) time steps spanning 1–6 years show that well-designed 100RElec systems, in which most electricity is generated by VRElec, can be as reliable as conventional systems without the presence of any baseload power stations and without vast amounts of storage. Reliability is achieved primarily by balancing the variations in the bulk electricity supply generated by VRElec with dispatchable RElec and other forms of storage. Reliability can be increased by:

- demand response, especially during periods of low supply and potentially high demand;
- having a wide variety of RElec technologies, if available;
- geographical dispersion of wind and solar farms;

- building a few key transmission links within the region; and
- interconnection with neighbouring regions where possible.

However, even in regions where external interconnection is infeasible, a high penetration of VRElec can be achieved while maintaining reliability. In the Australian National Electricity Market (NEM), which has no transmission connection to an external grid, simulation modelling by Elliston *et al.* [39] found that an optimal economic mix with 78% of annual electricity generation from VRElec (wind plus solar PV) together with 22% of annual generation from dispatchable RElec (CST, existing hydro and biofuelled gas turbines) can still meet the reliability criterion [39, Figure 2(c) and Table B.12] with additional storage equivalent to 2.2% of annual NEM generation. So, in an isolated grid with high solar and wind potential, relatively little storage is needed to main system reliability.

Independent simulation modelling of the NEM with 100RElec by Blakers *et al.* [40] achieved reliability with 90% VRElec balanced by 10% dispatchable RElec (comprising mostly new pumped hydro and existing bioenergy). This very high penetration of VRElec requires additional storage of only 0.22% of annual NEM generation.

As mentioned earlier, several of the simulation and other studies from around the world found that reliability can be achieved with zero or negligible baseload capacity, e.g. [39–45]. Furthermore, baseload power stations are poor partners for variable RElec, because of the former's relative inflexibility in operation [46,47]. Coal and nuclear stations are very slow to start from cold, taking a day or more, and most of them cannot respond rapidly enough to follow the daily variations in demand. That is why, we have peak-load power stations. Appropriate partners for VRElec are flexible, dispatchable power stations and other storage technologies [41,42], together with demand response [48,49].

Next we address the question of how a 100RElec system based predominantly on VRElec can maintain supply during rare periods of several days when there is simultaneously very little wind and solar power over a wide geographical region. Such periods have been named *Dunkelflaute* (dark doldrums) in Germany [50]. With reflection, one realises that the challenge is *not* to substitute for variable RElec continuously over several days because:

1. Solar PV and solar hot water systems with flat-plate collectors still operate during overcast conditions, although well below their rated or peak power.
2. The critical periods are generally of extent 1–3 h during the peaks in demand.

This entails that 100% backup of variable RElec with dispatchable RE or FF is unnecessary for maintaining reliability; the backup that is required only has to operate for several hours at a time and hence baseload power stations are not suited to the task. However, if the *Dunkelflaute* lasts for (say) 1 week in winter/summer, up to 14 demand peaks may have to be met, exhausting batteries and the small dams of off-river pumped hydro. In such rare events, OCGTs, reciprocating engines and contracted demand management can play a vital role. In the immediate future, OCGTs and reciprocating engines may have to operate on FF, but in the longer term they can run on renewable fuels such as biofuels, or hydrogen and ammonia produced by using renewable electricity.

2.3.3 Security

In electric power engineering, 'security' is a technical term denoting the ability of the power system to tolerate disturbances and hence maintain electricity supply to consumers [51]. Security is achieved by operating the system in a stable state and within the required bounds of a number of technical parameters such as frequency and voltage of alternating current, fault current levels and the operation of equipment within its design limits.

A disturbance in the system can result from, for example, the unexpected failure of a power station, the physical collapse or overloading of a major transmission line, or a sudden big change in demand. This in turn may cause an imbalance between supply and demand, resulting in a change in frequency and voltage of the alternating current. If supply is greater than demand, the frequency is greater than the specified level, which is 60 Hz (cycles per second) in the United States and 50 Hz in most other countries. Conversely, if supply is less than demand, the frequency is less than the specified level. In the past, frequency was generally maintained by the inertia of the heavy rotating turbines driven by boilers in conventional baseload power stations. With the gradual phase-out of baseload stations, new ways of controlling frequency are implemented. The following technologies and measures are commercially available for maintaining fixed frequency, but they require different time periods for widespread dissemination:

1. Dispatchable RElec technologies (e.g. CST, hydro with dam, OCGTs using renewable fuels).
2. VRElec technologies are 'dispatchable downwards' and so can help reduce frequency.
3. Contracted rapid demand response for critical periods can be expanded. It already exists for a few large industrial interruptible loads (e.g. aluminium smelters) and could be readily extended to households (e.g. water heating, air conditioners and refrigerator), commercial and small industrial consumers by rolling out 'smart' software, switches and other devices [49,51,52]. It is dispatchable both 'upwards' and 'downwards'.
4. At present, most inverters follow the grid frequency and so cannot in general contribute to frequency control or response. However, in the future, the capability may emerge for inverter technologies to set and constantly maintain frequency. Thus, wind, solar PV (including rooftop) and batteries will be able to provide both resilience and 'synthetic inertia' as opposed to the physical inertia provided by heavy rotating machinery [51,53].
5. If variable RElec technologies are operated slightly below their maximum possible output for prevailing weather conditions during periods when a disturbance appears likely, they can also increase their output almost instantly and maintain that increase for a brief period [53]. In that limited sense, they are briefly 'dispatchable upwards'. A small economic penalty results for operating VRElec in this manner.

6. Synchronous condensers can be installed in the grid. This is a well-established technology for providing reactive power and adjusting the phase difference between current and voltage (i.e. power factor) in industrial settings and on a transmission grid. It can control voltage and frequency [51].
7. Additional hydro generators can be installed on existing dams. The potential for this is limited in most regions.
8. Some regions with negligible potential for conventional once-through hydro may have considerable potential for off-river pumped hydro, especially at coastal sites. These dispatchable technologies will have small dams and large elevation differences, so that they can deliver high power for short periods (hours to days), the key periods for handling the variability of wind and solar PV [39,54,55].
9. Improved transmission interconnection between regions.

With the exception of batteries, the first five of these measures are inexpensive and the first six can be rolled out rapidly (0.5–2 years) as required. New off-river pumped hydro projects may take 3–5 years to plan and build. Major long-distance high-voltage transmission lines are expensive, very expensive where undergrounded, and can take much longer (7–10 years) to plan, acquire land easements, obtain environmental approvals and construct. However, increasing the capacity of existing transmission lines can generally be done much more rapidly.

2.3.4 System cost

Section 2.1 has cited several highly credible studies concluding that bulk electricity supply from wind and solar PV is much less expensive than from new FF and nuclear. For low penetrations of VRElec, even up to 45%–50% in some cases, existing balancing capacity in the form of peak-load plant (hydro and OCGT) can balance the fluctuations in VRElec and additional storage may not be needed [38,56]. However, as the penetration of VRElec grows, additional dispatchable RElec and other forms of storage will be increasingly required to maintain reliability and frequency of alternating current [38], and these technologies are more expensive than wind and solar PV [9,12]. The additional cost depends on the quantity and types of storage required, which in turn depend on the penetration of VRElec into the grid, the storage already present in the grid, e.g. in the form of reserve generation, the reliability and security criteria of the grid and the weather patterns and RE resources in the region of interest.

As discussed in Section 2.3.2, the quantity of storage to maintain high levels of reliability can be 2–3 orders of magnitude less than annual generation of the grid for high penetrations of VRElec. In a RElec grid, storage will be provided by dispatchable renewables and other forms of storage charged by VRElec. Although CST with thermal storage is still expensive, its LCOE has been almost halved since 2010 and continuing reductions in cost are expected as the market increases in size [12]. However, CST is limited to regions with an abundance of direct sunshine. OCGTs and reciprocating engines have low capital cost, typically US$800–1,000/kW, and if operated infrequently and for short periods, low annual fuel costs. In theory, they can operate as reliability insurance with a low premium [57,58]. However, their fuel costs in US$/MWh are relatively high and so, in an electricity market with bidding, they can push up the wholesale price of electricity [59, p. 8]. Batteries are expensive for storage capacities of more than a few hours; however, their costs will decline as their market grows. Meanwhile, even with just a few hours of storage, they are already playing a valuable role; because of their very fast response time, they can easily cover peaks in demand and contribute to frequency control [60–62]. In grids with high penetrations of VRElec, batteries will be part of a mix of different storage technologies with different response times, storage capacities and economic values.

One approach to calculate the LCOE of the whole generation system is to assume it to be the sum of the LCOEs of individual generation technologies. Then, in a system with a high penetration of VRElec, the cost of storage can be added either as a separate technology set or to each individual VRElec technology. In the view of the present author, the latter approach is likely to overestimate the amount of storage required and hence its cost, because the same storage technology at grid level can be used to manage the variability of different VRElec technologies at different times of the day, for instance, during the loss of solar PV generation at night and variations in wind power in daytime. Furthermore, given the method based on individual technologies, adjustments to individual LCOEs are needed to take account of the different climate risks of different technologies, in particular FF versus RE, and their different roles in the grid (e.g. bulk supply versus peak-load and different capacity factors of FF versus VRElec technologies) [11].

To take account of climate policy risk faced by FF generators, the GenCost study by CSIRO Australia and the AEMO calculates LCOE of FF technologies in three ways: with no adjustment, with a 5% premium added to their weighted average cost of capital (WACC) and with no WACC but with a carbon price added [11]. To take account of different capacity factors, the study adds to individual VRElec technologies a quantity of storage. The study also takes account of the declining cost of RElec technologies over time. The results are exhibited for 2020, 2030, 2040 and 2050 in Figures 4.2–4.5 by Graham *et al.* [11], respectively. One of the results is that in 2050 the LCOEs of wind and solar PV, with either 2 h of battery storage or 6 h of pumped hydro storage, are slightly cheaper than the *unadjusted* LCOEs of all the FF technologies with similar capacity factors and are much cheaper than both types of adjusted FF LCOEs. CST with 8 h of thermal storage has a similar cost to unadjusted FF and is cheaper than both types of adjusted FF costs [11].

2.3.5 *Timescale of transition*

Vaclav Smil has authored several articles and a book [63] on energy transitions. Based on his extensive, detailed, historical research, he stated [63, Chapter 5] that

> the process of restructuring the modern high-energy industrial and postindustrial civilisation on the basis of non-fossil, that is, overwhelmingly renewable, energy flows will be much more challenging that [sic] was replacing wood by coal and then coal by hydrocarbons.

To support his position, Smil made a number of assumptions, seven of which are critiqued in detail in the study by Diesendorf and Elliston [29]. Drawing upon the latter, this section summarises the refutations of Smil's two principal assumptions:

(1) Smil [63] asserted that 'changing the sources of electricity is much easier than changing the makeup of primary fuel supply' and 'most of the renewable targets defined by more than 160 countries apply only to electricity generation'. However, his focus on transitioning primary fuel supply put the cart before the horse. When RElec and energy efficiency reduce end-use electricity by a certain amount, they can substitute for approximately three times (range 2–4) that amount of energy in primary FF combusted for electricity generation, as discussed in Section 2.1. Transforming electricity is the key that unlocks the door to cutting primary energy and hence emissions.

(2) Smil also argued that the transition to RE is likely to be slow because it faces challenges similar to those of the slow historical energy transitions, namely that:
   1. The rate of increase in efficiencies of RE technologies is slow.
   2. The rapid decline in cost of RE technologies and their high annual growth rates are nothing special for the early stages of dissemination of new technologies.
   3. The transition to FF was faced with huge infrastructure requirements and the world is locked into them.

None of these reasons is convincing. Although RE technologies bring huge increases to the energy efficiency of the economy [3,4], their own increasing efficiencies are currently one of the several factors driving cost reductions. According to IRENA [12, p. 18], for wind turbines these factors include 'continued improvements in turbine design and manufacturing; more competitive global supply chains; and an increasing range of turbines designed to minimise LCOE in a range of operating conditions'; for solar PV, 'lower solar PV module prices and ongoing reductions in balance of system costs remain the main drivers of reductions in the cost of electricity' [12, p. 21]. In addition, both wind and solar PV have increasing capacity factors [12]. These improvements result from intensive, competitive research and rapid growth in market size. Note that the two key VRElec technologies, wind and solar PV, are economically competitive (Section 2.1), they have passed a tipping point and they are experiencing rapid market growth (Section 2.2). Rapidly expanding markets lead in turn to substantial economic savings in manufacturing processes and supply chains. Increasing experience in operation and maintenance leads to reductions in that cost. Reliability and security also have several low-cost options that are already being implemented as penetrations of VRElec increase (Section 2.3). The extraordinary speed of transition of electronic and artificial intelligence technologies offers a precedent. More than a precedent, smart technologies and systems are becoming an integral part of the transition [64]. For EVs, a tipping point is expected within several years as their capital costs approach those of equivalent internal combustion engine (ICE) vehicles and then they will rapidly take over the new car market, as expressed eloquently in a presentation by Seba [65]. EV operating costs are at present much less than those of ICE vehicles, making them already economically competitive for some fleets.

It is true that new infrastructure – especially transmission lines and a network of rapid charging stations for EVs – will be needed in a RE future, and that transmission lines take longer to build than VRElec power stations. However, much existing infrastructure can be used or adapted for the new energy future, e.g. existing transmission lines can be upgraded rapidly, without the need to acquire additional land, and trackless trams can use existing roads [66]. Later, the new essential infrastructure is already rolled out in some regions. Like it or not, some major FF infrastructure (e.g. coal mines, power stations and oil refineries) will become stranded assets [67].

Thus, the barriers to a rapid transition that Smil envisaged, based on historical experience, are unproven. Nevertheless, other barriers, identified by political science, must be overcome.

2.4 Conclusion

This chapter has shown that well-designed 100RElec grids, with very high penetrations of VRElec, can provide reliability and security of supply, are economically competitive and can be implemented rapidly provided governments cooperate with the wishes of the community at large. These grids do not need baseload power stations or, in several regions, large amounts of storage. Driven by favourable economics and the environmental crisis, the transition to 100RElec, and more generally to 100RE, is under way in many countries, states/provinces and towns and cities. It appears to be unstoppable. Many households, communities, corporations and utilities are already on the pathway to an ecologically and economically sustainable energy system based on RElec and energy efficiency. Several countries and states, and many towns and cities, are already aiming for 100RElec. The main barriers to a rapid transition are neither technical feasibility nor economic viability, but one of political will [4,29,68].

To leave the transition to an unguided market is a recipe for disruption. Government action is needed to address the barriers in physical infrastructure and institutions (laws, regulations and standards, organisations, financial mechanisms). To achieve a smooth transition that maintains reliability of supply and employment, strategic planning and effective government policies are needed. In practical terms, dispatchable renewables and other storage technologies, which are at present more expensive than wind and solar PV, must be encouraged with financial mechanisms and reformed rules for electricity market operation. Markets must include climate mitigation as an objective. They must be restructured to facilitate distributed generation and to avoid market control by a few powerful utilities. Key transmission links must be built. Overcoming some of these barriers is discussed in subsequent chapters.

While it is beyond the scope of this chapter to recommend specific policies, it does suggest the following broad strategy for transforming the socio-economy to 100RE:

- Rapidly replace FF electricity technologies with RElec technologies, including storage.
- Greatly increase energy efficiency of the built environment, appliances and equipment.
- Replace FF in gas and oil heating with electric heat pumps and direct electrical heating.
- Transition most transport from ICE to EV.
- Foster new cleaner industries in regions that were previously dependent on FF.
- Retrain and redeploy workers who are disadvantaged by the transition.
- In countries that have substantial RE resources, develop an industry to export RE, either in the form of RElec by transmission lines, or as renewable hydrogen or ammonia in tanker ships adapted from carrying liquefied natural gas.

While the first two of these goals can be completed before the others, all must be addressed together. Earth's climate is changing rapidly and science cannot predict when the tipping point into irreversible change will occur.

References

[1] Australian PV Institute (APVI). *Solar Map*. Available from https://pv-map.apvi.org.au/historical#4/-26.67/134.12 [Accessed February 2020].

[2] IEA. *Energy Technology Perspectives 2017*. Paris: OECD/IEA; 2017.

[3] Jacobson M.Z., Delucchi M.A., Cameron M.A., and Frew B.A. 'Low-cost solution to the grid reliability problem with 100% penetration of intermittent wind, water, and solar for all purposes'. *PNAS* 2015;112:15060–065.

[4] Ram M., Bogdanov D., Aghahosseini A., *et al. Global Energy System based on 100% Renewable Energy – Power, Heat, Transport and Desalination Sectors*. Lappeenranta & Berlin: Lappeenranta University of Technology and Energy Watch Group; 2019.

[5] Licht S., Cui B., Wang B., Li F-F., Lau J., and Liu S. 'Ammonia synthesis by N2 and steam electrolysis in molten hydroxide suspensions of nanoscale Fe_2O_3'. *Science* 2014;345:637–40.

[6] Bruce S., Temminghoff M., Hayward J., *et al. National Hydrogen Roadmap* [online]. Australia; CSIRO; 2018. Available from https://www.csiro.au/en/Do-business/Futures/Reports/Hydrogen-Roadmap/ [Accessed July 2019].

[7] REN21. *Renewables 2018 Global Status Report* [online]. 2018. Available from http://www.ren21.net/status-of-renewables/global-status-report/ [Accessed July 2019].

[8] Wiser R., and Bolinger M. 2018 *Wind Technologies Market Report*. US Department of Energy; 2019. Available from https://www.energy.gov/sites/prod/files/2019/08/f65/2018%20Wind%20Technologies%20Market%20Report%20FINAL.pdf/ [Accessed February 2020].

[9] Lazard. *Lazard's Levelized Cost of Energy Analysis, Version 13.0* [online]. 2019. Available from https://www.lazard.com/media/451086/lazards-levelized-cost-of-energy-version-130-vf.pdf [Accessed February 2020].

[10] Fu R., Feldman D., and Margolis R. *U.S. Solar Photovoltaic System Cost Benchmark: Q1 2018*. NREL/TP-6A20-72399; Golden, CO: National Renewable Energy Laboratory; 2018. Available from https://www.nrel.gov/docs/fy19osti/72399.pdf [Accessed June 2019].

[11] Graham P.W., Hayward J., Foster J., Story O., and Havas L. *GenCost 2018*. Australia: CSIRO; 2018.

[12] IRENA. *Renewable Power Generation Costs in 2018* [online]. Abu Dhabi: International Renewable Energy Agency; 2019. Available from https://www.irena.org/publications/2019/May/Renewable-power-generation-costs-in-2018/ [Accessed July 2019].

[13] Parkinson G. 'More stunning falls in solar and battery storage costs put fossil fuels on notice'. *RenewEconomy* 27 March 2019. Available from https://reneweconomy.com.au/more-stunning-falls-in-solar-and-battery-storage-costs-put-fossil-fuels-on-notice-31119/ [Accessed July 2019].

[14] Danish Energy Agency. *Energy Statistics 2018* [online]. Copenhagen: Danish Energy Agency; 2018. Available from https://ens.dk/en/our-services/statistics-data-key-figures-and-energy-maps/annual-and-monthly-statistics [Accessed February 2020].

[15] Scottish Renewables. *Renewables in Numbers* [online]. Available from https://www.scottishrenewables.com/forums/renewables-in-numbers/ [Accessed February 2020].

[16] Hein F., and Peter F. *The Energy Transition in the Power Sector: State of Affairs in 2019*. Berlin: Agora Energiewende; 2020. Available from https://www.agora-energiewende.de/en/events/energy-transition-in-the-european-power-sector-state-of-affairs-in-2019/ [Accessed February 2020].

[17] Clean Energy Council. *Clean Energy Australia Report 2019* [online]. Australia: Clean Energy Council; 2019. Available from https://www.cleanenergycouncil.org.au/resources/resources-hub/clean-energy-australia-report/ [Accessed July 2019].

[18] American Wind Energy Association. *Wind facts at a* glance [online]. Available from https://www.awea.org/wind-101/basics-of-wind-energy/wind-facts-at-a-glance [Accessed February 2020].

[19] Danish Energy Agency. *Security of Electricity Supply in Denmark* [online]. Danish Energy Agency; 2015. Available from https://ens.dk/sites/ens.dk/files/Globalcooperation/security_of_electricity_supply_in_denmark.pdf [Accessed July 2019].

[20] Jørgensen P. 'The Danish electricity grid is a European grid' in State of Green (ed.), *Wind Energy Moving Ahead: How Wind Energy has Changed the Danish Energy System* [online]. Denmark: State of Green; 2017. p. 8.

[21] Johnston I. 'Scotland's wind turbines provided more electricity than the country needed four days in a row' [online]. *The Independent* 9 January

2017. Available from http://www.independent.co.uk/environment/scotland-wind-turbines-more-electricity-country-needs-four-daysrenewable-energy-power-national-grid-a7517066.html/ [Accessed July 2019].

[22] Parkinson G. 'AEMO further relaxes limits on wind generation in South Australia' [online]. *RenewEconomy* 19 February 2019. Available from https://reneweconomy.com.au/aemo-further-relaxes-limits-on-wind-generation-in-south-australia-73634/ [Accessed July 2019].

[23] Parkinson G., and Vorrath S. 'South Australia has 10 GW wind and solar in pipeline as it heads to 100% renewables'. *RenewEconomy* 23 July 2019. Available from https://reneweconomy.com.au/south-australia-has-10gw-wind-and-solar-in-pipeline-as-it-heads-to-100-renewables-65116/?utm_source=RE+Dai00ly+Newsletter&utm_campaign=68faeb120f-EMAIL_CAMPAIGN_2019_07_23_05_04&utm_medium=email&utm_term=0_46a1943223-68faeb120f-24195345/ [Accessed July 2019].

[24] Go 100% Renewable Energy. http://www.go100percent.org/cms/index.php?id=21 [Accessed June 2019].

[25] ACT Government. *How do the ACT's reverse auctions work?* [online]. Available from https://www.environment.act.gov.au/energy/cleaner-energy/how-do-the-acts-renewable-energy-reverse-auctions-work [Accessed July 2019].

[26] Osmond D. 'Deep dive into the ACT's 100% renewable energy target' [online]. *RenewEconomy* 19 June 2019, https://reneweconomy.com.au/deep-dive-into-the-acts-100-renewable-energy-target-41592/ [Accessed July 2019].

[27] RE100. 'The world's most influential companies, committed to 100% renewable power'. Available from http://re100.org [Accessed July 2019].

[28] Jacobson M.Z., Delucchi M.A., Cameron M.A., and Mathiesen B.V. 'Matching demand with supply at low cost in 139 countries among 20 world regions with 100% intermittent wind, water, and sunlight (WWS) for all purposes'. *Renewable Energy* 2018;123:236–48.

[29] Diesendorf M., and Elliston B. 'The feasibility of 100% renewable electricity systems: a response to critics'. *Renewable & Sustainable Energy Reviews* 2018;93:318–30.

[30] Brown T.W., Bischof-Niemz T., Blok K., Breyer C., Lund H., and Mathiesen B.V. 'Response to "Burden of proof: A comprehensive review of the feasibility of 100% renewable-electricity systems"'. *Renewable & Sustainable Energy Reviews* 2018;92:834–47.

[31] Clack C.T.M., Qvist S.A., Apt J., *et al.* 'Evaluation of a proposal for reliable low-cost grid power with 100% wind, water, and solar'. *PNAS* 2017:201610381. doi:10.1073/PNAS.1610381114

[32] Jacobson M.Z., Delucchi M.A., Cameron M.A., and Frew B.A. 'The United States can keep the grid stable at low cost with 100% clean, renewable energy in all sectors despite inaccurate claims'. *PNAS* 2017;114:5021–3. doi:10.1073/pnas.1708069114

[33] Jacobson M.Z., Delucchi M.A., Bauer Z.A.F., *et al.* '100% clean and renewable wind, water, and sunlight all-sector energy roadmaps for 139 countries of the world'. *Joule* 2017;1:108–21. doi:10.1016/j.joule.2017.07.005

[34] Ram M., Bogdanov D., Aghahosseini A., *et al. Global Energy System based on 100% Renewable Energy – Power Sector*. Lappeenranta & Berlin: Lappeenranta University of Technology and Energy Watch Group; 2017.
[35] Čepin M. *Assessment of Power System Reliability: Methods and Applications*. Berlin: Springer; 2011.
[36] Brook B.W., and Bradshaw C.J.A. 'Key role for nuclear energy in global biodiversity conservation'. *Conservation Biology* 2015;29:702–12. doi:10.1111/cobi.12433
[37] Heard B.P., Brook B.W., Wigley T.M.L., and Bradshaw C.J.A. 'Burden of proof: A comprehensive review of the feasibility of 100% renewable-electricity systems'. *Renewable & Sustainable Energy Reviews* 2017;76:1122–33. doi:10.1016/j.rser.2017.03.114
[38] Diesendorf M. 'Subjective judgments in the nuclear energy debate'. *Conservation Biology* 2016;30:666–9. doi:10.1111/cobi.12692
[39] Elliston B., Riesz J., and MacGill I. 'What cost for more renewables? The incremental cost of renewable generation – An Australian National Electricity Market case study'. *Renewable Energy* 2016;95:127–39 doi:10.1016/j.renene.2016.03.080
[40] Blakers A., Lu B., and Stocks M. '100% renewable electricity in Australia'. *Energy* 2017;133:471–82.
[41] Heide D., Greiner M., von Bremen L., and Hoffmann C. 'Reduced storage and balancing needs in a fully renewable European power system with excess wind and solar power generation'. *Renewable Energy* 2011;36:2515–23. doi:10.1016/j.renene.2011.02.009
[42] Rasmussen M.G., Andresen G.B., and Greiner M. 'Storage and balancing synergies in a fully or highly renewable pan-European power system'. *Energy Policy* 2012;51:642–51. doi:10.1016/j.enpol.2012.09.009
[43] Elliston B., Diesendorf M., and MacGill I. 'Simulations of scenarios with 100% renewable electricity in the Australian National Electricity Market'. *Energy Policy* 2012;45:606–13. doi:10.1016/j.enpol.2012.03.011
[44] Elliston B., MacGill I., and Diesendorf M. 'Least cost 100% renewable electricity scenarios in the Australian National Electricity Market'. *Energy Policy* 2013;59:270–82. doi:10.1016/j.enpol.2013.03.038
[45] Elliston B., MacGill I., and Diesendorf M. 'Comparing least cost scenarios for 100% renewable electricity with low emission fossil fuel scenarios in the Australian National Electricity Market'. *Renewable Energy* 2014;66:196–204. doi:10.1016/j.renene.2013.12.010
[46] Troy N., Denny E., and O'Malley M. 'Base-load cycling on a system with significant wind penetration'. *IEEE Xplore Digital Library* 2010;25.
[47] Pfeifenberger J.P., Chang J., and Aydin M.G. *Advancing past 'Baseload' to a flexible grid*. Brattle Group; 2017. Available from http://www.brattle.com/system/publications/pdfs/000/005/456/original/Advancing_Past_Baseload_to_a_Flexible_Grid.pdf?1498246224/ [Accessed July 2019].
[48] Roscoe A., and Ault G. 'Supporting high penetrations of renewable generation via implementation of real-time electricity pricing and demand response'. *IET Digital Library* 2010;4:369–82.

[49] Aghaei J., and Alizadeh M.I. 'Demand response in smart electricity grids equipped with renewable energy sources: A review'. *Renewable & Sustainable Energy Reviews* 2013;18:64–72. doi:10.1016/j.rser.2012.09.019

[50] Huneke F., Linkenheil C.P., and Niggemeier M. *Kalte Dunkelflaute: Robustheit des Stromsystems bei Extremwetter (Cold Dark Doldrums: Robustness of the Current System in Extreme Weather)*. Berlin: Energy Brainpool; 2017. Available from https://www.energybrainpool.com/fileadmin/download/Studien/Studie_2017-06-26_GPE_Studie_Kalte-Dunkelflaute_Energy-Brainpool.pdf [Accessed July 2019].

[51] Finkel A., Moses K., Munro C., Effeney T., and O'Kane M. *Blueprint for the Future: Independent review into the future security of the National Electricity Market*. Canberra: Department of Environment & Energy; 2017. Available from https://www.energy.gov.au/government-priorities/energy-markets/independent-review-future-security-national-electricity-market [Accessed July 2019].

[52] Torriti J., Hassan M.G., and Leach M. 'Demand response experience in Europe: Policies, programmes and implementation'. *Energy* 2010;35:1575–83. doi:10.1016/j.energy.2009.05.021

[53] AEMO. *Wind Turbine Plant Capabilities Report*. Australian Energy Market Operator; 2013. Available from https://www.aemo.com.au/-/media/Files/PDF/Wind_Turbine_Plant_Capabilities_Report.pdf [Accessed July 2019].

[54] Lu B., Blakers A., and Stocks M. '90–100% renewable electricity for the South West Interconnected System of Western Australia'. *Energy* 2017;122:663–74. doi:10.1016/j.energy.2017.01.077

[55] MEI. *MEI and Partners Awarded ARENA Grant for Pumped Hydro Research* [online] http://energy.unimelb.edu.au/news-and-events/news/mei-and-partners-awarded-arena-grant-for-pumped-hydro-research/ [Accessed July 2019].

[56] Campey T., Bruce S., Yankos T., *et al. Low Emissions Technology Roadmap*. Australia: CSIRO; 2017. Available from https://www.csiro.au/en/Do-business/Futures/Reports/Low-Emissions-Technology-Roadmap [Accessed July 2017].

[57] Martin B., and Diesendorf M. 'Optimal thermal mix in electricity grids containing wind power'. *International Journal of Electrical Power and Energy Systems* 1982;4:155–61. http://dx.doi.org/10.1016/0142-0615(82)90044-8

[58] Martin B., and Diesendorf M. 'The economics of large-scale wind power in the UK: A model of an optimally mixed CEGB electricity grid'. *Energy Policy* 1983;11:259–66. http://dx.doi.org/10.1016/0301-4215(83)90082-4

[59] AEMO. *Quarterly Energy Dynamics Q1 2019*. Australian Energy Market Operator; 2019. Available from https://www.aemo.com.au/-/media/Files/Media_Centre/2019/QED-Q1-2019.pdf [Accessed July 2019].

[60] Parkinson G. 'Tesla big battery turns one, celebrates $50 million in grid savings'. *RenewEconomy* 30 November 2018. Available from https://reneweconomy.com.au/tesla-big-battery-turns-one-celebrates-50-million-in-grid-savings-95920/ [Accessed July 2018].

[61] Petkovic M. 'Tesla big battery at Hornsdale earns record revenue in September 2019'. *RenewEconomy* 21 October 2019. Available from

https://reneweconomy.com.au/tesla-big-battery-at-hornsdale-earns-record-revenue-in-september-2019/ [Accessed February 2020].

[62] Parkinson G. 'Wind and batteries saved the day when storm cut South Australia adrift'. *RenewEconomy* 14 February 2020. Available from https://reneweconomy.com.au/wind-and-batteries-saved-the-day-when-storm-cut-south-australia-adrift-22060/ [Accessed February 2020].

[63] Smil V. *Energy Transitions: Global and National Perspectives*. 2nd ed. Santa Barbara: Praeger; 2017.

[64] Andoni M., Robu V., Flynn D., *et al.* 'Blockchain technology in the energy sector: A systematic review of challenges and opportunities'. *Renewable & Sustainable Energy Reviews* 2019;100:143–74.

[65] Seba T. *Clean Disruption of Energy & Transportation* [online]. 2017. Available from https://www.youtube.com/watch?v=2b3ttqYDwF0 [Accessed July 2019].

[66] Newman P. 'Why trackless trams are ready to replace light rail'. *The Conversation* 26 September 2018. Available from https://theconversation.com/why-trackless-trams-are-ready-to-replace-light-rail-103690 [Accessed July 2019].

[67] Institute for Energy Economics & Financial Analysis. *Stranded Assets* [online]. Available from http://ieefa.org/tag/stranded-assets/ [Accessed July 2019].

[68] Hess D.J. 'Sustainability transitions: A political coalition perspective'. *Research Policy* 2014;43: 278–83.

Chapter 3

Sustainability perils and opportunities of clean electricity

Anders Arvesen[1]

Climate protection scenarios, such as those relied upon by the International Energy Agency and the IPCC, describe many possible future pathways to climate protection. While the pathways differ in many details, they share an emphasis on the importance of electricity. In the scenarios, it is the electricity sector that reduces its greenhouse gas emissions fastest and first; that is, before other sectors [1,2]. Furthermore, not only does the electricity sector cut its emissions first, it also helps to cut emissions in other sectors by 'electrifying' fuel uses in other sectors. One example of this is using electricity instead of petrol to fuel road transport vehicles. The increasing reliance of society on electricity in future climate protection scenarios is illustrated by Figure 3.1.

Given the scale and pace of the required transformation of electricity supply, it is important to understand the environmental consequences of different electricity generation options. With such understanding, we can better plan the selection and development of future electricity mixes. Later in this chapter, the author explores the results from life cycle assessments (LCAs) of electricity supply in the context of the UN Sustainable Development Goals (SDGs). The LCAs systematically consider the full life cycle – that is, the construction, operation and end of life – of electricity infrastructure, and activities occurring upstream in supply chains, such as mining, processing and transport of materials that go into electricity infrastructure. They also systematically translate effects of hundreds of pollutants and resource flows into a limited number of environmental impact indicator scores. In these ways, LCAs can provide a comprehensive picture of the environmental impacts associated with technology alternatives.

But before discussing LCA and environmental impacts of electricity production, the author deals with the subject of the increasing importance of electricity in future climate protection scenarios, and asks the question: Why? What are the fundamental reasons why, in the words of Williams *et al.* [5], electricity plays a

[1]Industrial Ecology Programme and Department of Energy and Process Engineering, Norwegian University of Science and Technology (NTNU), Trondheim, Norway

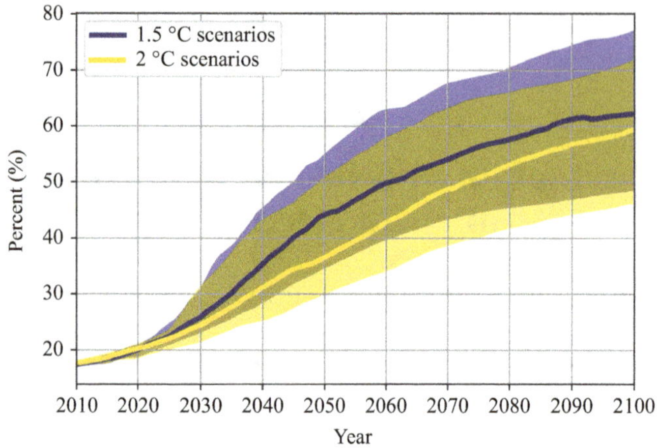

Figure 3.1 Share of electricity (%) in final energy (median and 10%–90% range) in 1.5 and 2 °C climate protection scenarios from the IPCC special report on global warming of 1.5 °C [3]. The figure is produced using data downloaded from the IAM 1.5 °C Scenario Explorer and Data [4]. It shows median values and 10%–90% ranges based on a total number of 84 scenarios at 1.5 °C and 131 scenarios at 2 °C

pivotal role in the path to deep greenhouse gas emissions cuts? The author attempts to answer this question in the section that follows.

3.1 Why the pivotal role of electricity in climate protection?

Here, the author tried to answer the question posed in the heading of this section: Why is electricity so important in climate protection scenarios? The answer has four parts, which are listed as A, B, C and D as follows:

(A) The physical world happens to be so that most climate-friendly energy options by their nature deliver electricity, not a combustible fuel. Wind, sunlight and natural uranium, as present in nature, are carriers of energy, but they cannot be burned, like fossil fuel resources can. In addition, the dominant technologies available for utilizing wind, solar and nuclear energy resources, by their design, generate electricity, not combustible fuels.* In this way, a push for climate-friendly energy in itself creates a push for increased use of electrical energy at the expense of combustible energy. This is one of

*Bioenergy is an exception. Unlike wind and solar energy, bioenergy is a combustible energy form relatively easily transformed into refined fuels (though depending on the type and quality of the resource).

the fundamental reasons why electricity is so important for the making of a climate-stable future.

(B) Renewable, nuclear or other climate-friendly electricity options are both diverse (i.e. different from each other and of various sorts), numerous (think about onshore wind, offshore wind, solar, hydro, nuclear, biomass, carbon capture) and largely based on energy resources that are geographically widespread (think about wind and solar energy in particular). In other words, there are both numerous choices and substantial flexibility in technology choice for generating electricity [2,6,7]. This makes climate-friendly electricity universally relevant despite variations in societal preferences and geographical resource availabilities.

(C) Then, besides the flexibility in choosing how to generate electricity, electricity itself is an energy carrier of extraordinary versatility and cleanliness at the point of use [8]. Just think how we can use electricity for any type of work – including charging your mobile phone – with zero emissions at the point of the use. Then imagine, as a contrast, a diesel-fuelled phone charger with exhaust fumes from a pipe. The versatility and cleanliness is the author's third reason why electricity is so important.

(D) Finally, of course there is also a cost dimension in understanding the growing importance of electricity: Electric energy is generally the cheapest to make climate-friendly. In comparison, emission reduction options that do not involve electric energy – such as introducing climate-friendly biofuels in transport, or capturing and storing carbon from fossil energy use in industry – tend to be more expensive [9,10]. Furthermore, costs of solar photovoltaic (PV) power and offshore wind power are declining rapidly and appear likely to continue to decline, as do the costs for batteries for electrical energy storage [11].

So, that is A, B, C – and D – on why the rise of electricity is happening and probably needs to happen to stabilize the climate. By these features, electricity as an energy carrier in itself can be regarded as a sustainability opportunity.

3.2 Environmental impacts of electricity generation

In this section, the author presents an overview of life cycle environmental impacts of different electricity production options, considering a broad set of impact types in the context of the UN SDGs. The impact results are derived from life cycle analyses conducted by Pehl *et al.* [12], Arvesen *et al.* [13], Luderer *et al.* [14] and Gibon *et al.* [15], in combination with sets of impact assessment methods developed for LCA (Recipe, 2012). The set of impact categories and application to the SDG context are original to this chapter. Table 3.1 lists the SDGs covered together with the related LCA impact categories and impact assessment methods used.

The analyses take into account the impacts not only caused directly by power stations (such as greenhouse gases emitted directly power stations), but also occurred over complete technology life cycles and in supply chain activities. In

Table 3.1 Overview of covered sustainable development goals (SDGs), related LCA impact categories and impact assessment methods

SDGs/subsection heading	LCA impact indicator	Impact assessment method
Climate action (SDG13)	Greenhouse gas emissions. Emissions covered: CO_2 (fossil and land use sources), CH_4, N_2O and other greenhouse gases	ReCiPe [48], midpoint hierarchist version (100-year global warming potential)
Good health and well-being (SDG3)	Pollution impacts (non-climate) on human health. Environmental mechanisms covered: stratospheric ozone depletion; particulate matter formation; photo-oxidant formation; human toxicity; ionizing radiation	ReCiPe [48], end point hierarchist version
Life below water (SDG14), life on land (SDG15)	Pollution impacts (non-climate) on ecosystems. Environmental mechanisms covered: freshwater eutrophication; terrestrial acidification; terrestrial, freshwater and marine ecotoxicity	ReCiPe [48], end point hierarchist version
	Land occupation impacts on ecosystems. Environmental mechanisms covered: urban and agricultural land occupation	ReCiPe [48], end point hierarchist version
Affordable and clean energy (SDG7), responsible consumption and production (SDG12)	Upstream investments of primary energy	Cumulative energy demand with losses excluded
	Mineral resource depletion	ReCiPe [48], midpoint hierarchist version

Note: In the following parts of this chapter, each set of 1–2 SDGs has its own subsection.

total, thousands of life cycle and supply chain activities and hundreds of pollutants and natural resource uses are considered. Results presented in this chapter are derived from future-oriented analyses of Pehl *et al.* [12], Arvesen *et al.* [13], Luderer *et al.* [14] and Gibon *et al.* [15], and are assumed to be representative for the year 2030 in a 2 °C scenario.† Note that for nuclear power, interim storage and final repositories of nuclear waste are considered based on Dones *et al.* [17], but with high uncertainty related to the extremely long-time horizon of nuclear waste disposal projects [18,19].

Some types of environmental or human health concerns are not captured or poorly captured by the LCA quantifications, as specific and/or agreed-upon methods for quantification do not exist. Examples of concerns that are not specifically treated are bird and bat mortality related to wind turbines, psychological effects of nuclear projects, local ecological effects of hydropower and visual intrusion of renewable energy projects in landscapes. Discussions of these and other factors are available elsewhere [20–22]. The quantifications presented in this chapter assume normal operation; that is, they do not attempt to quantify effects of accidents, such as nuclear meltdown or hydropower dam failures.

In the subsections that follow, life cycle environmental impact results representative for four sets of SDGs are discussed.

3.2.1 Climate action (SDG13)

The life cycle greenhouse gas emission results displayed in Figure 3.2 are dominated by the emissions of fossil fuel-based power, in particular coal power, and negative emissions for bioelectricity with carbon capture and storage (CCS). The contributions of electricity from solar, hydro, wind and nuclear energy sources are barely visible in the figure, as these contributions are 1–2 order of magnitude smaller than the contributions from coal and gas power. In fact, even the indirect emissions of the fossil fuel-based options – that is, the emissions occurring indirectly in the extraction, processing and transport of the fuels to power stations – are comparable and often higher than the complete life cycle emissions of the solar, hydro, wind and nuclear options.‡

A remote location of one of the hydropower cases investigated lead to a relatively high maximum value for hydropower (Figure 3.2). The characteristics of new hydropower installations in the future seem somewhat uncertain, as relatively high

†Scenarios for year 2030 for changes in electricity mixes and other technology changes are fed into the calculations. The electricity mixes originate from the IEA BLUE Map 2 °C scenario [15]. Besides electricity mixes, technology improvements are particularly significant for solar PV (based on [16]) and biopower [12,13] technologies.

‡This is largely owing to losses and leaks of methane to the atmosphere associated with natural gas extraction and supply and CO_2 emissions associated with coal transport. Opportunities exist for reducing methane releases from natural gas supply [23–25], but they are not considered in the results of this chapter.

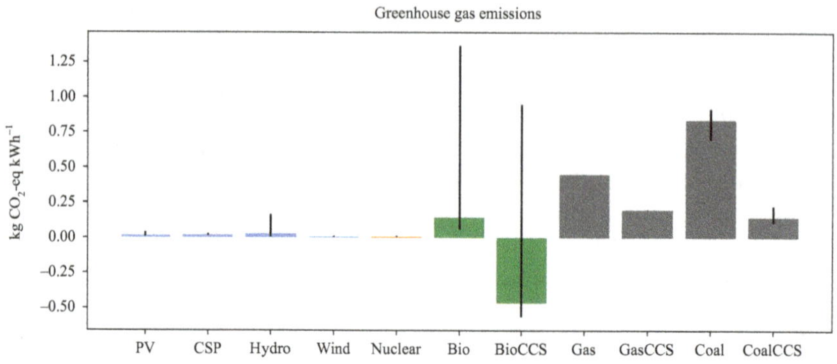

Figure 3.2 Life cycle greenhouse gas emissions for 11 main types of electricity production for the year 2030. Derived from the analyses of Pehl et al. [12] and Arvesen et al. [13]. Averages represent weighted average for specific technology variants (e.g. six different types of PV) and world regions, adopting assumed detailed technology market shares of Arvesen et al. [13]. Average values for biopower assume a 50–50 split of biomass from forest residues and lignocellulosic biomass from purpose-grown energy plants. The ranges reflect variation in results for different specific technologies and regions investigated. Note that a wider set of specific technologies were investigated for some main types of electricity production (e.g. six types of PV electricity) than for others (e.g. only one type of gas power). For biopower, the ranges additionally reflect different assumptions about land management [12]. Remote location of one of the hydropower cases investigated lead to a relatively high maximum value for hydropower. CSP: concentrated solar power; CCS: carbon capture and storage

shares of hydropower potentials have already been exploited and more difficult-to-access sites may have to be used in the future [8,26].

The use of CCS for coal and gas power eliminates most of the CO_2 that is emitted directly by power stations without CCS, but leaves residual direct CO_2 emissions because the carbon capture is incomplete. Besides, CCS does not reduce the greenhouse gas emissions occurring in the fuel supply chains.

For electricity derived from biomass, life cycle greenhouse gas emissions can be broadly categorized into two main types of sources: (1) fossil fuel-burning in supply chain activities, including use of machinery, irrigation and production of fertilizers and (2) non-fossil CO_2 emissions related to direct and indirect land use changes. In addition, nitrous oxide emissions stemming from nitrogen fertilizer use in bioenergy crops also contribute. The large ranges for biopower (Figure 3.2) reflect high uncertainty as well as geographical variability in model estimates of

CO_2 emissions from land use changes [12]. Uniquely, bioenergy – when combined with CCS – presents an opportunity for negative greenhouse gas emissions.

Finally, some hydropower dams cause substantial greenhouse gas emissions in the form of methane produced in biomass decomposition [27,28]. Such emissions are omitted in Figure 3.2 owing to high variability of emissions between dams and difficulty in establishing a global average.

3.2.2 Good health and well-being (SDG3)

Figure 3.3 shows estimated pollution-related human health impacts of electricity production, considering negative effects on human health from particulate matter formation, human toxicity and three other environmental mechanisms (but excluding climate change impacts on human health; see Table 3.1). The figure shows that coal power, irrespective of whether CCS is used or not, causes the highest non-climate impacts on human health. Next follows biopower with or without CCS, causing about half the non-climate human health impacts as coal (but note also the wide ranges for biopower, indicating high uncertainty and variability across regions). The main culprits behind these impacts for coal and biopower are emissions of particulate matter, and, for coal power, releases of toxic heavy metals from coal mine waste disposal.

According to the estimates, other electricity production options than coal and biopower are less damaging to human health (Figure 3.3). Toxic releases from copper mining waste are responsible for large parts of the non-climate human health impacts for solar and wind power, while toxic releases and ionizing radiation in connection to uranium mining are the main causes for nuclear power. In general, toxicity impact estimates have high uncertainty, owing to large number of substances contributing to impacts (much larger numbers than for air pollution impacts), and relatively immature impact assessment methods.

Notably, Figure 3.3 does not capture risks of accidents. For example, hydropower carries risks of dam failures; coal power carries risks of coal-mining accidents; wind and solar power carry risk of metals-mining accidents and explosions can occur in natural gas distribution systems [29–31]. The maximum possible severity of accidents seems particularly high for nuclear and perhaps hydropower. Besides accidents, concerns not covered by Figure 3.3 are reductions in well-being due to the noise or visual intrusion in landscapes of wind turbines [32], psychological effects of nuclear power [33], displacements of settlements in connection to new hydropower projects [21] and competition over land between food crops and bioenergy crops [34,35]. Overall, these issues range from largely subjective issues (e.g. perceived visual aesthetics) to more objective issues (e.g. number of people relocated).

3.2.3 Life below water (SDG14) and life on land (SDG15)

Analogously to the case of human health impacts shown in Figure 3.3, Figure 3.4(a) shows estimated impacts on ecosystem quality due to various forms of pollution (excluding effects of climate change; see Table 3.1). Coal power with or without

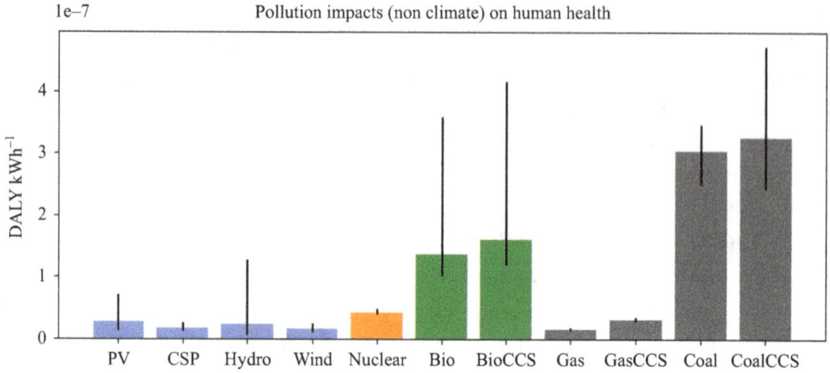

Figure 3.3 Life cycle pollution impacts on human health (excluding climate change impacts on human health) for 11 main types of electricity production for the year 2030. Derived from the analyses of Arvesen et al. [13] and Luderer et al. [14]. For interpretation of averages and ranges, see caption for Figure 3.2. CSP: Concentrated solar power

CCS exhibits the highest impacts in Figure 3.4(a), followed by biopower with or without CCS (but note here also the wide ranges for biopower). PV, CSP, hydro, wind and nuclear energy sources show lower impacts. It is noteworthy that Figure 3.3 (pollution impacts on human health) and Figure 3.4(a) (pollution impacts on ecosystems) look very similar, despite that different environmental mechanisms are modelled in both cases (see Table 3.1).

Pollution-related ecosystem impacts of coal power (Figure 3.4(a)) are mainly caused by two mechanisms: release of phosphorous nutrients from coal mining waste to freshwater bodies (causing freshwater eutrophication) and release of toxic heavy metals from coal mining waste (causing toxic effects in terrestrial, freshwater and marine ecosystems). Similarly, for biopower, release of phosphorous nutrients from bioenergy crops (causing freshwater eutrophication)[§] and use of pesticides in bioenergy crops (causing toxic effects on ecosystems) are the chief causes.

Land use-related impacts on ecosystem quality are shown in Figure 3.4(b). The most striking feature of Figure 3.4(b) is probably the high maximum values for biopower, which are attributable to bioenergy crops with low yields (i.e. low production of biomass per unit of crop land) for certain modelled scenarios or world regions (see the original analysis of Pehl *et al.* [12]). These maximums are highly

[§]Effects of releases of nitrogen nutrients from bioenergy crops are omitted as no suitable impact assessment method was available at the time the analysis was conducted. For this reason, pollution-related ecosystem impacts of bioenergy are likely underestimated.

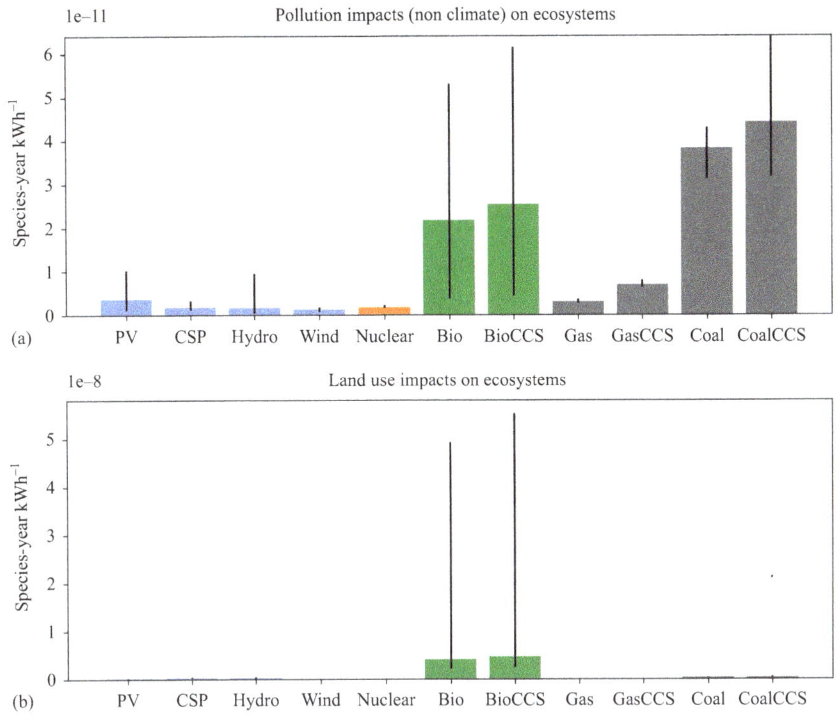

Figure 3.4 Life cycle pollution impacts on ecosystems (excluding climate change impacts on ecosystems) (a) and land use impacts on ecosystems (b) for 11 main types of electricity production for the year 2030. Derived from the analysis of Arvesen et al. [13] and Luderer et al. [14]. For interpretation of averages and ranges, see caption for Figure 3.2. CSP: Concentrated solar power

uncertain and are also to some extent artefacts of modelling choices. But even if we look only at the average values in Figure 3.4(b), which can be taken as rough 'default' estimates, the results indicate that land use impacts of biopower exceed that of any other power generation options by 1–3 order of magnitude. In the literature, concerns about the potential land use and biodiversity impacts associated with future increases in bioenergy use have been raised by many authors (e.g. Luderer *et al.* [14], Humpenöder *et al.* [35], Smith *et al.* [36], Gibon *et al.* [37], van Vuuren *et al.* [38]).

Land uses such as reservoirs for hydropower projects, open-pit coal mines and ground-mounted solar power plant areas are taken into consideration in these results. For wind power located onshore, only the areas used exclusively by turbine foundations and access roads are counted. The reason for this choice is that the spacing between turbines can be used by humans for other purposes or by

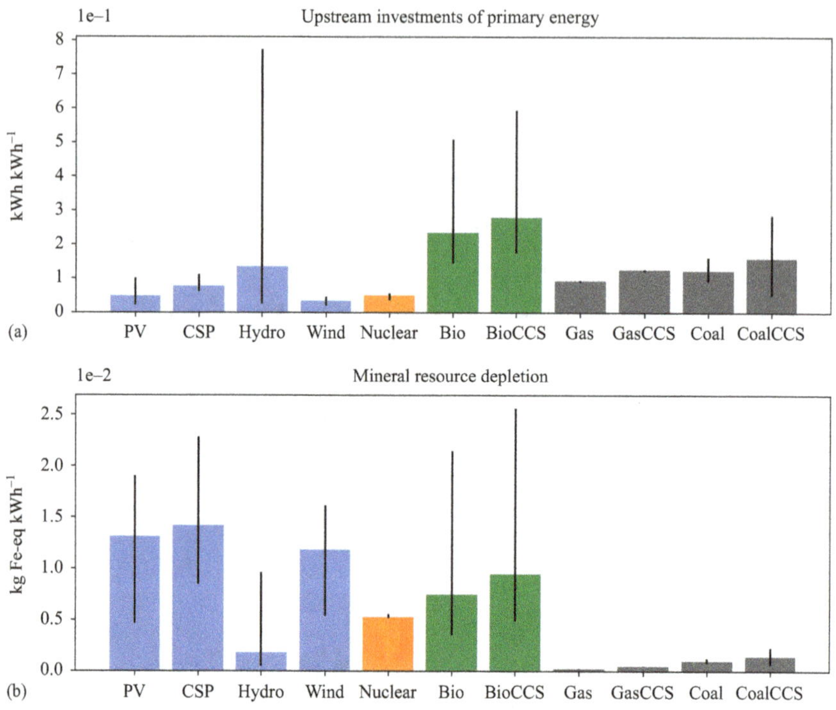

Figure 3.5 Life cycle investments of primary energy (excluding energy losses in the final conversion to electricity) (a) and mineral resource depletion impacts (b) for 11 main types of electricity production for the year 2030. Derived from the analysis of Arvesen et al. [13]. For interpretation of averages and ranges, see caption for Figure 3.2. CSP, Concentrated solar power

terrestrial wildlife. A wind farm area can, within some limits, coexist with agricultural crops, animal grazing or wildlife. The same cannot be said for the land used by open-pit coal mines or bioenergy crops. At the same time, a much larger wind farm area could also be regarded as impacted for some species, including bird and bat species [32,39].

3.2.4 *Affordable and clean energy (SDG7) and responsible consumption and production (SDG12)*

In Figure 3.5, upstream investments of primary energy (Figure 3.5(a)) and mineral resource depletion (Figure 3.5(b)) are used as two indicators loosely representative of SDG7, affordable and clean energy, and SDG12, responsible consumption and production.

The primary energy indicator shown in Figure 3.5(a) measures the ratio between life cycle primary energy invested by society to produce electricity and the actual electricity produced. Here, energy conversion losses occurring directly in the studied power plants are not counted as energy investments, based on the logic that such losses are, exactly, losses of energy, and not investment of energy.¶ Low energy investments per unit of energy supply are often thought to be positive for economic and human development [40,42–44].

As is evident from Figure 3.5(a), the magnitudes of total upstream investments of energy are comparable across the different electricity production options, but with a tendency for electricity from combustible energy sources (i.e. bio, gas and coal) to require more energy than electricity from solar, hydro, wind and nuclear energy sources. In other words, the energy requirements to extract, process and distribute the combustible fuels tend to exceed the energy requirements associated with building wind and solar power infrastructure. For the case of biopower, irrigation and production of fertilizers can also cause substantial energy use. Broadly similar conclusions – though with some variations in results, in particular for gas power – are drawn by Raugei and Leccisi [41].

The mineral resource depletion indicator (Figure 3.5(b)) aggregates contributions of 20 important metals to resource depletion, based on considerations of marginal cost increases per unit of metal extracted. Among all impact categories investigated in this chapter (Figures 3.2–3.5), mineral resource depletion is the only category where PV, CSP and wind power show the highest impacts, and the fossil fuel-based options show the lowest impacts. This is in large part attributable to substantial copper requirements of solar and wind power plants, as well as iron, nickel and other elements used for steel materials.

There is no widely agreed-upon method for measuring resource depletion, and quantifications of resource depletion reported here should be regarded as having high uncertainty. Nonetheless, it is reasonable to conclude that solar and wind power require more metals than fossil fuel-based power; this conclusion is also supported by literature quantifying life cycle individual metal requirements of power generation options [45,46]. It is also worth mentioning that electricity systems have been found to require more metals than combustible fuel energy systems [47].

3.3 Concluding remarks

In this chapter, the author has noted the crucial role of electricity in climate protection scenarios and attempted to answer why electricity possess this crucial role. The author has argued that electricity in itself can be seen as a climate protection opportunity – and thus a sustainability opportunity.

¶The upstream investments of primary energy indicator (Figure 3.5(a)) is the inverse of the energy return on investment (EROI) indicator used in literature (e.g. [40,41]). In the current chapter, the direct equivalent method [8] is used for primary energy accounting.

Then, based on environmental impact results from LCAs, the author has discussed sustainability merits of alternative ways of producing electricity. Key insights from this discussion are:

- Electricity from solar, wind and nuclear energy show (very) low pollution-related impacts for all investigated pollution impact categories – that is, greenhouse gas emissions, pollution impacts on human health and pollution impacts on ecosystems. Electricity from solar, wind and nuclear thus represents a major opportunity to reduce not only greenhouse gas emissions but also pollution-related impacts generally.
- Fossil fuel-based electricity combined with CCS reduces greenhouse emissions compared with straight fossil fuel-based electricity, but it does not offer similar opportunities as solar, wind and nuclear electricity for reducing other forms of pollution.
- Renewable energy-based electricity systems tend to require more metals than their fossil counterparts. Constrained supply of the required metal resources is thus a possible risk factor. The overall significance of the issue in the context of global climate protection and sustainability efforts seems difficult to ascertain.
- Uniquely, bioenergy combined with CCS presents an opportunity to produce energy with negative CO_2 emissions, a tremendously attractive prospect from a climate protection perspective. However, this attractive prospect is mirrored by two serious risks: ecological damages associated with the land use of bioenergy crops and greenhouse gas emissions stemming from the land use of bioenergy crops.

Finally, no electricity production is without environmental concerns. This statement is confirmed by Figures 3.2–3.5 to some degree, and further confirmation can be obtained if we also consider concerns not addressed by Figures 3.2–3.5. Examples of such concerns not addressed by Figures 3.2–3.5 include water use of thermal power generation or bioenergy crops, geological storage space requirements associated with CO_2 storage, reductions in the visual quality of landscapes due to wind turbines or negative effects of hydropower projects on ecosystems [14,20,21,47]. Thus, despite the substantial flexibility in choosing technologies for producing electricity, there is a danger that future transformations to climate-friendly electricity are constrained by other environmental objectives than climate.

Acknowledgement

The author acknowledged support from the Research Council of Norway (Grant No. 209697).

Appendix A

Table A1 Numerical values for average results presented in Figures 3.2–3.5

Figure:	**Figure 3.2**	**Figure 3.3**	**Figure 3.4 (a)**	**Figure 3.4 (b)**	**Figure 3.5 (a)**	**Figure 3.5 (b)**
Unit:	kg CO_2-eq.	DALY	species-year		kWh	kg Fe-eq.
PV	1.7E−02	2.8E−08	3.6E−12	1.1E−10	4.8E−02	1.3E−02
CSP	2.0E−02	1.8E−08	1.7E−12	2.9E−10	7.7E−02	1.4E−02
Hydro	2.9E−02	2.4E−08	1.7E−12	2.7E−10	1.3E−01	1.8E−03
Wind	8.5E−03	1.7E−08	1.2E−12	6.8E−12	3.4E−02	1.2E−02
Nuclear	7.7E−03	4.3E−08	1.7E−12	1.2E−11	5.0E−02	5.3E−03
Bio	1.4E−01	1.4E−07	2.2E−11	4.1E−09	2.3E−01	7.5E−03
BioCCS	−4.7E−01	1.6E−07	2.5E−11	4.7E−09	2.8E−01	9.5E−03
Gas	4.5E−01	1.6E−08	3.0E−12	6.4E−12	9.2E−02	2.3E−04
GasCCS	2.0E−01	3.2E−08	6.9E−12	9.7E−12	1.2E−01	4.7E−04
Coal	8.4E−01	3.0E−07	3.8E−11	3.5E−10	1.2E−01	9.8E−04
CoalCCS	1.5E−01	3.3E−07	4.4E−11	3.8E−10	1.6E−01	1.4E−03

Note: All values are measured per kWh of electricity produced (e.g. for Figure 3.2, values are in kg CO_2-equivalents/kWh).

References

[1] Luderer, G., Kriegler, E., Delsa, L., *et al.*, 2016. *Deep Decarbonisation towards 1.5 °C–2 °C Stabilisation. Policy Findings from the ADVANCE Project (first edition)*. Available from: http://www.fp7-advance.eu/sites/default/files/documents/WP7/ADVANCE-Synthesis-Report.pdf [accessed February 10, 2020].

[2] Rogelj, J., Shindell, D., Jiang, K., *et al.*, 2018. *Mitigation Pathways Compatible with 1.5 °C in the Context of Sustainable Development*, Special Report on the Impacts of Global Warming of 1.5 °C. Intergovernmental Panel on Climate Change, Geneva.

[3] IPCC, 2018. *Global Warming of 1.5 °C*. An IPCC Special Report on the Impacts of Global Warming of 1.5 °C Above Pre-Industrial Levels and Related Global Greenhouse Gas Emission Pathways, in the Context of Strengthening the Global Response to the Threat of Climate Change, Sustainable Development, and Efforts to Eradicate Poverty.

[4] Huppmann, D., Kriegler, E., Krey, V., *et al.*, 2018. *IAMC 1.5 °C Scenario Explorer and Data hosted by IIASA. Version 1.1.* Integrated Assessment Modeling Consortium & International Institute for Applied Systems Analysis.

[5] Williams, J.H., DeBenedictis, A., Ghanadan, R., *et al.*, 2012. The technology path to deep greenhouse gas emissions cuts by 2050: The pivotal role of electricity. *Science* 335, 53–59.

[6] Jenkins, J.D., Luke, M., and Thernstrom, S., 2018. Getting to zero carbon emissions in the electric power sector. *Joule* 2, 2498–2510.

[7] Luderer, G., Krey, V., Calvin, K., *et al.*, 2014. The role of renewable energy in climate stabilization: results from the EMF27 scenarios. *Climatic Change* 123, 427–441.

[8] Grübler, A., Johansson, T.B., Mundaca, L., *et al.*, 2012. Chapter 1: Energy primer. In: *Global Energy Assessment – Toward a Sustainable Future*. Cambridge University Press, Cambridge, UK and New York, NY, USA and the International Institute for Applied Systems Analysis, Laxenburg, Austria. Available from: http://www.iiasa.ac.at/web/home/research/researchPrograms/Energy/Home-GEA.en.html [accessed June 6, 2018].

[9] Kriegler, E., Weyant, J.P., Blanford, G.J., *et al.*, 2014. The role of technology for achieving climate policy objectives: overview of the EMF 27 study on global technology and climate policy strategies. *Climatic Change* 123, 353–367.

[10] Luderer, G., Vrontisi, Z., Bertram, C., *et al.*, 2018. Residual fossil CO_2 emissions in 1.5–2 °C pathways. *Nature Climate Change* 8, 626–633.

[11] Nykvist, B., and Nilsson, M., 2015. Rapidly falling costs of battery packs for electric vehicles. *Nature Clim. Change* 5, 329–332.

[12] Pehl, M., Arvesen, A., Humpenöder, F., Popp, A., Hertwich, E.G., and Luderer, G., 2017. Understanding future emissions from low-carbon power systems by integration of life cycle assessment and integrated energy modelling. *Nature Energy* 2, 939–945.

[13] Arvesen, A., Luderer, G., Pehl, M., Bodirsky, B.L., and Hertwich, E.G., 2018. Deriving life cycle assessment coefficients for application in integrated assessment modelling. *Environmental Modelling & Software* 99, 111–125.

[14] Luderer, G., Pehl, M., Arvesen, A., *et al.*, 2019. Environmental co-benefits and adverse side-effects of alternative power sector decarbonization strategies. *Nature Communications* 10, 5229.

[15] Gibon, T., Wood, R., Arvesen, A., Bergesen, J.D., Suh, S., and Hertwich, E.G., 2015. A methodology for integrated, multiregional life cycle assessment scenarios under large-scale technological change. *Environmental Science & Technology* 49, 11218–11226.

[16] Bergesen, J.D., Heath, G.A., Gibon, T., and Suh, S., 2014. Thin-film photovoltaic power generation offers decreasing greenhouse gas emissions and increasing environmental co-benefits in the long term. *Environmental Science & Technology* 48, 9834–9843.

[17] Dones, R., Heck, T., Faist Emmenegger, M., and Jungbluth, N., 2005. Life cycle inventories for the nuclear and natural gas energy systems, and examples of uncertainty analysis (14 pp). *The International Journal of Life Cycle Assessment* 10, 10–23.

[18] Warner, E.S., and Heath, G.A., 2012. Life cycle greenhouse gas emissions of nuclear electricity generation. *Journal of Industrial Ecology* 16, S73–S92.

[19] Barron, R.W., and Hill, M.C., 2019. A wedge or a weight? Critically examining nuclear power's viability as a low carbon energy source from an intergenerational perspective. *Energy Research & Social Science* 50, 7–17.

[20] Hertwich, E.G., Aloisi de Larderel, J., Arvesen, A., *et al.* (Eds.), 2016. *Green Energy Choices. The Benefits, Risks, and Trade-Offs of Low-Carbon Technologies for Electricity Production*. United Nations Environment Programme (UNEP), https://www.resourcepanel.org/reports/green-energy-choices-benefits-risks-and-trade-offs-low-carbon-technologies-electricity [accessed October 12, 2018].

[21] Sathaye, J., Lucon, O., Rahman, A., *et al.*, 2011. Renewable energy in the context of sustainable energy. In *IPCC Special Report on Renewable Energy Sources and Climate Change Mitigation* [O. Edenhofer, R. Pichs-Madruga, Y. Sokona, et al. (Eds.)]. New York, NY: Cambridge University Press.

[22] Volkart, K., Bauer, C., Burgherr, P., Hirschberg, S., Schenler, W., and Spada, M., 2016. Interdisciplinary assessment of renewable, nuclear and fossil power generation with and without carbon capture and storage in view of the new Swiss energy policy. *International Journal of Greenhouse Gas Control* 54, 1–14.

[23] Schwietzke, S., Sherwood, O.A., Bruhwiler, L.M.P., *et al.*, 2016. Upward revision of global fossil fuel methane emissions based on isotope database. *Nature* 538, 88–91.

[24] Bouman, E.A., Ramirez, A., and Hertwich, E.G., 2015. Multiregional environmental comparison of fossil fuel power generation – Assessment of the contribution of fugitive emissions from conventional and unconventional fossil resources. *International Journal of Greenhouse Gas Control* 33, 1–9.

[25] Burnham, A., Han, J., Clark, C.E., Wang, M., Dunn, J.B., and Palou-Rivera, I., 2012. Life-cycle greenhouse gas emissions of shale gas, natural gas, coal, and petroleum. *Environmental Science & Technology* 46, 619–627.

[26] Gernaat, D.E.H.J., Bogaart, P.W., Vuuren, D.P.v., Biemans, H., and Niessink, R., 2017. High-resolution assessment of global technical and economic hydropower potential. *Nature Energy* 2, 821–828.

[27] Hertwich, E.G., 2013. Addressing biogenic greenhouse gas emissions from hydropower in LCA. *Environmental Science & Technology* 47, 9604–9611.

[28] Scherer, L., Pfister, S., 2016. Hydropower's biogenic carbon footprint. *PLOS ONE* 11, e0161947.

[29] Hirschberg, S., Bauer, C., Burgherr, P., *et al.*, 2016. Health effects of technologies for power generation: Contributions from normal operation, severe accidents and terrorist threat. *Reliability Engineering & System Safety* 145, 373–387.

[30] Sovacool, B.K., Andersen, R., Sorensen, S., *et al.*, 2016. Balancing safety with sustainability: Assessing the risk of accidents for modern low-carbon energy systems. *Journal of Cleaner Production* 112, 3952–3965.

[31] Burgherr, P., and Hirschberg, S., 2014. Comparative risk assessment of severe accidents in the energy sector. *Energy Policy* 74, S45–S56.

[32] Arvesen, A., Zhongying, W., Jingting, Y., Mingliang, L., and Hertwich, E. G., 2016. Chapter 5: Wind power, in Hertwich, E.G., Aloisi de Larderel, J.,

and Arvesen, A. (Eds.), *Green Energy Choices. The Benefits, Risks, and Trade-Offs of Low-Carbon Technologies for Electricity Production*. Report of the International Resource Panel. United Nations Environment Programme (UNEP). Available from: http://www.resourcepanel.org/reports/green-energy-choices-benefits-risks-and-trade-offs-low-carbon-technologies-electricity [accessed October 12, 2018].

[33] Edwards, M.W., Schweitzer, R.D., Shakespeare-Finch, J., Byrne, A., and Gordon-King, K., 2019. Living with nuclear energy: A systematic review of the psychological consequences of nuclear power. *Energy Research & Social Science* 47, 1–15.

[34] Creutzig, F., Ravindranath, N.H., Berndes, G., *et al.*, 2015. Bioenergy and climate change mitigation: an assessment. *GCB Bioenergy* 7, 916–944.

[35] Humpenöder, F., Popp, A., Bodirsky, B.L., *et al.*, 2018. Large-scale bio-energy production: How to resolve sustainability trade-offs? *Environmental Research Letters* 13, 024011.

[36] Smith, P., Davis, S.J., Creutzig, F., *et al.*, 2015. Biophysical and economic limits to negative CO_2 emissions. *Nature Climate Change* 6, 42.

[37] Gibon, T., Hertwich, E.G., Arvesen, A., Singh, B., and Verones, F., 2017. Health benefits, ecological threats of low-carbon electricity. *Environmental Research Letters* 12, 034023.

[38] van Vuuren, D.P., Hof, A.F., van Sluisveld, M.A.E., and Riahi, K., 2017. Open discussion of negative emissions is urgently needed. *Nature Energy* 2, 902–904.

[39] Dai, K., Bergot, A., Liang, C., Xiang, W.-N., and Huang, Z., 2015. Environmental issues associated with wind energy – A review. *Renewable Energy* 75, 911–921.

[40] Hall, C.A.S., Lambert, J.G., and Balogh, S.B., 2014. EROI of different fuels and the implications for society. *Energy Policy* 64, 141–152.

[41] Raugei, M., and Leccisi, E., 2016. A comprehensive assessment of the energy performance of the full range of electricity generation technologies deployed in the United Kingdom. *Energy Policy* 90, 46–59.

[42] Carbajales-Dale, M., Barnhart, C.J., Brandt, A.R., and Benson, S.M., 2014. A better currency for investing in a sustainable future. *Nature Climate Change* 4, 524–527.

[43] Lambert, J.G., Hall, C.A.S., Balogh, S., Gupta, A., and Arnold, M., 2014. Energy, EROI and quality of life. *Energy Policy* 64, 153–167.

[44] Ayres, R., and Voudouris, V., 2014. The economic growth enigma: Capital, labour and useful energy? *Energy Policy* 64, 16–28.

[45] de Koning, A., Kleijn, R., Huppes, G., Sprecher, B., van Engelen, G., and Tukker, A., 2018. Metal supply constraints for a low-carbon economy? *Resources, Conservation and Recycling* 129, 202–208.

[46] Hertwich, E.G., Gibon, T., Bouman, E.A., et al., 2015. Integrated life-cycle assessment of electricity-supply scenarios confirms global environmental benefit of low-carbon technologies. *Proceedings of the National Academy of Sciences* 112, 6277–6282.

[47] Volkart, K., Weidmann, N., Bauer, C., and Hirschberg, S., 2017. Multi-criteria decision analysis of energy system transformation pathways: A case study for Switzerland. *Energy Policy* 106, 155–168.
[48] ReCiPe, 2012. *ReCiPe Method, Version 1.08*. http://www.lcia-recipe.net [accessed May 21, 2016].
[49] Jungbluth, N., and Frischknecht, R., 2007. Cumulative energy demand, in Frischknecht, R., and Jungbluth, N. (Eds.), *Implementation of Life Cycle Impact Assessment Methods*. Ecoinvent Report No. 3. Ecoinvent. http://www.ecoinvent.org/fileadmin/documents/en/03_LCIA-Implementation-v2.2.pdf [accessed March 13, 2017].
[50] Arvesen, A., and Hertwich, E.G., 2015. More caution is needed when using life cycle assessment to determine energy return on investment (EROI). *Energy Policy* 76, 1–6.

Part II

Tools for renewable energy grid integration

Chapter 4

The role of transmission for renewable energy integration and clean exports

Pedro Crespo del Granado[1], Christian Skar[1] and Raquel Alonso Pedrero[1]

A large deployment of renewable energy sources (RES) in the power system will not only require the development of key technologies (e.g. batteries or hydrogen), but also ensure that the transmission infrastructure provides the means to share flexibility across regions and countries. This chapter describes challenges and opportunities for the energy transmission infrastructure in Europe in terms of: (i) planning the energy transition of the power sector under a long-term perspective while considering short-term challenges, (ii) the role of transmission as an enabler to transnational balancing source for renewables and its cost–benefit implications, (iii) gas as a key flexible provider to balance RES but with challenges on maintaining the gas infrastructure relevant under climate targets and (iv) the perspective of balancing exporter countries on its policy and strategies to deliver clean exports.

4.1 Introduction

4.1.1 Background

The integration of large amounts of renewable energy sources (RES) in the power system creates supply uncertainty and hence balancing challenges to cope with peak demand. Flexibility options in the electricity and natural gas transmission system can absorb short-term demand fluctuations and smooth intermittent RES (mostly solar and wind). The ability of gas power plants to provide flexibility in a short-time period makes them especially suitable for fast balancing. Similarly, hydropower production, pumped storage, electricity storage and demand response are alternatives for flexibility services and balancing. Also, stronger grid transmission between countries with abundant hydropower reservoirs can open the inherent flexibility to other countries or regions hence making transmission expansion an important enabler of supply-side flexibility options [1].

[1]Department of Industrial Economics and Technology Management, Norwegian University of Science and Technology, Trondheim, Norway

Therefore, to consider a 100% renewable-based energy system, it requires a transformation supported by policies and a careful understanding of techno-economic challenges occurring in the short, medium and long terms. For example, the study by Papaefthymiou and Dragoon [2] recommended not only to focus on long-term solutions but also to complement it with mid-term initiatives such as the development of smart grid frameworks. The study advocated for the need of transmission system expansion to access flexibility and to design nodal markets and prices for the management of congestion [3]. These insights and perspectives on transmission expansion and clean exports are further explored in the study by Heard *et al.* [4]. This review explores how 24 global, regional and national studies have analysed the implications and the viability of achieving a 100% RES energy system. The review concludes that none of these studies achieved the minimum criteria to guarantee a reliable and secure power system. The authors noted that 'in general' energy studies tend to ignore the requirements of transmission and distribution systems required for 100% renewable power system (also supported by Reedman [5]). To support this argument, the study noted that this could be due to the underestimation of network expansion decisions by focusing only on short-term operations (only a typical day). In a counter paper, critiquing several key aspects of the study by Heard *et al.* [4], Brown *et al.* [6] noted that most of the technical feasibility challenges of high RES scenarios can be met easily at low economic cost. Moreover, the authors noted that grid upgrade costs represent a small percentage of the total costs in the energy transition.

In line with the literature, in this chapter we start with the premise that the cost-optimal transition to a larger share of RES in the electricity system relies extensively on a combination of flexibility measures to balance RES. In particular, a massive expansion of transmission (in our case, cross border) capacity is found optimal in Europe. A lack of transmission expansion (maybe due to public resistance to overhead lines) might imply cost increases if there are no alternative flexibility options. The viability of gas carbon capture and storage (CCS) could play a major role in providing flexibility from gas power while capturing CO_2 and hence meeting EU climate targets. The analysis shows that designing a system with limited transmission and mostly relying on renewables, energy storage and flexible demand will likely be a very expensive way to transform the power sector. We also analyse the role of the natural gas infrastructure in the energy transition and how clean export countries (like Norway) contribute to providing flexibility.

The next two subsections provide a general overview of the challenges on modelling the transformation of the power system towards larger shares of RES and related studies on modelling and analysing transmission expansion in Europe.

4.1.2 Modelling flexibility: time and spatial dimensions

Integrating RES requires flexibility to be considered in long-term planning and short-term operations. According to Gracceva and Zeniewski [7], flexibility is defined as '[t]he ability of a system to cope with the short-term uncertainty of energy system variables'. Flexibility analysis requires high time resolution as well as coverage of several timescales, detailed power system modelling, detailed unit commitment models, uncertainty representation of generation and demand, and integration of energy sectors and carriers [8].

This leads to two main areas of discussion: (i) how to properly quantify flexibility in terms of cost, products and value and (ii) how to efficiently model flexibility in terms of detail of results, model complexity and software capabilities [1]. For this reason, the difference among models is usually on the level of technical detail, representation of uncertainty (e.g. wind variations), temporal considerations, spatial aggregation and planning horizons [9]. Regarding time resolution, RES temporal dynamics and flexibility assets such as demand response and energy storage require representation of hourly characteristics [44]. In terms of technical details, such as voltage, frequency, etc., Koljonen *et al.* [45] argued that they are important in valuing and planning for flexibility, although issues with frequency and voltage control in systems with high shares of renewables are manageable as pointed out by Brown *et al.* [6]. In summary, the trade-offs between computational tractability and the level of modelling details face new challenges when assessing high RES share scenarios. For example, there are other important aspects: (i) to consider modelling capacity expansion by incorporating a long-term perspective where investments in grids and generation are included and (ii) to ensure a good coverage on spatial resolution. All of these are recurring challenges in modelling flexibility and capacity expansion. In the following section, some of existing studies and models focusing on the European energy transition have been analysed and some insights on capacity expansion are provided.

4.1.3 Related literature on the value of transmission

Recent studies have focused on analysing the role of transmission in the European energy transition. Most studies have partially included long- and short-term modelling features on their approaches. Brown *et al.* [6], for example, analysed various energy sectors beyond the power system and concluded that offshore wind will be a major actor in northern European countries if there are strong transmission capacity reinforcements in that region. This is compared with a scenario of less transmission expansion ('isolated systems') in which wind power still has a prominent role, but with more distributed national investments across Europe, and photovoltaics (PV) requires more expansion. Comparing both cases, the transmission expansion case helps reduce the average electricity price by 10% compared with the no-expansion scenario. This is partially due to the lesser need of stationary batteries or hydrogen storage infrastructure. Similarly, in the study by Schlachtberger *et al.* [10], the infrastructure requirements for the implementation of storage technologies to provide the necessary flexibility in a scenario with no grid expansion will raise costs by around 30% compared with the cost-optimal solution. The paper noted the importance of having a higher utilization factor for storage which is facilitated by increased transmission connections and hence sharing long-term storage options among countries. In line with these studies, Child *et al.* [11] analysed transmission expansion by including another flexibility

Table 4.1 Literature studies on transmission lines scenarios with high shares of RES

Paper and objective	Method and approach	Insights to transmission expansion and RES share
Paper: Brown *et al.* [6] **Objective:** The model couples electricity, transport and heat sectors, and considers transmission grid expansion to balance the variability of renewables.	Linear optimization model to minimize operational and investment costs. The model is subject to supply–demand balance, generator and storage operations constraints, flows capacities and CO_2 emissions.	Transmission grid expansion is the most cost-optimal solution in all scenarios. Higher levels of energy sector coupling reduce the benefits of interconnection reinforcements. The combination of flexibility options with transmission expansions can reduce the total system costs by 37%.
Paper: Schlachtberger *et al.* [10] **Objective:** Optimal system for balancing wind and solar power generation using storage units and the transmission grid.	Linear optimization minimizing fixed costs (generation, storage and transmission) and variable costs (generation and storage dispatch). The model is subject to power balance, generators and storage operations, transmission and CO_2 emissions caps.	The results indicate that a ninefold increase in transmission grid capacity would be optimal. However, the transmission investment costs represent only 12% of the total cost in the optimal scenario. If transmission is maintained at current levels, there is a reduction of the share of wind as it will require hydrogen or extended battery storage. In contrast, the share of solar PV systems with batteries for daily load-shifting and grid services is increased by 131% for current interconnecting line volume.
Paper: Child *et al.* [11] **Objective:** Compares two different pathways to achieve 100% renewables for the power sector in Europe for 2050: the first considering 20 countries as independent energy systems and the second considering the interconnection between them.	Linear model minimizing total annualized system and end-user consumption costs. The model considers generation and installation costs as well as costs of prosumer electricity consumption. The model is operationally constrained regarding: RES installed capacity, biogas deployment, prosumer demand and nuclear or fossil fuels plants.	The capacity of the interconnection facilities in Europe goes from 63 GW (in 2015) to 262 GW in 2050. Most of it should be installed before 2025 (7 GW/km installed between 2020 and 2025). Most developments occur between areas with wind (e.g. Great Britain, Ireland, Denmark), solar (e.g. Turkey, Italy, Iberia), hydropower resources (e.g. Norway) and densely populated countries (e.g. Germany and France).
Paper: Zappa *et al.* [46] **Objective**: Compares several scenarios with different future demands and technology availability in 2050 with focus on Concentrating Solar Power (CSP).	Mixed-integer linear programming model that allows to couple investments and unit commitment calculations to minimize the total net present value of both operational and strategic costs.	By 2050, the power system may rely just on domestic European resources. The installation of CSP capacity has a significant impact in the configuration of the transmission network. Utility-scale daily storage units are less cost effective than large-scale transmission extensions.

(Continues)

Paper: Pleßmann and Blechinger [13] **Objective:** Show pathways needed to reach European climate targets by 2050 ensuring techno-economic feasibility focusing on the power supply system.	Linear model to optimize the power supply system considering generation, storage and transmission technologies. The model responds to short-term power requirements and includes long-term investments at 5-year intervals.	The best economic pathway has wind generation capacity at around 1,485 GW and PV at 909 GW. The transmission line capacity increases from 79.5 GW (in 2016) to 362 GW. The greatest transmission expansion is between the Western and Southern Balkan countries. The Denmark-to-Germany link shows a large capacity increase.
Paper: Scholz *et al.* [12] **Objective:** Analyses the role of CSP in the European energy transition. The study looks at curtailed energy and system costs.	Model capable to set optimal power generation and transmission capacities while hourly dispatch is simulated. The study runs 66 different scenarios differing in RES share, solar-to-wind ratios and CO_2 certificate price.	The increase of Variable Renewable Energy Sources (VRES) shares drives the installation of DC connections between regions. Even at value 0, VRES penetration transmission capacity between regions is installed. The maximum grid extension occurs with high penetrations of wind power capacity across all CO_2 certificate price scenarios.
Paper: Egerer *et al.* [14] **Objective:** Provides a pathway to transform the power system with special attention on network and infrastructure development until year 2050. Presents three scenarios based on CO_2 reductions.	Mixed-integer problem model to calculate the need for transmission grid extension. Minimizes total costs for dispatch operations and grid investments. Good representation of the power infrastructure (e.g. HV network with nodal energy balance and DC Load Flow). The model invests in both AC network and DC grids.	Differing from most of the literature, results show that a strong HVDC network is unlikely to be established in the European context. Instead, the model suggests that the most cost-optimal solution is to upgrade and extend national and transnational AC networks.

option for congestion management: the use of demand-side measures. The study emphasized and demonstrated that peak transmission capacity can be lowered up to 6% given the right mix of investments.

Other studies looked at specific RES technology aspects together with transmission expansion. Zappa *et al.* [46] and Scholz *et al.* [12] explored the role of concentrating solar panels and how they contribute to the energy mix in 2050. These and other studies, focusing on Europe, are summarized in Table 4.1 providing some more detail in terms of their perspectives and insights into the pathways for achieving higher shares of RES in the system by 2050. In line with the European targets for 2050, most of the studies achieve a 60%–75% RES share, with some studies even including a 100% scenario. It is interesting to note that to some degree these studies include a good representation of short-term and long-term decisions. All their modelling approaches have at least an hourly resolution, full investment analyses with a 2050 horizon and similar European spatial coverage.

4.2 Modelling large-scale energy systems

To analyse the impact of transmission expansions in the decarbonization of the power system in Europe, we use the EMPIRE model (**E**uropean **M**odel for **P**ower System **I**nvestment with high shares of **R**enewable **E**nergy) [9,15–19]. This section details the EMPIRE modelling features and introduces a similar type of model for gas infrastructure assessments.

4.2.1 The EMPIRE model

EMPIRE is formulated as a multi-horizon stochastic programme, an approach based on the study by Kaut *et al.* [20] in which the calculation of optimal technology portfolios (or energy mixes) is subjected to capacity expansion decisions and short-term uncertainty (scenarios for solar and wind realizations). EMPIRE provides a unique representation by merging the short- and long-term dynamical aspects of the power system, in contrast to most methodologies that adopt just one timescale. That is, the model incorporates strategic decisions for the long term considering operational uncertainties. The integration of the two planning horizons is a key contribution as short-term decisions need proper generation capacity to ensure the supply–demand balance while investment decisions are determined by the adequacy of balancing in the short term. In other words, the model provides the optimal capacity investments (generation and transmission) along with system operation details over long-term planning horizons. With the objective of minimizing the total system costs, EMPIRE considers common elements of technical and investment models such as generation, storage and transmission capacity expansions, net transfer capacity (NTC), up-ramping constraints for generation and investment and operation costs of storage technologies. The model can therefore be adapted to different technological advances (e.g. CSS, demand response, EV charging) and policy changes (e.g. carbon price or RES targets). To represent the operational decisions and features of the power system, EMPIRE has the following elements:

Short- and long-term horizons: A main constraint when integrating operations in long-term horizons is the large dimension of the problem. Running hourly operations compromise 8,760 periods over a year which makes it an intractable problem when we span it for years or decades ahead. An approach for overcoming this issue is to sample representative weeks that serve as parameters for incoming investment periods. In EMPIRE, these are specific weeks representing the seasons and setting days with extreme scenarios (e.g. high peak load, low RES production). With investment windows* of 5 years, the operational decisions from the representative weeks are scaled up to illustrate the short-term decisions during 5 years. EMPIRE considers a planning horizon of 35 years (from 2015 to 2050).

Spatial aggregation: Thirty-one European countries from the European Economic Area along with 55 interconnectors are covered by the model. Each node is an abstraction (or aggregation) of each country and is associated with a certain generation capacity and estimated demand. To represent the transmission system, each node is modelled to resemble as a country connected to its neighbouring countries.

Technical details: For the short-term operations, EMPIRE defines hourly operations of transmission among countries and the scheduling of generators. In the long term, the model considers production costs of all generators in combination with the investment costs of generators and transmission lines.

Policy changes and economic approaches: EMPIRE can test different policies by changing the parameters related to technology evolution (learning curves), climate targets (e.g. CO_2 caps) or different assumptions regarding transmission expansions. In this way, the model outputs provide the usage factor of power plants, transmission and generation expansion costs, load shedding, curtailment and trade between countries to achieve decarbonization goals.

In summary, EMPIRE computes the optimal investment strategy, with 5-year increments of transmission and generating capacities, while considering the unit commitment operations for the European system. Figure 4.1 depicts the inputs, model and output details of EMPIRE while Figure 4.2 shows the schematics.

4.2.1.1 Mathematical formulation

The EMPIRE model is designed as a two-stage stochastic optimization problem. In line with typical unit commitment models, EMPIRE is formulated as a multi-period optimization model with hourly resolution. This allows to consider technical features of generations such as the ramp-up times, power rates of the units and their associated operational costs. The objective of the model is to minimize total system costs, comprising investment costs and the expected operational costs (4.1). The decision

*Not to be confused with *planning horizon*. We refer to investment windows as the time steps in which long-term investment decisions are modelled.

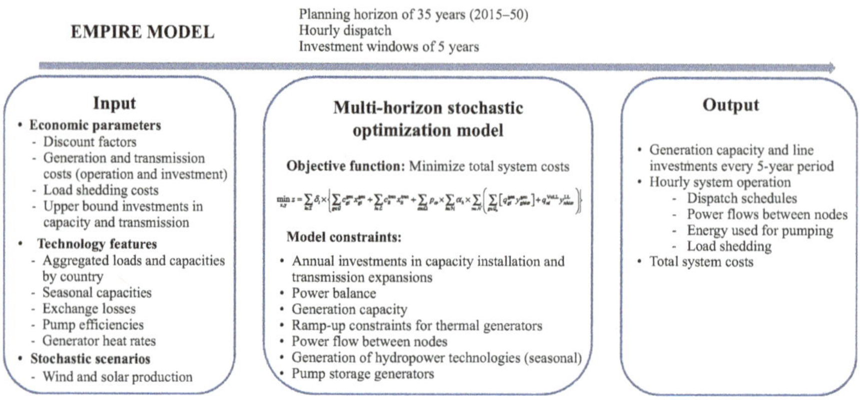

Figure 4.1 EMPIRE model components and structure

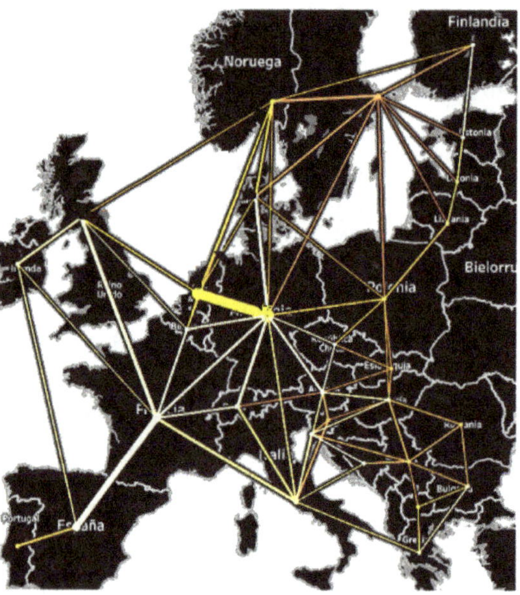

Figure 4.2 Geographical representation of the EMPIRE model and coverage

variables of the model are investment decisions and the schedule of generation units. In addition, the model includes a discount rate, δ_i, per year i to calculate the net present value (NPV) of investments. We use an interest rate of 5% in the discount factor.

$$\min_{x,y}\ z = \sum_{i\in\mathcal{I}} \delta_i \times \left\{ \sum_{g\in\mathcal{G}} c_{gi}^{\text{gen}} x_{gi}^{\text{gen}} + \sum_{l\in\mathcal{L}} c_{gi}^{\text{tran}} x_{li}^{\text{tran}} + \sum_{\omega\in\Omega} p_\omega \times \sum_{h\in\mathcal{H}} \alpha_h \right.$$
$$\left. \times \sum_{n\in\mathcal{N}} \left(\sum_{g\in\mathcal{G}_n} \left[q_{gi}^{\text{gen}} y_{ghi\omega}^{\text{gen}} \right] + q_{ni}^{\text{VoLL}} y_{nhi\omega}^{\text{LL}} \right) \right\} \tag{4.1}$$

Total investment costs result from the sum of the installation of generation capacity, x_{gi}^{gen}, for each unit g at year i with costs c_{gi}^{gen} and transmission grid extensions, x_{li}^{tran}, in line l at year i with costs c_{li}^{tran}. On the other hand, the operational costs from power generation will be defined by the production $y_{ghi\omega}^{\text{gen}}$ of each supply unit g at hour h in year i under the stochastic scenario w (different wind and solar hourly profiles). Similarly, the operation costs also include the load shedding $y_{nhi\omega}^{\text{VoLL}}$ for node (country) n under operational hour h in year i and the stochastic scenario ω. Both the production and load shedding are associated with their costs q_{gi}^{gen} and q_{ni}^{VoLL}, respectively. The factor α_h represents the total actual hours in a year aggregated as the operational hour h. This is required to scale up and match the number of hours of the investment periods (investment window of 5 years). Since the uncertainty affects the load profiles and the renewable generation (wind and solar), we assign the probability p_ω for that scenario. Note that the model has the perspective for economic social surplus maximization under perfectly competitive markets and predetermined consumer decisions (no demand elasticity).

4.2.1.2 Model constraints

The model must ensure realistic investments in terms of time and spatial adequacy. For instance, it will not be realistic to make a single big investment in just 1 year or invest in technologies with low potential in certain countries due to meteorological conditions or resource availability. Therefore, the amount of investments is constraint by an upper bound $\bar{x}_{nt*}^{\text{gen}}$ for specific node n and aggregations of generation technologies t. The constraint is for specific years i (4.2) and for the cumulative generation capacity over the planning horizon (4.3).

$$\sum_{g\in\mathcal{G}_{nt}} x_{gj}^{\text{gen}} \le \bar{x}_{nti}^{\text{gen,Period}}, \quad n\in\mathcal{N}, t\in\mathcal{T}, i\in\mathcal{J}. \tag{4.2}$$

$$\sum_{j=1}^{i}\sum_{g\in\mathcal{G}_{nt}} \left(x_{gj}^{\text{gen}} \le \bar{x}_{nt}^{\text{gen,Cumulative}} - \left(1-\rho_{gi}\right)\bar{x}_{g0}^{\text{gen}}, \quad n\in\mathcal{N}, t\in\mathcal{T}, i\in\mathcal{J}. \tag{4.3}$$

A similar approach is applied to investments in interconnection expansion (4.4).

$$x_{li}^{\text{tran}} \le \bar{x}_{li}^{\text{tran,Period}}, \quad l\in\mathcal{L}, i\in\mathcal{I}. \tag{4.4}$$

While minimizing the costs and considering realistic investments, the model is also subject to operational constraints. Equation (4.5) ensures the demand–supply balance for each node considering internal production $y_{ghi\omega}^{\text{gen}}$, interaction flows with neighbouring nodes $y_{ahi\omega}^{\text{flow}}$, pumping $y_{nhi\omega}^{\text{pump}}$, load shedding $y_{nhi\omega}^{\text{LL}}$ and the aggregated load of the country $\xi_{nhi\omega}^{\text{load}}$. Exchange losses η_a^{line} are given as a share of the total flow.

$$\sum_{g\in\mathcal{G}_n} y_{ghi\omega}^{\text{gen}} + \sum_{a\in\mathcal{A}_n^{\text{in}}} (1-\eta_a^{\text{line}}) y_{ahi\omega}^{\text{flow}} - \sum_{a\in\mathcal{A}_n^{\text{out}}} y_{ahi\omega}^{\text{flow}} - y_{nhi\omega}^{\text{pump}} + y_{nhi\omega}^{\text{LL}} = \xi_{nhi\omega}^{\text{load}},$$

$$n\in\mathcal{N}, h\in\mathcal{H}, \omega\in\Omega, i\in\mathcal{I}. \tag{4.5}$$

In addition, the generation capacity for each generation unit will be constrained by the available capacity from prior installations plus the new capacity resulting from the investments in the corresponding window (4.6).

$$y^{\text{gen}}_{ghi\omega} \leq \xi^{\text{gen}}_{ghi\omega} \times \left(\left(1 - \rho_{gi}\right)\bar{x}^{\text{gen}}_{g0} + \sum_{j=1}^{i} x^{\text{gen}}_{gj}\right), \quad g \in \mathcal{G}, h \in \mathcal{H}, i \in \mathcal{I}, \omega \in \Omega. \tag{4.6}$$

Ramp-up constraints of thermal generators are also included as part of the operational constraints (4.7). The limit on total upward ramping γ^{gen}_{g} is expressed as a fraction of total installed capacity.

$$y^{\text{gen}}_{ghi\omega} - y^{\text{gen}}_{g(h-1)i\omega} \leq \gamma^{\text{gen}}_{g} \times \left(\left(1 - \rho_{gi}\right)\bar{x}^{\text{gen}}_{g0} + \sum_{j=1}^{i} x^{\text{gen}}_{gj}\right),$$
$$g \in \mathcal{G}^{\text{Thermal}}, s \in \mathcal{S}, h \in \mathcal{H}^{-}_{s}, i \in \mathcal{I}, \omega \in \Omega. \tag{4.7}$$

The flow between nodes is also bounded by the previous interconnection capacity plus the transmission investments for year i (4.8).

$$y^{\text{flow}}_{ahi\omega} \leq \bar{x}^{\text{tran}}_{l0} + \sum_{j=1}^{i} x^{\text{tran}}_{lj}, \quad l \in \mathcal{L}_n, a \in \mathcal{A}_l, h \in \mathcal{H}, i \in \mathcal{I}, \omega \in \Omega. \tag{4.8}$$

Generation from hydro technologies have inter-temporal or seasonal constraints due to water availability. Equation (4.9) expresses the limit on hydropower production within a season (S).

$$\sum_{h \in \mathcal{H}_s} y^{\text{gen}}_{ghi\omega} \leq \xi^{RegHydroLim}_{gsi\omega}, \quad g \in \mathcal{G}^{\text{RegHydro}}, \quad s \in \mathcal{S}, i \in \mathcal{I}, \omega \in \Omega. \tag{4.9}$$

Another dispatch constraint is for the pumped-storage generators. The model ensures the inter-temporal balance and also guarantees that the reservoir level at the end of a season is equal to the initial season level (4.10).

$$\left.\begin{aligned} w^{\text{upper}}_{n(h-1)i\omega} + \eta^{pump}_{n} y^{\text{pump}}_{nhi\omega} - y^{\text{gen,pump}}_{nhi\omega} &= w^{\text{upper}}_{nhi\omega} \\ w^{\text{upper}}_{nhi\omega} &\leq \bar{w}^{\text{upper}}_{n} \end{aligned}\right. , n \in \mathcal{N}, h \in \mathcal{H}_s, i \in \mathcal{I}, \omega \in \Omega. \tag{4.10}$$

In summary, the EMPIRE model has been widely used in other studies for the European Commission, national research projects, industry and others. For example, a more comprehensive model formulation, data sources and other details are found in the studies by Skar *et al.* [17], Skar *et al.* [21] and ZEP [22]. Note that the model is openly available for download and use. Guidelines and details on how to use it are available at the EMPIRE website.†

†https://www.ntnu.edu/web/iot/energy/energy-models-hub/empire [accessed date: 13 March 2020].

4.2.2 Analysing gas network flows: The Global Gas Model

Energy system modelling approaches applied to (natural) gas infrastructure systems shares similarities to the framework of the EMPIRE model. However, gas systems have its own technological features. For example, some gas models represent supply-to-demand decisions in detail by modelling pressure–flow relationships (e.g. in the gas compression process) and other engineering aspects of gas infrastructure operations. Further gas modelling details might include reassurance constraints, gas quality and multi-commodity flows, features of reservoirs, technicalities of compressors and processing plants, and others. However, similarly as in the EMPIRE modelling framework, representing technological qualities in detail creates computational challenges that are more pronounced when considering long-term aspects (scenarios) of the gas system. Indeed, similar to power system models, (natural) gas models require the implementation of a strategic long-term perspective to represent infrastructure decisions such as capacity expansion and network design [23]. In gas systems, this feature allows analysing the impact of gas infrastructure projects in the market. For example, if a country requires natural gas, this demand must be satisfied based on the availability of indigenous production (reservoirs, offshore gas platforms or overall national production) or by acquiring it on gas markets. The latter implies importing gas from neighbouring countries or installing Liquefied Natural Gas (LNG) terminals on designed industrial maritime ports to receive the imported gas. Both instances require major investment decisions with long-lasting consequences. Building gas pipelines among countries involves a delicate decision-making process that involves key issues such as: (i) the economic benefits of such investments, (ii) the geopolitical circumstances and consequences, (iii) the technical feasibility and logistics of the gas infrastructure project and (iv) public perceptions on the impact of the gas infrastructure in the region and environmental concerns.

From a modelling perspective, the assessment of capacity expansion and deciding on where to invest in the gas infrastructure fall into the following main categories: (i) analyse which pipelines (and interconnectors) to build, (ii) weight the benefits of creating a production site (or reservoir) along with the surrounding infrastructure (e.g. storage or liquefication) and (iii) develop a LNG terminal that is open to multiple sellers. To illustrate an example of gas infrastructure analyses, the following text box illustrates the assessment of expanding the gas infrastructure in Europe based on pipeline and LNG projects proposed in 2016 by the European Commission.

Europe's Gas Network – Illustrative example [30]

Strengthening and diversifying the connections to major suppliers as well as investing in LNG capacity is key to reducing energy supply risk in Europe. Therefore, the relative location of gas demand plays a key role in planning gas infrastructure upgrades. The European Union proposes Projects of Common Interests to start a decision process on infrastructure priorities.

This is a list of pipelines and LNG projects which are evaluated altogether in one model as options to invest in. The figure on the right shows the European natural gas map infrastructure and the projects (orange) were assessed by Kotek *et al.* [30]. Here, the gas model, representing the whole gas network flow of supply–demand systems, is capable of determining the cost-effective projects based on production and demand in Europe in the next decades, the capacity of existing pipelines and interconnectors, the effect of LNG terminals and other technical aspects.

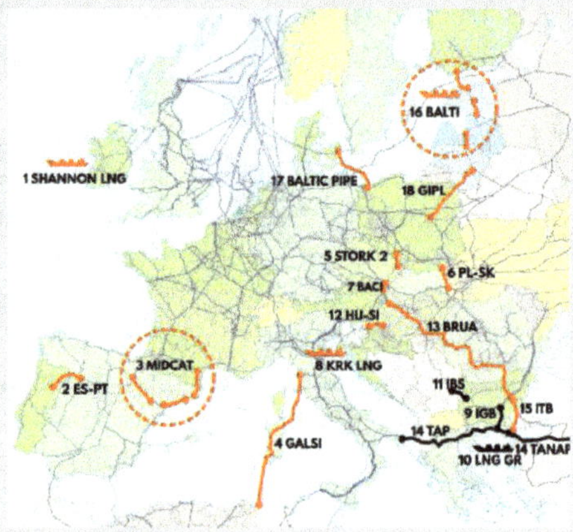

The main recommendations of the study [30] are: (i) to build an LNG terminal in Ireland due to declining production in Great Britain, (ii) to consider investing the LNG terminal in Croatia which would trigger investment in the Hungary–Slovenia interconnector, (iii) to establish one interconnector for Poland–Lithuania and one for Bulgaria–Turkey. An interesting underlying assumption of the study is the decision to partially limit the dependency on Russian gas. In this case, modellers blocked Russian gas coming through Ukraine, which triggered the investments mentioned in central Europe. In this study, the authors concluded that the current European gas infrastructure is robust as gas demand is declining in most European scenarios.

Modelling a gas network requires the representation of production, pipelines, liquefaction, regasification, shipping and storage. As some investment decisions entail a yes-or-not approach, they are represented as a binary variable. This entails formulating a model based on mixed-integer optimization. Similarly to the EMPIRE model, gas modelling includes both investment decisions and operational features of the gas network. Figure 4.3 depicts the way of representing two nodes (countries) and the gas infrastructure elements and their interactions.

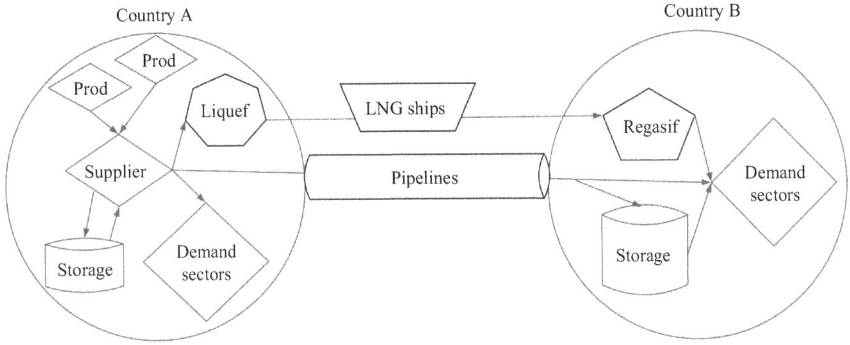

Figure 4.3 Schematic overview of the types of actors along the natural gas value chain which are modelled in GGM

In this chapter, we invited the reader to use the Global Gas Model (GGM). The GGM has country-level reference projections for production and consumption values and end-user prices, various infrastructure investment and operational costs, capacities, depreciation and loss rates, seasonal demand loads, sector shares and price-demand elasticities, production costs, and market power assumptions. This model is particularly useful as it is openly available and downloadable. The website[‡] of GGM provides a user guide and directions to use the model. GGM was developed by Egging [23] based on the development and application of the European Gas Model [24] and the World Gas Model [25,26]. GGM was applied in studies of the future global [27] and European gas markets [28,29]. GGM has been used to analyse climate policy and supply security (e.g. [27]). GGM has contributed to the analyses of projects of common interest (PCI) noted in the earlier example [30]. In a nutshell, the GGM is multi-period partial equilibrium model accounting for market power in world natural gas markets. GGM endogenously determines country-level seasonal production, consumption and trade patterns and prices, pipeline, liquefaction, regasification and storage infrastructure expansions and utilization, price variations in scenarios, and sector profits, costs, consumer surplus and social welfare impacts.

4.3 Large-scale RES share scenarios in Europe

The starting point of our EMPIRE analysis of the role transmission is two low-carbon scenarios previously presented in the report 'Norway's role as a flexibility provider in a renewable Europe' [19]. These scenarios are labelled *Baseline* and *NoCCS*. In the following sections, we present the underlying data assumptions for these scenarios along with the results from the previous analysis, focusing on the cost-optimal energy capacity and generation mix in Europe and the transmission

[‡]https://www.ntnu.edu/web/iot/energy/energy-models-hub/ggm [accessed date: 13 March 2020].

infrastructure investments. We accompany this analysis with an additional set of scenarios without transmission expansion which have not been presented earlier. These scenarios allow us to compare the effect of transmission on cost-optimal decarbonization strategies and on renewable energy integration.

4.3.1 Baseline scenario of a low-carbon European power system

A decarbonization scenario in EMPIRE is a complete data set describing parameters that are exogenous to the model. The most important drivers for the dynamics of power system investments are projections for future demand for electricity, fossil fuel prices, costs and characteristics of power generation, energy storage, demand response and transmission technologies. In the following sections, we briefly discuss the data sources used for our Baseline scenario. The NoCCS scenario use the same data set, however with the additional assumption that capture, transport and storage of carbon is not available as a technology for power generation.

The decarbonization in all the scenarios is enforced in a set of constraints that ensures a linear reduction of direct annual CO_2 emissions from power generation of 90% in the period from 2010 to 2050. That is, we impose a carbon budget restriction that gradually decreases as follows: in 2030 to 700MtCO_2, in 2040 to 420MtCO_2 and in 2050 to 130MtCO_2.

4.3.1.1 Data sources

Electricity demand projects for individual European countries were obtained based on the 'EU Reference Scenario 2016' report by the European Commission [31]. The annual demand data are used as inputs to generate hourly load profiles using a transformation routine described by Skar *et al.* [21]. Future prices for fossil fuels are taken from IEA's 2-degree scenario (2DS) found in the 'Energy Technology Perspectives 2016' [32]. Figure 4.4 shows these demand and fuel price data used for the Baseline.

The data sources used for power generation technologies, transmission, energy storage and demand response costs and characteristics are discussed in Skar *et al.* [19]. The investment cost data used for generation technologies in our analysis is shown in Table 4.2. The data sources used are summarized in Table 4.3. Data for battery energy storage and renewable investment costs are shown in Figure 4.5. For the value of lost load, we use an estimate of 22,000 €/MWh which has been derived from London Economics [38].

When it comes to modelling of carbon capture, transport and storage, we add a heat rate penalty to the CCS power generation (to account for the capture process) and cost terms to account for transport and storage (data found in ZEP [22]). The estimated storage capacity in Europe is 94 GtCO_2 [39]. Although the carbon storage capacity is not modelled explicitly, a post-optimization check is done to verify that the CCS use in a solution does not imply exceeding the estimated European limit.

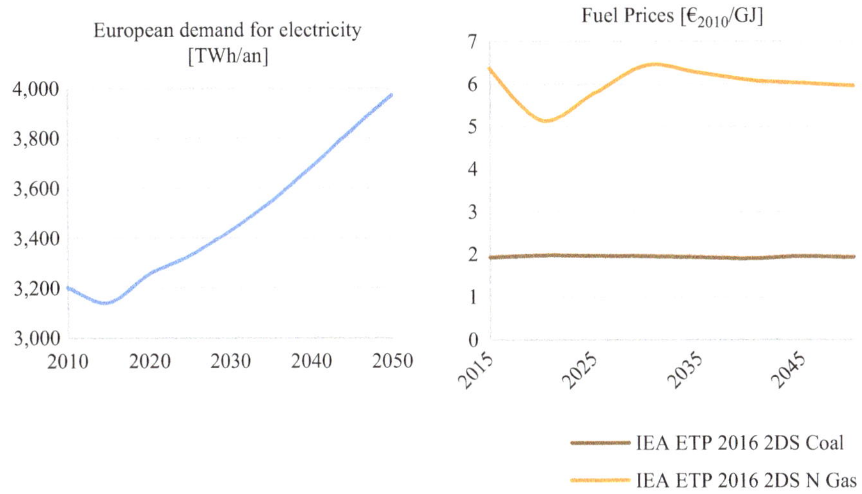

Figure 4.4 Demand for electricity and fuel prices used for the EMPIRE analysis

Table 4.2 Technology investment cost used in EMPIRE

Technology	2010	2015	2020	2025	2030	2035	2040	2045	2050
Lignite	1,600	1,600	1,600	1,600	1,600	1,600	1,600	1,600	1,600
Lignite CCS				2,600	2,530	2,470	2,400	2,330	2,250
Coal	1,500	1,500	1,500	1,500	1,500	1,500	1,500	1,500	1,500
Coal CCS				2,500	2,430	2,370	2,300	2,230	2,150
Gas OCGT	400	400	400	400	400	400	400	400	400
Gas CCGT	800	800	800	800	800	800	800	800	800
Gas CCS				1,350	1,330	1,310	1,290	1,270	1,250
Bio	2,250	2,250	2,250	2,250	2,250	2,250	2,250	2,250	2,250
Bio 10pcnt cofir.	1,600	1,600	1,600	1,600	1,600	1,600	1,600	1,600	1,600
Bio 10pcnt cofir. CCS				2,600	2,530	2,470	2,400	2,330	2,250
Nuclear	4,500	4,500	4,500	4,500	4,500	4,500	4,500	4,500	4,500
Hydro reservoir	3,000	3,000	3,000	3,000	3,000	3,000	3,000	3,000	3,000
Hydro run-of-the-river	4,000	4,000	4,000	4,000	4,000	4,000	4,000	4,000	4,000
Wind onshore	1,200	1,063	1,033	1,002	972	942	912	881	851
Wind offshore	3,500	3,500	2,625	2,190	1,930	1,795	1,710	1,642	1,592
Solar PV	1,900	998	826	653	481	463	445	427	409

Note: All costs in €$_{2010}$/kW (for offshore wind an additional 500 €$_{2010}$/kW is added to the values here in the model for grid connection cost).

4.3.1.2 EMPIRE results from the Baseline and NoCCS scenarios

In Figure 4.6, the cost-optimal developments of power generation capacity and energy mix in Europe are shown for the Baseline and the NoCCS scenarios. In the medium term (until 2040), these scenarios are indistinguishable, as CCS is only

Table 4.3 Data sources used in the EMPIRE analysis (all scenarios)

Technology	Reference	Comment
Fossil fuel technologies including CCS	[21,22]	Natural gas data updated [19]
Nuclear power	6,000 €/kW for all investment periods	Investments limited to ENTOS-E's visions 1 and 2 of the 2016 TYNDP market modelling data
Solar PV	[33]	Medium scenario used. National investment potential from Gils *et al.* [34]
Wind onshore/offshore	[35]	Capacity potentials from IEA's 'Nordic Energy Technology Perspectives 2016' [36]
Utility-scale battery storage	Cole *et al.* [47]	Medium scenario used
Demand response	Gils [34]	Further documentation found in Marañón-Ledesma and Tomasgard [16]
Transmission infrastructure	[21]	By 2020, we assume the capacities from ENTSO-E's TYNDP from 2016 are in operation [37]. Beyond 2020, each interconnector expansion has a 4 GW limit every fifth year.

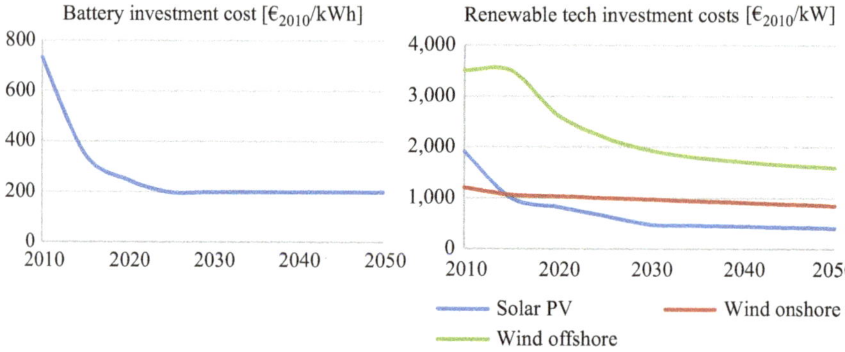

Figure 4.5 Battery energy storage and renewable generation investment cost assumptions. Sources are found in Table 4.3

deployed in the later part of the analysis period. This follows from the model's ability to anticipate CCS technology development (and resulting cost reductions) which is exogenous to the model. The steps 2010 and 2015 are used for model calibration against historical data, and investments are not possible for this period.

In the EMPIRE results from 2020 and beyond, we see power production from coal being displaced by power produced with natural gas, wind and solar, following the gradual decrease of the emission limit. Nuclear power capacity is steadily

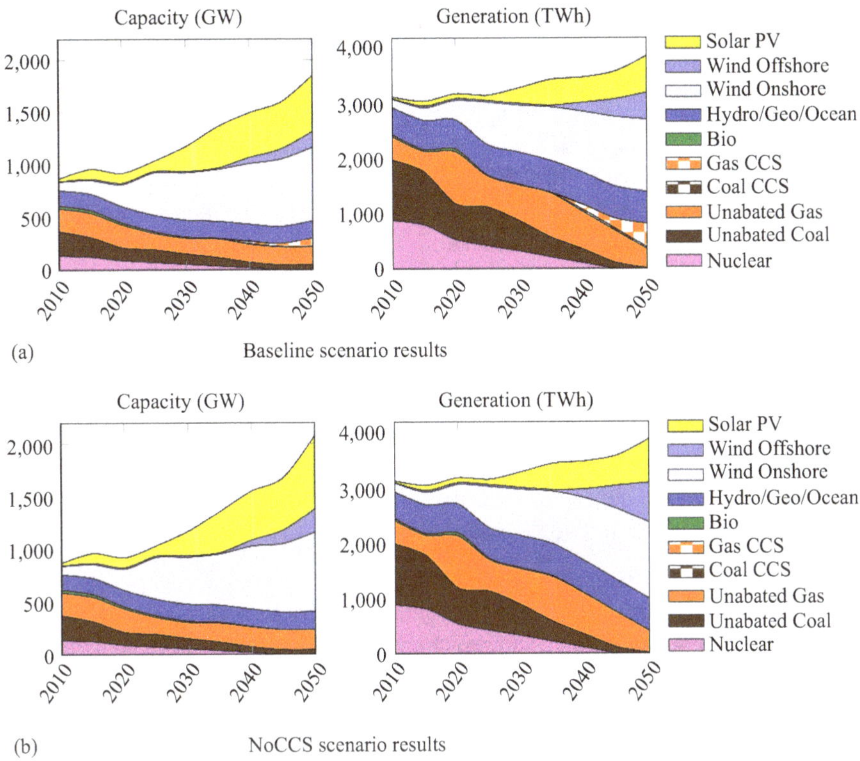

Figure 4.6 European generation capacity and energy mix from EMIPRE 2010–2050 for the Baseline and the NoCCS scenarios

phased out from 2015, and not replaced by reinvestments due to the prohibitively high costs. From 2025 and onwards, wind energy quickly becomes the technology with the dominant share of the energy mix. When it comes to offshore wind, the first significant deployment starts in 2040. By 2050, almost half of the total power generation is wind based, either from onshore or offshore wind. Solar PV investments in this analysis are delayed until after 2030 when a decrease in PV investment costs is sufficient to give the technology a competitive edge.[§]

Natural gas comes out as the major fossil fuel technology in this analysis. The peak production occurs in 2020 when the total power generation from natural gas is close to 945 TWh/year. This is caused by the low gas price for this period found in the IEA ETP 2016 2DS data (see Figure 4.4). The role of natural gas towards 2050

[§]The delayed solar PV expansion may come as a surprise, as even with current cost level the technology is deployed throughout Europe. The modelling result may be an artefact of several effects: we do not: model direct renewable support schemes, include distribution grid costs (one thing that can be reduced by distributed energy production), account for difference in land cost or grid connection costs (can potentially be different for solar than other technologies).

relies heavily on the availability of CCS. In 2040, natural gas with CCS for the Baseline scenario is starting to be deployed, and by 2050 there is a total of close to 800 TWh/year power generation using natural gas (with a bit more than half using CCS). In the NoCCS scenario, the total power generation from natural gas is almost exactly half of this amount. In both scenarios, conventional coal and nuclear power are completely phased out.

In 2050, the natural gas CCS generation is offset by increased renewable energy generation for the NoCCS scenario when compared with the Baseline scenario. The difference is about 300 GW in installed capacity, which is mostly solar PV and offshore wind. Hydropower remains a significant contributor to the power generation share in Europe, however the total installed capacity and generation is stable over the analysis period.

Table 4.4 shows the European capacity and generation mix results in 2050. In the Baseline scenario, CCS power generation accounts for 12% of the total mix. Comparing the total intermittent renewable generation in these scenarios, we see that wind (onshore/offshore) and solar PV have a total share of the generation mix of 65% in the Baseline and 75% in the NoCCS.

Transmission

The results of the transmission investments by 2050 for our two scenarios are shown in Figure 4.7 (alongside the initial capacities in the European system as of 2010). The total capacity in the system (sum of all interconnector capacities) by 2050 is 456 GW in the Baseline scenario and 527 GW in the NoCCS scenario. Compared with the 2010 system capacities (which was about 65 GW in Europe in 2010), there is an increase of 701% and 811%, respectively.

It is evident from these results that an optimal decarbonization, in which large shares of RES are deployed, relies on a significant transmission system expansion. When CCS is available and deployed, the need for such infrastructure investments is

Table 4.4 European generation capacity and energy mix by technology in 2050 from EMPIRE for the Baseline and NoCCS scenarios

Technology/fuel (2050)	**Baseline**				**NoCCS**			
	Capacity		**Generation**		**Capacity**		**Generation**	
	GW	**Share**	**TWh**	**Share**	**GW**	**Share**	**TWh**	**Share**
Solar	536	(29%)	665	(17%)	690	(33%)	788	(20%)
Wind onshore	698	(38%)	1,314	(34%)	751	(36%)	1,381	(36%)
Wind offshore	149	(8%)	492	(13%)	222	(11%)	730	(19%)
Gas CCS	81	(4%)	436	(11%)				
Coal CCS	6	(0%)	33	(1%)				
Gas unabated	173	(9%)	346	(9%)	190	(9%)	393	(10%)
Coal unabated	45	(2%)	9	(0%)	43	(2%)	11	(0%)
Others (hydro, geo, etc.)	164	(9%)	577	(15%)	173	(8%)	580	(15%)

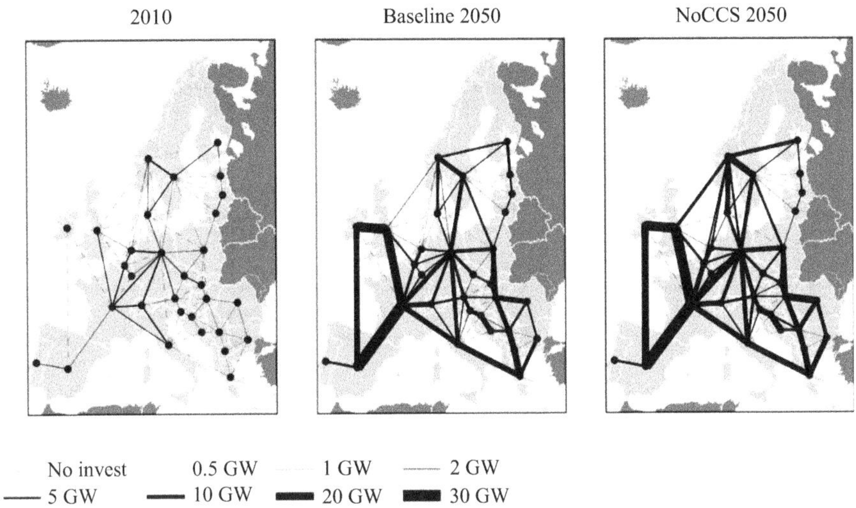

Figure 4.7 European cross-border interconnector capacities in 2050 from EMPIRE in the Baseline and the NoCCS scenarios

reduced somewhat as the CCS plants can be dispatched when needed to balance the fluctuating renewable generation. Therefore, there is less need for spatial balancing.

4.3.2 A deeper look at the role of transmission in decarbonization

The common cost-optimal energy transition strategy in both the Baseline and the NoCCS scenarios clearly involves a significant expansion of the interconnector capacities in Europe. This result is supported by other similar modelling studies (see Section 4.1.3); however, to better understand the role of transmission in high renewable scenarios two additional sensitivity scenarios are presented here: NoTrans and NoTransNoCCS. These scenarios are variations of the Baseline and the NoCCS scenarios, where the only difference is that interconnector capacities in Europe are locked to the planned expansions by 2020 found in ENTSO-E's Ten-Year Network Development Plan (TYNDP) 2016 [37]. Although such an assumption does not reflect any reasonable future for a European power system with high shares of renewable energy, the results from these transmission constrained scenarios can illustrate how the availability of grid expansion options (or lack thereof) impacts aspects of a cost-optimal system design and operation. In this analysis, the focus is on the following 2050 European aggregate results from EMPIRE: (i) deployment of wind and solar generation capacities, demand response and battery energy storage capacities, (ii) the use of natural gas for power generation, (iii) curtailed renewable energy generation and (iv) electricity prices. These results from both of our main scenarios and their transmission-constrained counterparts are shown in Table 4.5.

Table 4.5 Results from EMPIRE for 2050: overall generation mix for Europe

	iRES [GW]	DR [GW]	BESS [GWh]	Network [GW]	Natural gas [TWh]	Curtailed RES [TWh]	Avg. 'price' €$_{2010}$/MWh
Baseline	1,444	176	0	456	781	44	59
NoCCS	1,736	215	0	591	396	134	65
NoTrans	1,172	168	0	107	1,260	35	60
NoTrans NoCCS	1,646	220	71	107	423	107	80

Column legend: (i) iRES (intermittent RES): The combined installed capacity of wind (onshore + offshore) and solar PV. (ii) DR (demand response): The combined capacity of all demand response measures. (iii) BESS (battery energy storage system): The combined utility scale Li-ion battery storage capacity. (iv) Network: The total cross-border interconnector capacity (overhead AC and subsea DC). (v) Natural gas: The total power generation using natural gas (unabated + CCS). (vi) Curtailed RES: The total amount of (potential) renewable power generation that could have been used to cover load, utilized by demand response measures or used for battery charging, but had to be curtailed to preserve system power balance. (vii) Avg 'price': The average long-run marginal cost in 2050, found as the average of all the shadow prices for the load balance constraints in 2050. This reflects the marginal increase of system cost resulting from a marginal increase in load.

The apparent importance of natural gas CCS in the absence of grid investments for keeping the system costs down is of course highly dependent on the input data assumptions used in the modelling. By using other cost assumptions, (higher) for CCS and (lower) for batteries these results would likely change. However, the importance of local flexibility options and the impact of lack of such options on prices and costs would still remain valid.

As discussed in the previous section, the NoCCS scenario sees a higher installed capacity of renewable energy production, transmission and demand response than the Baseline. The total power generated using natural gas is twice as high in the Baseline scenario, which has been made possible within the emission constraint using CCS on half the production. Neither of these scenarios has any deployment of dedicated battery energy storage capacity; however, this is a result of temporal flexibility being made available through demand response measures. For the data used in this modelling analysis, demand response comes out as more cost competitive than batteries. The total average price is 10% higher in the NoCCS scenario.

Contrasting the main scenarios to those without additional transmission expansion reveals some notable details. When CCS is still available as a technology for power generation, the absence of grid expansion results in far less renewable energy capacity deployment than the Baseline (reduced demand response capacity as well). With the increased investments in CCS capacity, more natural gas is used, and the total production is three times higher than the NoCCS scenario. The underlying mechanism here is that the limited spatial balancing capability in the system necessitates far more national deployment of generation capacity. In this setting, a lot of renewables are still deployed; however, natural gas with CCS becomes an economic way of supplementing the renewable generation in each country as resource sharing is no longer as effective to mitigate local variations in renewable output. In terms of prices, this scenario is almost identical to the Baseline even though the system designs are quite different.

When both transmission expansion and CCS investments are unavailable, natural gas can no longer play the same role to cover demand for electricity. Similarly, as for the previously discussed scenario, the renewable resource sharing that was possible with a reinforced grid in our main scenarios has been reduced. The result is that a significant deployment of storage is needed to keep the system balanced. Neither of the scenarios has any significant curtailment of load (although voluntary load reduction as a demand response measure is used); however, the NoTransNoCCS scenario still shows signs of having a constrained resource situation as can be seen from the average price (reported as the long-run marginal cost). It is about 36% higher than for the Baseline and shows that any increase in the demand would be significantly more expensive to cover than for the other cases. As the average price is reported as the long-run marginal cost, we are considering that it may be covered by generation (or storage) capacity yet to be built. Therefore, the possible necessity of developing less favourable renewable generation sites, or deploying more batteries, to cover additional demand can impact this metric. As the total flexible power generation (i.e. natural gas) is not possible to increase further due to the carbon constraint, this is a likely explanation for the high price difference (Table 4.5).

Considering all the scenarios together, we can derive the following insights:

1. In power systems, covering wide regions with heterogeneous weather conditions for wind and solar power generation large-scale transmission expansion is part of a cost-optimal high renewable system design.
2. Without transmission investments accompanying renewable deployment 'local' generation and demand flexibility options become more important. If these are available at a reasonably low cost, the total system cost may still be similar to that of a solution with transmission expansion.
3. By constraining availability of local flexibility options, prices and costs quickly rise.

4.4 Clean energy exports in 2050 scenarios: the case of Norway

4.4.1 Norway power system perspective in 2050

This section focuses on the Norwegian results** in the EMPIRE analysis introduced in Section 4.3. The case of Norway is interesting in light of the discussion on the role of transmission in clean energy scenarios. Over the last decade, Norway has had an annual net power surplus every year except 2010 and 2019 (ranging between 5 TWh and 16 TWh in the period 2009–2019) [48]. Almost all power generation comes from hydropower (close to 95%), most of which with a high degree of

**A more extensive analysis of this case can be found in the report 'Norway's role as a flexibility provider in a renewable Europe' [19].

Table 4.6 *Energy balances for Norway in 2050 from EMPIRE for the Baseline and NoCCS scenarios (in TWh)*

Type	Baseline	NoCCS	NoTransNoCCS
Demand	152	152	152
Generation	266	459	191
Hydro regulated	120	121	120
Hydro run-of-the-river	30	28	31
Wind onshore	51	50	18
Wind offshore	65	258	23
Export	128	320	52
Import	16	16	14
Net export	112	304	38

possible regulation due to large mountain reservoirs [19]. Still, there is a vast potential in Norway for further increased power generation both from onshore and offshore wind. A development of wind resources in Norway can be used to support the energy transition of European power; however, this can only be achieved with strong interconnections between Norway and neighbouring countries. In the following, we analyse the role of Norway in the Baseline, NoCCS and NoTransNoCCS scenarios.

Table 4.6 shows EMPIRE results for energy generation by technology along with exchange across interconnectors for Norway in 2050. In all three scenarios considered Norway has a net positive power balance over the year, but in the NoCCS scenario the surplus is almost 200 TWh higher than in the Baseline. The reason for this difference in surplus can almost entirely be explained by a significant increase in offshore wind deployment in the NoCCS scenario compared with the Baseline. The effect of not having the possibility to reinforce interconnectors between Norway and the neighbouring countries is quite evident from the NoTransNoCCS scenario, as the wind energy deployment is far smaller than in other scenarios.

In Table 4.7, the interconnector capacities for both the Baseline and the NoCCS scenarios are shown for all connections between Norway and the neighbouring countries. Also, the capacities for 2020 from the ENTSO-E TYNDP 2016 are shown in the table, which are used in the NoTransNoCCS scenario. This table effectively shows what is required in terms of interconnector capacity expansion going from today's situation of about 6.5 GW[††] into a future where large amounts of wind generation can be potentially developed for export of energy to Europe. It is possible to export a significant amount of power with the planned capacities as of 2020; however, to realize a net export of 300 TWh/year (as in the NoCCS scenario) based on a massive offshore wind deployment the total exchange capacity must increase by a factor of 6.

[††]Both the NordLink cable, connecting Norway and Germany, and the North Sea Link, connecting Norway and Great Britain, will come into operation in 2021.

Table 4.7 Interconnector capacity between Norway and connected European countries

Connection to	2020 (ENTSO-E TYNDP 2016) [MW]	2050 Baseline (EMPIRE) [MW]	2050 NoCCS (EMPIRE) [MW]
Sweden	4,000	11,800	16,900
Denmark	1,600	7,600	12,900
Finland	100	4,900	9,800
Germany	1,400	1,400	5,400
Great Britain	1,400	1,400	8,100
The Netherlands	700	700	4,900
Belgium			4,000
Total	*9,200*	*27,800*	*62,000*

Note: Numbers rounded to nearest 100 MW.

This case study of Norway illustrates how transmission investments can facilitate development of renewable projects at resourceful locations (here especially offshore wind outside the Norwegian coast), and how lack of further grid reinforcements can provide a barrier for such projects.

4.4.2 Natural gas infrastructure in the energy transition

As in the scenarios applied to the EMPIRE model, most of the European decarbonization scenarios expect that gas demand projections could decrease from current level to at least 50% in 2050. Both Kotek *et al.* [40] and SET-Nav [41], a GGM-based study, analyse two gas demand projections in Europe. The gas demand in 2050 will be affected by the EU carbon budget goals (90%–95% decarbonization of the energy sector by 2050) targeted in different energy sectors (see Section 4.3.1). For example, this could include reducing gas consumption in buildings, industry and transport. Hence, the demand of gas will not only diminish in the power sector (as illustrated in the EMPIRE model results) but might have a wider impact on its utilization in other sectors and hence on the overall European gas infrastructure. To understand the effect of the energy transition on the gas infrastructure, let us consider two scenarios similar to the ones analysed in the EMPIRE model. These two scenarios are illustrated in Figure 4.7. One scenario is based on combining a projection study for the European Union, the PRIMES Reference scenario [31] and the 2017 new policies scenario (NPS) of the World Energy Outlook (WEO), which is called NPS-Ref [42]. The other scenario is based on the project SET-Nav,[‡‡] funded by the European Commission, in which a scenario called 'Directed Vision' [43] is combined with the 2017 Sustainable Development Scenario (SDS) of the WEO. These scenarios narrate two contrasting stories: on the one hand, the NPS-Ref does not achieve climate goals and hence shows a still significant gas demand projection. On

[‡‡]http://www.set-nav.eu/content/pages/results [accessed date: 13 March 2020].

the other hand, the SDS Vision scenario presents a strong decarbonization case in line with the 1.5°C Paris agreements (Figure 4.8).

Under these gas scenarios, we observed that in Norway gas export declines to 67 bcm (around 30% from current levels). According to the projections of the 'directed vision' scenario in SET-Nav, the average gas consumption in Europe in 2050 will be composed of 25%–30% for power generation and the remaining 70%–75% for buildings, industry, transport and others. By looking at the possible price of electricity in 2050, EMPIRE results and SET-Nav analyses report an average price (wholesale) for electricity around 80 EUR/MWh (in real 2010 value). The analyses also showed that the carbon price of electricity will be above 100 EUR/tCO_2. These carbon and electricity prices will mean that 25%–30% of the gas usage will likely be unprofitable and incentives will be needed to maintain the gas infrastructure. This emphasized the point that relying on gas infrastructure is economically uncertain and that transmission expansion in the power system is even more important if gas will not be a viable technology in 2050. The following subsection better exemplifies the gas utilization conditions from the perspective of a power system.

4.4.3 RES fluctuations and natural gas capacity (utilization) factor

On top of having a decreasing demand in various gas sectors and the consequences to the economic viability of gas, there is another challenge to use gas as a flexibility provider. As described in the EMPIRE results, gas power plays an important role in the energy mix in 2050, but the usage of gas is highly influenced by the hourly dynamics of supply–demand operations. In this regard, Figure 4.9 illustrates the operations of the top 200 peak time hours under a scenario where gas is a key

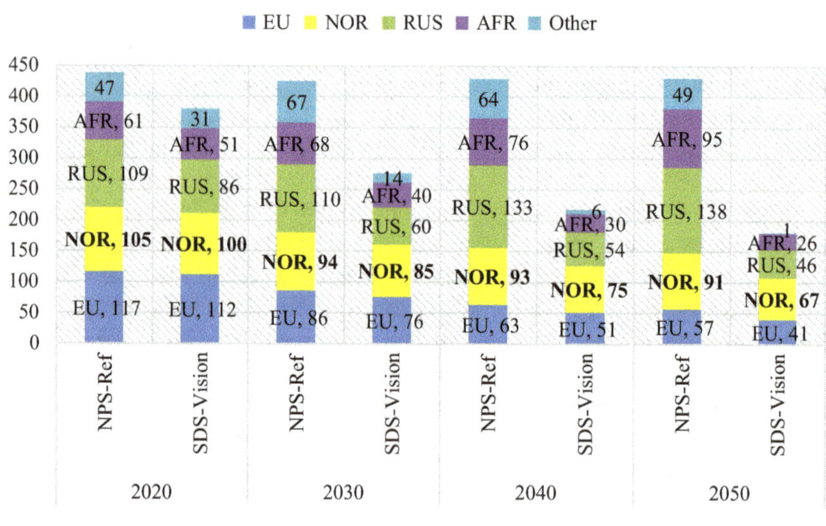

Figure 4.8 Total natural gas consumption in Europe and the contributing supply regions (BCMA/Year) (see Crespo del Granado et al., [9])

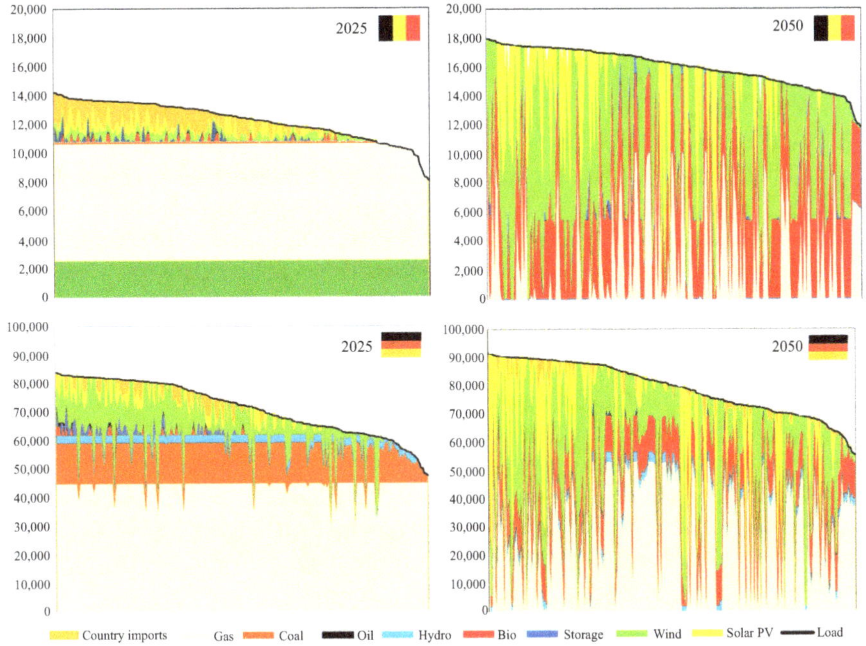

Figure 4.9 Load curves (highest peaks only, in MWh) for Belgium and Germany under a transmission expansion restricted scenario in 2025 vs 2050

flexibility source to a decarbonization case performed in EMPIRE (note that this case has restrictions imposed to transmission expansion, see [1]). This illustrative result focuses on the load curves for Belgium and Germany, where the hours are sorted from the largest to the smallest peak hour for 2025 and 2050. The following insights can be gained from the figure:

- Gas power plants play a strong transitional role: gas usage is primarily base-load in 2025 while the main use is for balancing support in 2050. This diminishes its utilization factor greatly (with the gas capacity factor decreasing from 65% in 2025 to 15% in 2050), but its synergy and flexibility with other technologies in 2050 contribute to integrating a large RES share.
- Storage and hydro become key flexibility providers when there is lack of RES generation. The load curves clearly note their involvement in periods of low solar availability.
- Towards 2050, biomass emerges as a key flexibility source, in part due to the high emission costs of other technologies and due to its synergy with gas power plants.
- Interconnection support and curtailment: in 2025, interdependency on importing generation for medium-to-high peak demand among countries is notoriously different compared with 2050. In addition, curtailments increase from practically nothing (2025) to 11% (2050).

4.5 Conclusions and highlights

Transmission expansion is a key enabler for the grid integration of wind and solar. Also, it is a fundamental tool to ensure stronger cross-sector integration to deliver affordable, secure and clean exports among countries. As illustrated in this chapter, having limitations in transmission expansion strongly undermines the prospects of reaching a 100% RES share in the energy system. Coupling extensive renewable deployment with a strong transmission infrastructure has several benefits. Regions with favourable renewable generation potentials can supply other parts of the system where RES prospects are limited. Also, weather patterns vary extensively across wide areas (such as the whole of Europe), but to be able to use spatial variations in production to balance the system a sufficiently strong grid is required. As shown in this chapter, if transmission investments are limited the reliance on local flexibility measures becomes more important. This can quickly become costly if such options are either expensive or limited. This is in line with a wide range of European energy transition studies analysing transmission which conclude that without transmission expansion other technologies with higher investment costs (batteries, hydrogen, etc.) will need to come into play. Planning for a strong grid therefore seems a robust strategy for achieving a deep decarbonization of the power sector with high shares of RES in an affordable way.

Another take away of this chapter is the approach on how to analyse the long-term aspects of capacity expansion in relation to the short-term dynamics of the power system. The demonstration and analysis of the EMPIRE model shows the complexity of analysing and integrating a large spatial and a long-term horizon in a single modelling framework,[§§] which also includes a stochastic representation of RES. This state-of-the-art modelling suite is unique in its ability to model a large number of countries (30 countries in Europe, in the studies presented in this chapter) and to study possible decarbonization strategies for the continent. It is a robust modelling method that balances the representation of technological details and timescale resolution. Despite of this, it does not give a full picture of the effects of the transition of the power system on other sectors. For example, the gas infrastructure under EMPIRE analyses is likely to become a stranded asset. Even then, the 2050 scenarios analysed show that gas is a key flexibility provider in Europe. This has contrasting messages when performing a long-term analysis of the gas infrastructure. Therefore, it is shown that gas demand is likely to decline, not only on the power system sector but also in other sectors (buildings, industry, transport, etc.). This means that not only gas power plants will change from providing baseload to regulation services, but also the gas demand market will decline, leading to a low utilization of the network infrastructure. Hence, a key conclusion is the fact that gas systems will likely be very expensive and could become stranded assets without new market designs (e.g. capacity markets and well-functioning markets for operational reserves) or subsidies. This should include reforming regulation markets and analyse in detail the valuation of gas flexibility

[§§]EMPIRE's computational dimension includes approximately 15 million variables and 22 million constraints.

participation in ancillary services. Modelling these services are usually overlooked in most studies as well as in the models discussed in this chapter.

Acknowledgements

The authors acknowledge the Research Council of Norway for its support to the project 'Planning Clean Energy Export from Norway to Europe', CleanExport. Furthermore, we are especially grateful for the funding of the CleanExport project provided by Agder Energi, Air Liquid, Equinor Energy, NEL Hydrogen Electrolyser, Gassco, Total E&P Norge and The Research Council of Norway under grant 295078. Authors are also thankful to colleagues of the FME CenSES centre and the H2020 SET-Nav project for fruitful discussions and contribution to EMPIRE and GGM models results.

References

[1] SET-Nav. (2018). *Unlocking Unused Flexibility and Synergy in Electric Power and Gas Supply Systems*. Retrieved from www.set-nav.eu [accessed: 13 March 2020].

[2] Papaefthymiou, G., and Dragoon, K. (2016). Towards 100% renewable energy systems: Uncapping power system flexibility. *Energy Policy*, *92*, 69–82. doi:https://doi.org/10.1016/j.enpol.2016.01.025

[3] Weigt, H., Jeske, T., Leuthold, F., and von Hirschhausen, C. (2010). "Take the long way down": Integration of large-scale North Sea wind using HVDC transmission. *Energy Policy*, *38*(7), 3164–3173. doi:10.1016/j.enpol.2009.07.041

[4] Heard, B. P., Brook, B. W., Wigley, T. M. L., and Bradshaw, C. J. A. (2017). Burden of proof: A comprehensive review of the feasibility of 100% renewable-electricity systems. *Renewable and Sustainable Energy Reviews*, *76*, 1122–1133. doi:https://doi.org/10.1016/j.rser.2017.03.114

[5] Reedman, L. (2012). *High Penetration Renewables Studies: A Review of the Literature*. Report prepared for the Australian Energy Market Operator (AEMO). Retrieved from https://publications.csiro.au/rpr/download?pid=csiro:EP127113&dsid=DS2 [accessed: 13 March 2020].

[6] Brown, T., Schlachtberger, D., Kies, A., Schramm, S., and Greiner, M. (2018). Synergies of sector coupling and transmission reinforcement in a cost-optimised, highly renewable European energy system. *Energy*, *160*, 720–739. doi:10.1016/j.energy.2018.06.222

[7] Gracceva, F., and Zeniewski, P. (2014). A systemic approach to assessing energy security in a low-carbon EU energy system. *Applied Energy*, *123*, 335–348. doi:10.1016/j.apenergy.2013.12.018

[8] González, I. H., Ruiz, P., Sgobbi, A., *et al.* (2015). *Addressing Flexibility in Energy System Models*. Retrieved from https://op.europa.eu/en/publication-detail/-/publication/f45e2db6-54c2-44a8-a937-03cbfe204ee8/language-en [accessed: 13 March 2020].

[9] Crespo del Granado, P., Skar, C., Doukas, H., and Trachanas, G. P. (2019). Investments in the EU power system: A stress test analysis on the effectiveness of decarbonisation policies. In H. Doukas, A. Flamos, and J. Lieu (Eds.), *Understanding Risks and Uncertainties in Energy and Climate Policy: Multidisciplinary Methods and Tools for a Low Carbon Society* (pp. 97–122). Cham: Springer International Publishing.

[10] Schlachtberger, D. P., Brown, T., Schramm, S., and Greiner, M. (2017). The benefits of cooperation in a highly renewable European electricity network. *Energy*, *134*, 469–481. doi:10.1016/j.energy.2017.06.004

[11] Child, M., Kemfert, C., Bogdanov, D., and Breyer, C. (2019). Flexible electricity generation, grid exchange and storage for the transition to a 100% renewable energy system in Europe. *Renewable Energy*, *139*, 80–101. doi: https://doi.org/10.1016/j.renene.2019.02.077

[12] Scholz, Y., Gils, H. C., and Pietzcker, R. C. (2017). Application of a high-detail energy system model to derive power sector characteristics at high wind and solar shares. *Energy Economics*, *64*, 568–582. doi:10.1016/j.eneco.2016.06.021

[13] Pleßmann, G., and Blechinger, P. (2017). How to meet EU GHG emission reduction targets? A model based decarbonization pathway for Europe's electricity supply system until 2050. *Energy Strategy Reviews*, *15*, 19–32. doi:10.1016/j.esr.2016.11.003

[14] Egerer, J., Gerbaulet, C., and Lorenz, C. (2016). European electricity grid infrastructure expansion in a 2050 context. *The Energy Journal*, 1–7. doi:10.5547/01956574.37.SI3.jege

[15] Backe, S., Crespo del Granado, P., Tomasgard, A., Pinel, D., Korpast, M., and Lindberg, K. B. (2018). Towards zero emission neighbourhoods: Implications for the power system. Paper presented at the *2018 15th International Conference on the European Energy Market* (EEM).

[16] Marañón-Ledesma, H., and Tomasgard, A. (2019). Analyzing demand response in a dynamic capacity expansion model for the European power market. *Energies*, *12*(15), 2976. doi:10.3390/en12152976

[17] Skar, C., Doorman, G., and Tomasgard, A. (May 2014). The future European power system under a climate policy regime. Paper presented at the *2014 IEEE International Energy Conference (ENERGYCON).*

[18] Skar, C., Egging, R., and Tomasgard, A. (2016). The role of transmission and energy storage for integrating large shares of renewables in Europe. *IAEE Energy Forum*, *Q1*, 35–36.

[19] Skar, C., Jaehnert, S., Tomasgard, A., Midthun, K. T., and Fodstad, M. (2018). *Norway's Role as a Flexibility Provider in a Renewable Europe*. Retrieved from NTNU: https://www.ntnu.no/documents/7414984/1281984692/Norway%E2%80%99s+role+as+a+flexibility+provider+in+a+renewable+Europe.pdf/a055c776-f2a5-468f-bce2-c2cf1943f1fc [accessed: 13 March 2020].

[20] Kaut, M., Midthun, K. T., Werner, A. S., Tomasgard, A., Hellemo, L., and Fodstad, M. (2014). Multi-horizon stochastic programming. *Computational Management Science*, *11*(1–2), 179–193. doi:10.1007/s10287-013-0182-6

[21] Skar, C., Doorman, G., Pérez-Valdés, G. A., and Tomasgard, A. (2016). *A Multi-Horizon Stochastic Programming Model for the European Power*

System. Retrieved from Trondheim: http://www.ntnu.no/documents/7414984/202064323/1_Skar_ferdig.pdf/855f0c3c-81db-440d-9f76-cfd91af0d6f0 [accessed: 13 March 2020].

[22] ZEP. (2013). *CO_2 Capture and Storage (CCS) – Recommendations for Transitional Measures to Drive Deployment in Europe*. Retrieved from https://zeroemissionsplatform.eu/ [accessed: 13 March 2020].

[23] Egging, R. (2013). Benders decomposition for multi-stage stochastic mixed complementarity problems – Applied to a global natural gas market model. *European Journal of Operational Research, 226*(2), 341–353. doi:10.1016/j.ejor.2012.11.024

[24] Egging, R., Gabriel, S. A., Holz, F., and Zhuang, J. (2008). A complementarity model for the European natural gas market. *Energy Policy, 36*(7), 2385–2414. doi:10.1016/j.enpol.2008.01.044

[25] Egging, R., Holz, F., Von Hirschhausen, C., and Gabriel, S. A. (2009). Representing gaspec with the world gas model. *The Energy Journal, 30* (Special Issue). Retrieved from https://www.jstor.org/stable/41323198

[26] Egging, R. (2010). *Multi-Period Natural Gas Market Modeling Applications, Stochastic Extensions and Solution Approaches*. University of Maryland.

[27] Holz, F., Richter, P. M., and Egging, R. (2015). A global perspective on the future of natural gas: Resources, trade, and climate constraints. *Review of Environmental Economics and Policy, 9*(1). doi:10.1093/reep/reu016

[28] Holz, F., Brauers, H., Richter, P. M., and Roobeek, T. (2017). Shaking Dutch grounds won't shatter the European gas market. *Energy Economics, 64*, 520–529. doi:10.1016/j.eneco.2016.03.028

[29] Holz, F., Richter, P. M., and Egging, R. (2016). The role of natural gas in a low-carbon Europe: Infrastructure and supply security. *The Energy Journal, 37*(1), 33–60. doi:10.5547/01956574.37.SI3.fhol

[30] Kotek, P., Tóth, B. T., Crespo del Granado, P., *et al.* (2018). *Case Study on Project of Common Interest and Gas Producers Pricing Strategy*. Retrieved from www.set-nav.eu [accessed: 13 March 2020].

[31] European Commission. (2016). *EU Energy Transport and GHG Emissions: Trends to 2050 – Reference Scenario 2016*. Brussels.

[32] IEA. (2016). *Energy Technology Perspectives 2016*. Retrieved from: www.iea.org.

[33] Fraunhofer ISE. (2015). *Current and Future Cost of Photovoltaics. Long-Term Scenarios for Market Development System Prices and LCOE of Utility-Scale PV Systems: Study on Behalf of Agora Energiewende*. Retrieved from https://www.ise.fraunhofer.de/content/dam/ise/de/documents/publications/studies/AgoraEnergiewende_Current_and_Future_Cost_of_PV_Feb2015_web.pdf [accessed: 13 March 2020].

[34] Gils, H. C., Scholz, Y., Pregger, T., Luca de Tena, D., and Heide, D. (2017). Integrated modelling of variable renewable energy-based power supply in Europe. *Energy, 123*, 173–188. doi:10.1016/j.energy.2017.01.115

[35] Gerbaulet, C., and Lorenz, C. (2017). *dynELMOD: A Dynamic Investment and Dispatch Model for the Future European Electricity Market*. Retrieved from https://www.diw.de/de/diw_01.c.558131.de/publikationen/data_documentation/2017_0088/dynelmod__a_dynamic_investment_and_dispatch_model_for_the_future_european_electricity_market.html [accessed: 13 March 2020].

[36] IEA. (2016). *Nordic Energy Technology Perspectives 2016*. Retrieved from: www.iea.org.

[37] ENTSO-E. (2016). *Ten Year Network Development Plan (TYNDP)*. Retrieved from https://www.entsoe.eu/publications/tyndp/tyndp-2016/ [accessed: 13 March 2020].

[38] London Economics. (2013). *The Value of Lost Load (VoLL) for Electricity in Great Britain*. Retrieved from https://www.ofgem.gov.uk/ofgem-publications/82293/london-economics-value-lost-load-electricity-gbpdf [accessed: 13 March 2020].

[39] Oei, P.-Y., Mendelevitch, R., and Herold, J. (2014). Modeling a carbon capture, transport, and storage infrastructure for Europe. *Environmental Modeling and Assessment*, *9*(6), 515–531. doi:10.1007/s10666-014-9409-3

[40] Kotek, P., Crespo del Granado, P., Egging, R., and Toth, B. T. (18–20 September 2019). *European Natural Gas Infrastructure in the Energy Transition*. Paper presented at the *2019 16th International Conference on the European Energy Market (EEM)*.

[41] SET-Nav. (2019). *Comparative Assessment and Analysis of SET-Nav Pathways*. Retrieved from www.set-nav.eu [accessed: 13 March 2020].

[42] IEA. (2017). *World Energy Outlook 2017*. Retrieved from Paris: https://www.iea.org/reports/world-energy-outlook-2017 [accessed: 13 March 2020].

[43] Crespo del Granado, P., Resch, G., Holz, F., *et al.* (2020). Energy transition pathways to a low-carbon Europe in 2050: The degree of cooperation and the level of decentralization. *Economics of Energy & Environmental Policy*, *9*(1), 121–135. doi:10.5547/2160-5890.9.1.pcre.

[44] Denholm, P. (2015). *Simplified flexibility parameters for evaluating renewable integration*. Addressing flexibility in energy system models, JRC Science and Policy Reports.

[45] Koljonen, T., Kiviluoma, J., Lehtilä, A., Ikäheimo, J., Koreneff, G., Holttinen, H., and Forsström, J. (2015). Holistic approaches in addressing flexibility in energy system modelling. In Hidalgo González, G., Ruiz Castello, P., Sgobbi, A., Nijs, W., Quoilin, S., Zucker, A., and Thiel, C. (Eds.), *JRC science and policy report: Addressing flexibility in energy system models* (pp. 27–29). Luxembourg: Publications Office of the European Union.

[46] Zappa, W., Junginger, M., and van den Broek, M. (2019). Is a 100% renewable European power system feasible by 2050?. *Appl. Energy*, 233–234, August 2018, 1027–1050. doi: 10.1016/j.apenergy.2018.08.109.

[47] Cole, W. J., Marcy, C., Krishnan, V. K., and Margolis, R. (2016). Utility-scale lithium-ion storage cost projections for use in capacity expansion models. In: *2016 North American Power Symposium (NAPS)*. doi:10.1109/NAPS.2016.7747866

[48] NVE. (2020). Quarterly report for the power market, fourth quarter 2019 [In Norwegian]. Retrieved from https://www.nve.no/media/9092/kraftsituasjonen_q4_2019.pdf [accessed: 13 March 2020].

Chapter 5

Integrating renewable energy into the distribution grid: general aspects and the case of Mexico

Ricardo Cruz[1], *Daniel Chacón*[1] *and Rodrigo Palacios*[1]

Distributed generation (DG) is traditionally identified with both off-grid applications at remote locations, often serving a minimal yet vital load, and on-grid facilities supporting the main grid, such as emergency generators, as well as niche applications located in communities with a strong commitment with sustainability. Decentralized cogeneration, for example, made its appearance in countries with a strong heating load long before climate change concerns made it into the minds of the general public. In the current society, distributed *renewable* energy generation[*] is rapidly moving out of the niche, both to provide *energy services* such as illumination, access to potable water, access to information through the Internet, and even formal education for those still living off the grid, and to make a sizeable contribution to the clean energy transition. Undoubtedly, the main momentum currently comes from the dramatically lower cost of solar photovoltaic (PV) systems, compared with less a decade ago, available in many parts of the world. As storage solutions are recently evolving with a similar pace as solar systems, firming up DG or making it *dispatchable* at the convenience of the grid, DG is rapidly becoming a feasible alternative to conventional centralized generation.

Evidently, electric utilities (in the traditional sense of a vertically integrated electric company with a regional monopoly) and grid operators (where the traditional regional monopolies have been broken up, at least partially, into market-based structures) are quick to point to the considerable challenges brought about by a myriad of generators distributed over their service areas, and they often have a technical case for it. Designing a robust yet flexible scheme for the coordination of protections, to name one major issue, is not a trivial issue, although modern approaches based on artificial intelligences and distributed sensors, feeding into the Internet-of-Things (IoT), are poised not only to solve the problems caused by

[1]Iniciativa Climática México, Ciudad de México, CDMX, Mexico

[*]Energy *generation*, as in most of the relevant literature, is used here, rather colloquially, instead of the physically correct term energy *conversion*.

distributed generators but also to provide a whole range of functional capabilities unthinkable in a traditional distribution grid. These new services range from the detection of electricity theft (a common problem in many parts of the world) to demand response services; these services are covered in greater detail in Chapter 8 on smart grids and Chapter 9 on demand response. Evidently, upgrades to the distribution infrastructure have to be paid for, and even though much of the concerns can be addressed by requiring distributed generation (DG) equipment (such as inverters) to comply with certain standards, thereby obligating the end user to foot the bill on his or her part of the grid upgrade, other upgrades have to be managed in a more centralized way. When the grid operator, the oversight authorities and the end users are all on board with this project, such a task seems manageable. More often than not, however, do grid operators or utilities stand in the way of such upgrades, driven by an understandable fear of uncertainty and loss of control. A revised role of the utility or grid operator as a smart administrator of resources, rather than a monolithic provider of a commodity, may bring about more opportunities than threats but evidently requires both individuals and institutions to change, an attitude often uncommon in institutions built to function in a predictable way. More importantly, a changing environment is likely to favor the spirited and motivated but may be detrimental to those grounded in continuity, and maybe even those whose technical expertise has been a cornerstone of the company over long periods of time. Addressing such challenges is at least as important as solving the technical issues of the transition.

In this chapter, an overview of the benefits and challenges of DG is given, followed by an overview of the status of DG in different parts of the world. The remainder of this chapter is dedicated to an in-depth study of particular country case, the one of Mexico.

5.1 Origins, benefits, and challenges of DG

DG, sometimes also referred to local or "retail" generation, is quite generally defined as electricity generation occurring near the load being served. A distributed generator will typically be a solar PV generator, a small wind turbine, a fuel cell, a biogas-fired generation plant, or a generator propelled by conventional fuels such as a diesel generator, a micro-turbine, or a cogeneration plant [1]. System sizes range from a few hundreds of watts to several hundred kilowatt, though the exact range will depend on the jurisdiction where the system is located in, and the corresponding regulations.

In Mexico, the residential distributed generator (using solar PV or wind generators) was limited to a nominal capacity of 10 kW and was managed on a net metering scheme. Distributed commercial generation was allowed up to a capacity of 100 kW, with other sale and marketing options available (see Section 5.3). In the last legislative reform, nominal capacity was increased to 500 kW; the three connection mechanisms were defined; and the residential–commercial separation was modified to a division based on voltage level and consumption.

The original of DG is often identified in the myriad of emergency generators and cogeneration installed by private users (generally businesses of all kinds) and providing backup to the system, without grid operators and utilities having to care for them. A 2007 study by the U.S. Department of Energy [2], commissioned by the U.S. Congress, found that about 12 million of emergency units existed at the time, for a total installed capacity of 200 GW, in addition to 84 GW of consumer-owned Combined Heat and Power (CHP) units. The same DOE study concluded that

> DG offers potential benefits to electric system planning and operations. On a local basis there are opportunities for electric utilities to use DG to reduce peak loads, to provide ancillary services such as reactive power and voltage support, and to improve power quality. Using DG to meet these local system needs can add up to improvements in overall electric system reliability.

The study also found that

> [distributed generation may] decrease the vulnerability of the electric system to threats from terrorist attacks, and other forms of potentially catastrophic disruptions, and to increase the resiliency of other critical infrastructure sectors as defined in the National Infrastructure Protection Plan (NIPP) issued by the Department of Homeland Security, such as telecommunications, chemicals, agriculture and food, and government facilities.

While many of the conclusions in the DOE study stem from a period when widespread solar PV and other renewable energy systems were yet uncommon in the distribution grid, most of the conclusions remain valid, especially if the advantages of modern inverter-based technologies are properly taken advantage of. Distributed renewable generation, particularly if equipped with a (modest amount of) storage, may significantly contribute to grid resiliency by keeping the lights on (through intentional islanding) and stabilize the distribution grid in the event of a power outage. As stated in the DOE study, "94% of all power outages are caused by faults and failures in the distribution system." Resiliency can be further increased by micro-grids [3] which, though normally operating interconnected to the main grid, can fully function on its own in case the surrounding grid goes into a cascade of disconnects. Such stability islands keep failures from spreading and represent effective insurance policies against large-scale blackouts known to cause billions in economic damages [4].

Challenges standing in the way of a comprehensive restructuring of the distribution grid in a way that allows to accommodate large amounts of clean energy abound as well, however. Oft-quoted examples [1] include (1) unintentional islanding in the event of service events, potentially exposing maintenance crews to safety hazards, (2) reversed power flow (from sections to the distribution grid traditionally expected to have a lower electric potential to sections with higher potential, and even from the low-voltage into the medium-voltage grid), obligating grid operators to rethink the type of relays used for protective services,

(3) protection coordination, i.e., the triggering of protective measures in response to certain geographic voltage patterns, and (4) reactive power flows. Concern (4), stemming mostly from the times when relatively small wind turbines (with rated capacities in the order of tens or a few hundred kilowatts) operating with asynchronous generators dominated the DG landscape, at least in some key areas such as California, Denmark, or Germany. With the event of modern power electronics, most distributed generators are capable of operating at unit power factory or may even inject or absorb reactive power at the convenience of the local grid, helping reduce losses and keep distribution lines and transformers within their capacity limits. Issues (1)–(3) can all be traced back to the fact that traditional distribution grids have a *radial* design, i.e., expect the power flow to occur from the feeder of the distribution grids toward the farthest branches (in an electrical sense). Accommodating large amounts of (renewable) DG will therefore undoubtedly require a major effort in regulations, technical norms, and investments in modernized infrastructure.

5.2 Overview of DG internationally

Across the world, renewable DG is showing two main areas of development: supplying energy to isolated communities and reducing the great cities' energy dependence. The first area is more notable in developing countries where collective distributed energy is a more economic and faster way to provide energy to distant communities than the development of a project for connecting the locality to the rest of the country—often, that kind of projects are not economically viable since the high costs of the project and the low demand of the community. The problem is such that the WEO estimated in 2017 that there were around 1.1 billion people—equivalent to 14% of the global population—without access to electricity around the world and that 95% of those people live in the sub-Saharan Africa or developing Asia [WEO, 2017] [5]. In countries, such as Kenya and Uganda, the number of isolated systems has overtaken the grid connections provided by the municipal or national electrical services [7]. It is foreseen that the cost reduction trends and the easier-to-operate systems will help more than 1,000 million people who currently lack energy access. In fact, Brazil, Cambodia, South Africa, China, Bangladesh, and India have promoted rural electrification by using a range from individual solar home systems to village-level microgrids [8].

One important option is the off-grid solar devices (OGS) such as pico solar PV [9] or the Plug-and-Play solar systems [10] since it is easy to install and can provide not only electricity but also energy for heat and cooking. It is thought that the OGS potential market is around 434 million houses in Africa and Asia. In Bangladesh, solar home systems are experimenting a growing trend and there are thousands of them. However, they are not connected to the grid although the country has a vast power distribution network [11].

For the matter of connected services, the potential is also high in developing countries. For instance, in Kenya, the Government has set the goal of 100%

electrification in 2020—while it was 47% in 2016—which is the main approach of the off-grid power stations in rural areas [5]. In Uganda, where the electricity access was 12% in 2010, the main source of electricity for the grid is based on diesel generators [12]. As in Kenya, distributed energy is more cost-effective and provides more reliable energy than grid extension [13].

The other area is more focused in developed countries where urbanization has made cities become, in energy sinks, dependent on the power generation from other regions. Despite DG—even with storage—is unlikely to be able to supply the whole demand of the cities, it certainly alleviates the issues that the power system faces in such energy sinks. For instance, it can reduce the congestion and losses in transmission, and it can discard the necessity of investment in new transmission lines.

One very interesting case, due to its similitude with Mexico's case—as it is seen in this work—is Brazil where DG based on solar PV has seen a significant increase since 2012, but it is thought that regulatory improvements are still required. Some proposals for that improvement are to (a) reduce taxes on PV equipment, (b) allow the consumers to sell their energy surplus provided by the systems, and (c) include the solar PV and energy efficiency in housing programs [14].

The DG is thriving. For instance, in the United States solar DG addition is comparable with the ones from utility scale solar or wind, accounting around 18% of the total new capacity in that country every year since 2016 [15,16].

California is an outstanding example of a state supporting DG, having more than 143,000 projects summing up to 8,854 MW installed [17]. Since 1998, California has supported these technologies and in 2006 the California Solar Initiative was created by Senate Bill 1 with a total budget of US$2.167 billion and a goal to install approximately 1,940 MW of new solar generation capacity between 2007 and 2016 [17]. Now, the state has specific sector programs to maintain the impulse on DG. This turns important since, as it is further discussed in this chapter, one of the needs in Mexico is to create programs dedicated to each sector. For example, California has solar programs aimed at determined types of housings. One is the Solar on the Multifamily Affordable Housing (SOMAH) Program that is directed to low-income families. Its aim is to install 300 MW of solar PV by 2030 [17]. Another program is the Multi-family Affordable Solar Housing (MASH), that is aimed at multifamily buildings and has the goal of installing 35 more MW by 2030 [17]. But programs are not exclusive to the residential sector. The RES-BCT Program allows local governments with generating facilities to export energy to the grid and receive credits that benefit the accounts of the same local government [17]. But programs are not exclusive to the residential sector. The RES-BCT Program allows local governments with generating facilities to export energy to the grid and receive credits that benefit the accounts of the same local government [17].

It is true that currently there is a debate all around the world about if the implications of distributed energy are all good for energy companies—and their centralized assets—transmission and distribution facilities, users, and the quality and security of the energy in the power systems.

For the technical aspects, the variability, intermittency, and uncertainty of DG are still to be studied to understand the impacts on the distribution circuits that may see voltage, frequency, and flow variations, energy losses, and tap changing operations depending on the specific circuit configuration [18]. Thus, the technical implications of DG depend on the site. Knowing that solar PV has greater impacts in the quality of energy than wind technologies, and low voltage systems are more affected than medium voltage ones [19], it is possible to use methodologies to assess the limit of insertion of DG technologies in the distribution systems, for instance using an indicator known as the hosting capacity [20].

Impacts on average retail rates may not be so high and, as well as the impacts on the profits of utilities, they are highly dependent on the regulatory environment [21]. Tariff structures and the decisions that the DG owners may take will affect deeply the attractiveness—payment period, energy costs savings, among others—of DG [22]. But tariff regulations are also an important piece of the puzzle. In Germany, for example, the allocation of recovery costs was discussed in 2018. Grid costs, connection charges, and other tariffs have to be very well regulated to avoid the risk of disproportionately burdening load customers in sparsely populated areas with favorable conditions for renewable energy production and, at the same time, to keep away customers from a real or perceived inequity that could negatively affect the public acceptance for the energy system transformation [23]. There are ways to optimize the use of grid-connected DG and its economic benefits, being the design of multipart tariffs with fixed components for the recurring grid costs and variable costs for the net effect of operating the distribution grids [24].

To conclude, DG has great potential in helping with accessible, low costs and clean energy, although there are technical and regulatory barriers that may be addressed. DG is transforming the whole electric industry around the world but its greatest contribution is how it is changing the energy paradigms by providing energy users a way that alleviates both their energy costs and needs while demanding from them more active participation and a better understanding of how the electricity sector works, and how do they get energy and their responsibility in its consumption.

5.3 DG in the Mexican electric sector

Mexico has recently evolved from a state monopoly, characterized by a vertically integrated utility, into a wholesale electricity market where generators compete freely, at least in principle. The control over the transmission and distribution grids remains in the hands of the state, through the National Energy Control Center, and the physical operation of both grids is currently limited to state companies as well. The creation of the electricity market is a direct result of the constitutional energy reform and specifically the Power Industry Law (PIL). The PIL heavily draws on international experience and explicitly allows for DG and demand control (DC). While the latter is still in its very early infancy, DG, mostly fueled by the low cost of solar PV panels, has experienced a considerable boom in the last few years. As opposed to other parts in the world (e.g., Germany) where solar PV installations have been heavily subsidized, the Mexican growth is completely self-sustained, gaining its momentum from both the abundant solar resource and the low module prices not available a few years ago. Unsurprisingly, PV systems are installed by businesses and middle- and upper-class households with mostly high electricity bills. Low-consumption households, representing most dwellings in Mexico, have been barely touched by these recent developments, both because of subsidized tariffs for the basic consumption block and the lack of financial resources or access to competitive financing conditions. In the current contribution, a detailed plan is laid out by which the current practice of subsidizing electricity tariffs could be replaced by an ambitious financed solar installation program that could both free considerable public resources and boost the transition to a cleaner electricity sector.

5.4 The Mexican power sector

The Mexican power industry, as in most of the countries, began with the building of the first power plants (PPs) for direct use in industries such as mining and textile manufacturing. In 1889, the country accounted for a total installed capacity of 838 kW with a 9.5% contribution from hydroelectric generation and the rest corresponding to thermal generation. The first hydroelectric PP in Mexico, with a capacity of 22 kW, had been built that same year. In 1937, the Federal Electric Commission (CFE, for its Spanish acronym) was created by the government with the goal of organizing and steering a national generation, transmission, and distribution power system. One of its commands, by law, was to procure power to every Mexican, looking not for the company benefit but the development of the country.

Figure 5.1 shows the evolution of the installed capacity in Mexico for the period 1889–1960 in which the generation grew around the main cities as isolated regions, driven by both the state and private investments. On September 27, 1960, the 27th article of the Mexican constitution was modified, establishing that the generation, transmission, and supplying of power were exclusive to the state. All these activities were then labeled as the Public Service, regulated by the subsequent Law of the Public Service of Electric Energy (LPSEE, for its Spanish acronym), which also had the strong ideal that the power industry had to be a development engine for the country. Later, the nationalization of the Mexican power industry began and CFE progressively acquired the private assets. In addition, a transmission grid began to be developed to connect the different generation regions. The grid was first was based on a 230 kV voltage level, but with the first great hydro projects—such as the *Infiernillo* hydro PP, currently in operation—the voltage for the larger transmission lines became 400 kV. Remarkably, it was not until 1971 that the standard frequency was set to 60 Hz for the whole of the country.

The last private participation in the Mexican power industry remained in the hands of a company founded with Canadian capital known as The Mexican Light and Power Company. This company, established in 1898, supplied power to

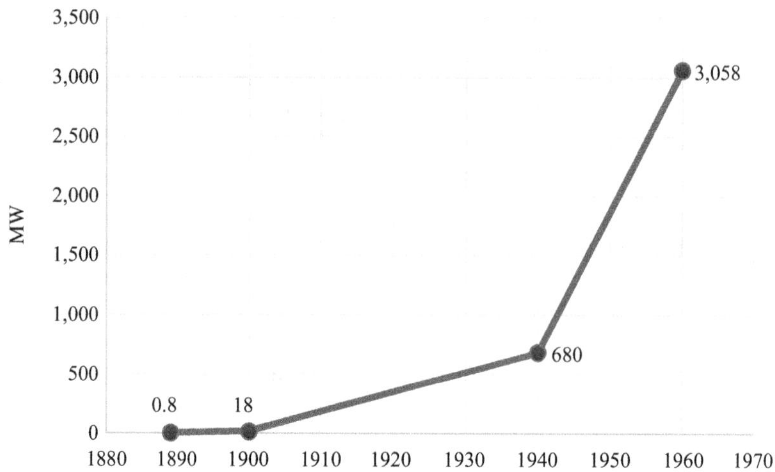

Figure 5.1 Evolution of the total installed capacity in Mexico from 1889 to 1960

Mexico City and its surroundings and was the main company of the region. As Mexican human resources became better trained and reached higher positions in the company, the private share decreased. In 1971, the state merged the different subsidiaries of this company to form Light and Power of the Center (LyFC for its Spanish acronym), the only power company in the center of Mexico, still with a small private share. The Mexican Government tried several times to liquidate the private share to extinguish LyFC, so CFE could be the only power company in Mexico but the intervention of the syndicate—the Mexican Electricians Syndicate, or SME for its Spanish acronym—made this impossible until 2009, when Felipe Calderon, the Mexican president declared the extinction of the LyFC. Previously, in 1999, the private assets in the company had been already settled. The next day after the president's decree CFE took on the operation of the whole of LyFC's system.

Even though the LSPEE restricted the power industry to the state only, some schemes were defined to allow for private investment. One is the still existing PIDIREGAS (the Spanish acronym for Productive Infrastructure Investment Project with Deferred Registration in Public Expenditure), result of a modification of the law for public expenditure. This scheme allows the federal institutions to develop infrastructure by bidding the project out to private investment; the government's payment obligations start once the private company delivers the operational facility. In this sense, a PIDIREGAS is a project in which a private party pays the construction costs to finance public infrastructure.

Another scheme emerged as a modification to the LSPEE itself. In 1992, a law amendment was passed decreeing that generation itself was not a public service anymore. Hence, any private party was allowed to generate energy but subsequent to a number of specific conditions. One of the most prominent conditions was self-supply, i.e., the self-consumption of the total output of the PP. Other schemes

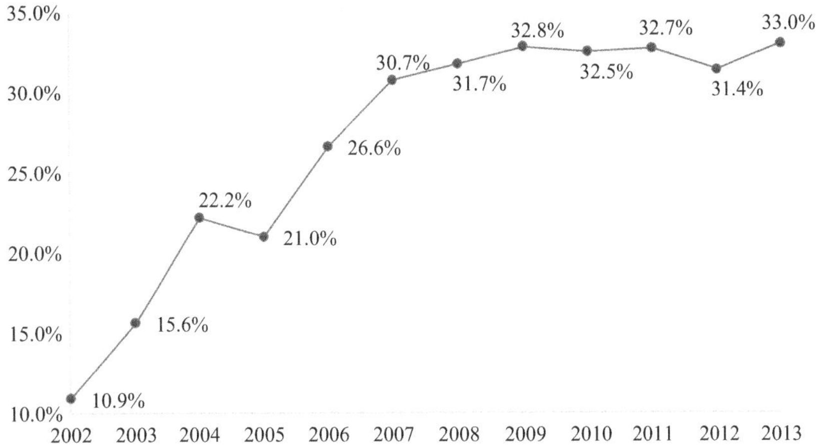

Figure 5.2 Share of the generation from PIEs in the Mexican power system

were subjected to the condition that the energy would be sold exclusively to CFE since transmission, distribution, and supplying were still considered a public service. As a result, the concept of the Independent Energy Producer (IPP or PIE, for its Spanish acronym) became highly important in the Mexican power industry. A PIE was a generation company that planned, financed, built, and operated PPs to supply CFE under a long-term contract, thus establishing the amount, price, and conditions in which the private would deliver its energy. This scheme became one of the main sources of power generation in the country reaching in 2013, the year when the energy reform abrogated the LSPEE, a 33% of the total generation (Figure 5.2).

5.4.1 The changes within the Mexican power industry law

In August 2014, Mexico approved the PIL where several changes in the electric legislation were made. For that purpose, the LPSEE—effective since 1975—was abrogated. That law allowed no other than the state to generate, transport, distribute, and control the Power System, with the specific exceptions in the generation sector (effective in 1992) pointed out earlier. Retail was state-controlled and the only differentiation between services was the tariff. The whole management of the industry was under the state-owned power company (CFE). In other words, there was a state-owned vertical integrated monopoly. Figure 5.3 shows the phases of this industry where all the processes to deliver energy to the Final Users—the blue arrow—belonged to CFE. Payments—the green arrow—of course went from the Final Users to CFE.

As already mentioned, after a 1992 reform to the LPSEE, private generators were allowed to produce energy for CFE—who would then sell it to the final users—or for their own use. In that sense, Mexican power industry was evolved in 1992 from a state-owned monopoly into a kind of market since there was

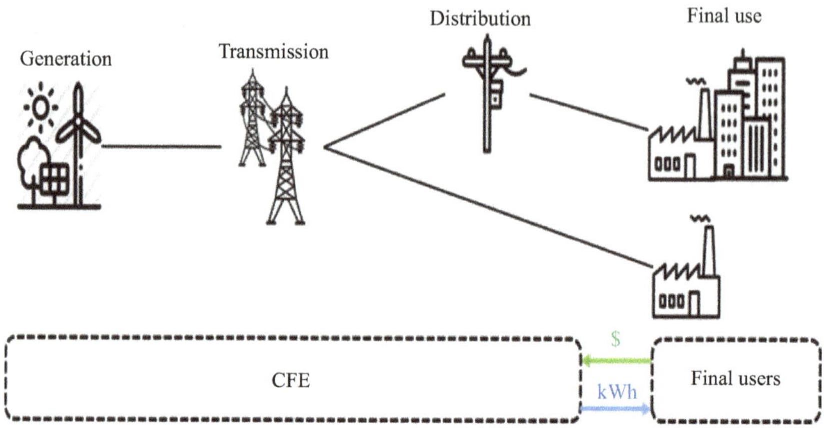

Figure 5.3 The Mexican power industry under the LSPEE with a state-owned monopoly (until 2014)

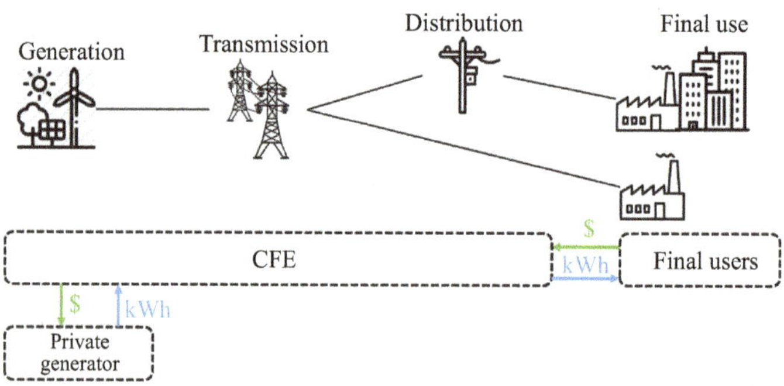

Figure 5.4 The Mexican power industry under the LSPEE after the 1992 modification. CFE retains the monopoly but private generation is allowed for exclusive sales to CFE or for self-generation

competition in generation. Another feature that was introduced with this reform was the self-generation concept in which a user—for example, an industry—could have a PP to generate its own energy without having this central in the same location of the consumption. Figure 5.4 illustrates how private generators were allowed to participate in the electricity sector of the 1992 reform to the LPSEE.

In 2013, Mexico changed its constitution to permit private companies to participate in various processes of the energy industry. One year later, the country approved the PIL that allowed for the creation of the Wholesale Electricity Market (MEM, for its Spanish acronym). In this new market, every generation company—called Generator—can produce and sell energy and other products to wholesale buyers (known as Certified Users). Yet distribution and transport are exclusive to the state with the assets belonging to the newly created state-owned companies

(CFE-Transmission and CFE-Distribution). However, both distribution and transport are considered activities of the MEM and its payments are under tariffs established by the state regulator of the market, the Energy Regulation Commission (CRE, for its Spanish acronym). Retail is regulated and currently the state-owned company, created specifically for that purpose, CFE-Basic Supply is the only one in the market. Nevertheless, every private company may participate as a figure known as a Basic Supplier since 2018. Basic Supply must retail at the tariffs established by the CRE. Certified Users can purchase energy and associated products directly in the market or from Certified Suppliers. Certified Suppliers and Basic Suppliers may buy and sell energy and associated products directly from the market or from Marketers who may sell and buy energy as well but do not represent physical assets. To be protected from the volatility of the spot market, Basic Suppliers are required to get those products through long-term contracts. For that purpose, Long Term Auctions are required to be executed by law. Finally, PPs may sell energy and associated products directly to the market—as Generators—or to a Marketer or Supplier.

The goods that may be bought and sold are energy (electricity) and the following associated products:

- *Capacity* ("Potencia"). Capacity can be roughly interpreted as the amount of energy provided by a PP to the system during the most critical hours of the system. The 100 yearly hours with the lowest reserve margins (originally, the highest load) are used to calculate that number.
- *Clean Energy Certificates* (CEL, for its acronym in Spanish). An accrediting title that certifies that 1 MWh has been generated by a Clean Energy Power Plant, roughly defined as a renewable energy, an "efficient" cogeneration, or a nuclear PP. Large hydro plants are subject to a power density criterion. Only plants or plant expansions built under the PIL terms are entitled to receive CEL credits.
- *Ancillary Services*. Certain products related to power system operations such as reserves, voltage, frequency regulation, and emergency start. While some products are traded in the market, such as (secondary) frequency regulation (see Chapters 6–8), others are a requirement for interconnection (such as primary frequency regulation, also known as frequency response) and are not remunerated.
- *Financial Transmission Rights*. Hedges on the price differences between the nodes of the transmission system.
- *Controllable Demand (CD)*. Demand that may be reduced upon request by the grid operator during certain critical times. Specific regulations for enabling a CD market are still missing.

All products may be bought and sold through bilateral contracts while energy and related services may be also bought and sold in any of the three short-term phases of the MEM:

- Day-ahead market (MDA, for its Spanish acronym). A market where all the Participants of the Market (Generators, Marketers, Suppliers, and Certified Users) make their offers for the next day of operation.

- Real-time market (MTR, for its Spanish acronym). The market where the differences between the forecast of the MDA and the real operation of the system are dealt with. The Participants of the Market make their offers for the next 15 min of operation.
- Hour-ahead market (MHA, for its Spanish acronym). This market deals with the differences between the forecast made 1 day ahead—in the MDA—and the one made 1 h before the execution. This market is not in operation yet.

The operation of the power system, as well as the management and execution of the short-term phases of the MEM, is managed by the state-owned system operator, the National Energy Control Center (CENACE, for its Spanish acronym).

In Figure 5.5, an overview of the topology of the Mexican electric system after the introduction of the PIL is shown, exhibiting the different market participants, as well as their physical connections, shown in black lines. The black dotted lines represent the limited physical connections that may be operating in special cases such as Isolated Supply (IS), Local Generation (LG), and certain types of DG. The green lines that connect the Participants of the Market—in the green boxes—depict the possible flux of money. Since Basic Users are not Participants of the Market, they can only buy energy from Basic Suppliers.

As seen in Figure 5.5, the MEM not only considers the traditional power system scheme where generation is centralized in large grid-connected PPs. To account for the possibility that the grid cannot supply energy to a given user, or the user wants to self-supply him/herself, the PIL created three different options: DG and two kinds of on-site generation—IS and LG.

IS and LG are schemes where a PP and a load are connected by a private transmission line that is not initially connected to the national grid. The PP and the load center may or may not belong to the same economic group—this is where the

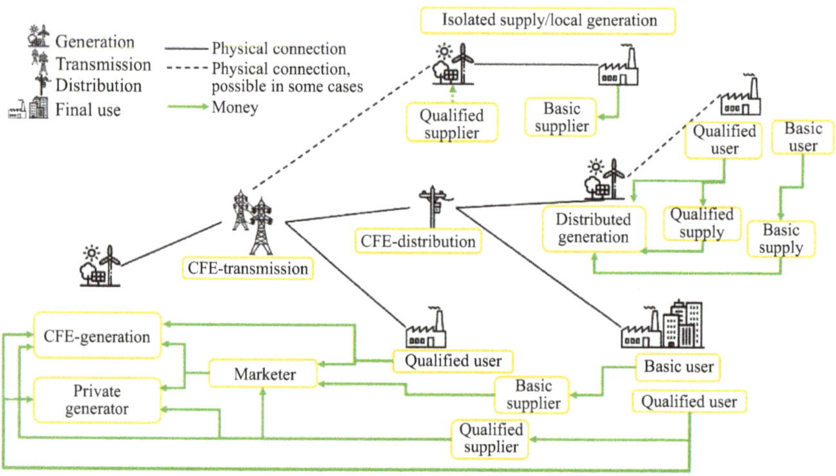

Figure 5.5 The Mexican power industry under the PIL (since 2014)

difference between both schemes lies—and the economic and energy exchange processes are not part of the MEM. Nonetheless, the PP may be connected to the transmission grid and the load to a distribution or transmission grid to sell and buy energy implying that when this happens the rules of the PIL and the MEM become the same as to any other PP and load.

5.4.2 Regulation of DG

A Distributed Generation Power Plant is recognized in the PIL as long as the facility fulfills some considerations; first among them is that it must be connected usually to the distribution grid or the transmission lines. Another consideration is made based on the electrical influence that the DGPP has on the grid in which it is connected. Such influence was depicted in the law as follows:

- The PP capacity must be lower than the nominal demand of the load centers of the distribution circuit to which the PP is connected, or
- The PP must lower or not to affect the loads of each element of the distribution circuit.

If the PP has a capacity lower than 500 kW, the law assumes that both criteria are satisfied. In other words, a PP smaller than 0.5 MW connected to a distribution grid is a DGPP, and therefore it does not require a Generation Permission.

Regarding energy trade, the owner of a DGPP needs to have a contract with a Supplier—Basic or Qualified, depending on the user's energy demand—not only to receive energy when needed, but also to sell the energy surplus that the DGPP may provide. In the case the DGPP is part of an IS scheme, that contract is not needed. If the contract is with a Basic Supplier, it is regulated by the CRE. If it is with a Qualified user, there is no regulation. Figure 5.6 shows both cases for a DGPP: the first where the plant is connected to a distribution grid, and the second where the plant under an IS scheme is also connected to a distribution grid. For the first case, there are two options: if the plant is the so-called Exempt Generator—the name the PIL gives to any DGPP—it has to sell its energy to a Basic Supplier but if not it may sell its energy and associated products to a Qualified User or directly into the Market. For both cases, the green dotted lines represent the money flow that occurs when the DGPP sells energy to the Supplier—paid at the Local Marginal Price (LMP) as is explained later.

In the case of a DGPP who sales its energy to a Basic Supplier, the CRE has developed a regulatory frame where three different schemes of economic considerations are defined with its specific cash flows.

1. *Net-metering*. The DGPP generates energy that can be both consumed at the facilities where the plant is located—a house, store, industry, etc.—or injected into the distribution circuit where the Basic Supplier can use it for its other consumers. When the DGPP is not producing energy, the facilities are provided with energy from the distribution grid. A bidirectional meter is needed to account for the energy imported from and exported to the grid. Every month, the Basic Supplier calculates the net energy delivered to its client. If this

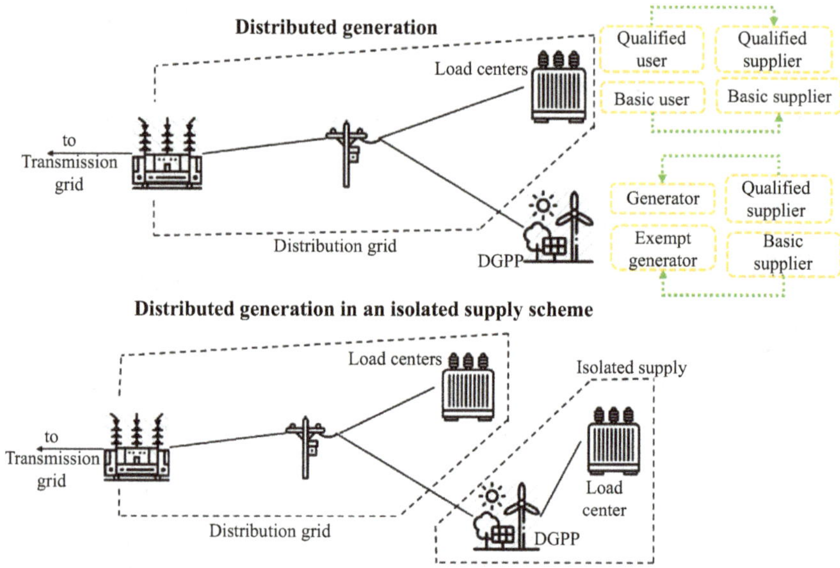

Figure 5.6 The two DG interconnection options

amount is positive, the client (the DGPP) has to pay for the net energy received at the cost of the tariff in the contract—that is the same tariff of any Basic Supply of the same kind. If the amount is negative, it means that the client exported more energy to the grid than the one it imported. Hence, the client gets credits that can pay for energy for months in future, if needed. At the end of one operational year, the supplier pays the client's remaining credits at the average LMP where the distribution grid is connected. It is worth to say that there are no limits to the energy sold to the Supplier (Figure 5.7).

2. *Net-billing*. While in the previous scheme, only the surplus energy—if any—is paid by the Basic Supplier to the consumer,[†] in this scheme all the energy produced by the DGPP is received and paid by the Basic Supplier; no energy is used by the load center with which the DGPP is associated. Conversely, the client consumes the energy delivered by the supplier at the cost of the applicable tariff. Under this scheme, the system must feature two distinct meters to account for energy delivery and reception separately, given that the rate for sales and purchase may differ. The energy generated by the DGPP is paid by the supplier at the LMP and the energy consumed from the grid is paid by the client at the tariff (Figure 5.8).
3. *Total energy sale*. In this scheme, no energy is consumed from the distribution grid. All energy generated by the DGPP is either consumed in the facilities of the user or exported to the distribution grid supplier. As in the other schemes, the energy is paid by the Basic Supplier at the LMP (Figure 5.9).

[†]Prosumer (= producer + consumer) is now a relatively common term coined for this type of users.

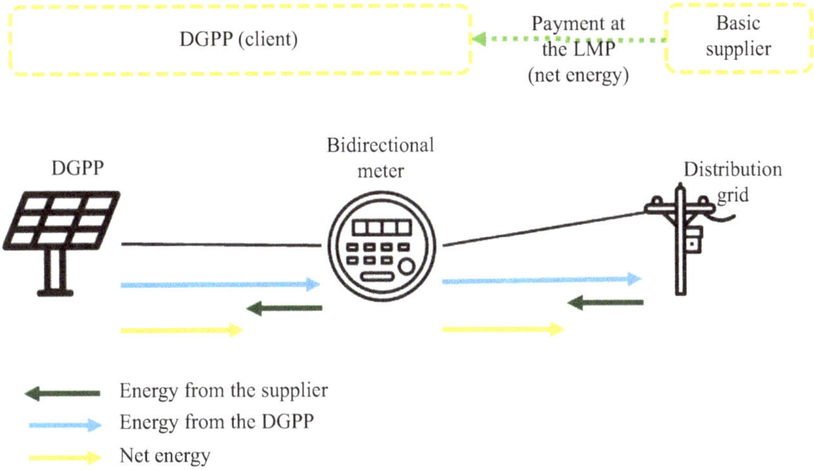

Figure 5.7 The DG net-metering scheme

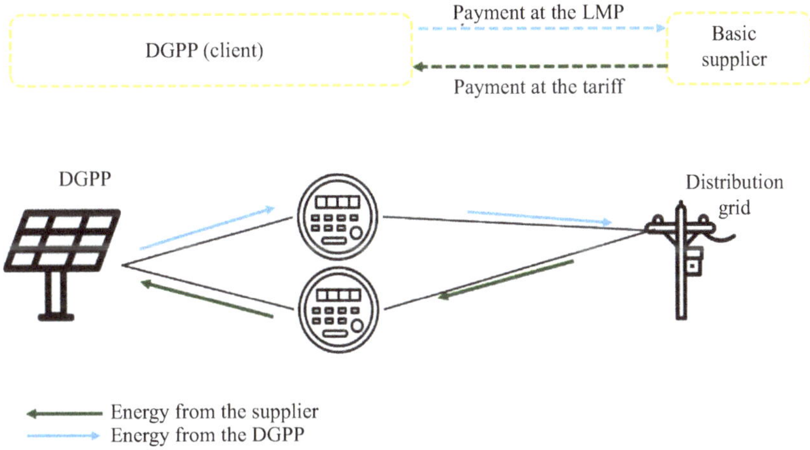

Figure 5.8 The DG net-billing scheme

The total energy sales scheme is thought to be used by the industry sector clients or by energy companies that identify potential places where DG could be a good strategic place for power generation—under 500 kW of capacity—and do not need self-consumption. For instance, it may be an interesting business for a landowner. In another example, a user identifies that the benefits of these schemes could be higher if he definitely separates energy use and energy sells since there is a

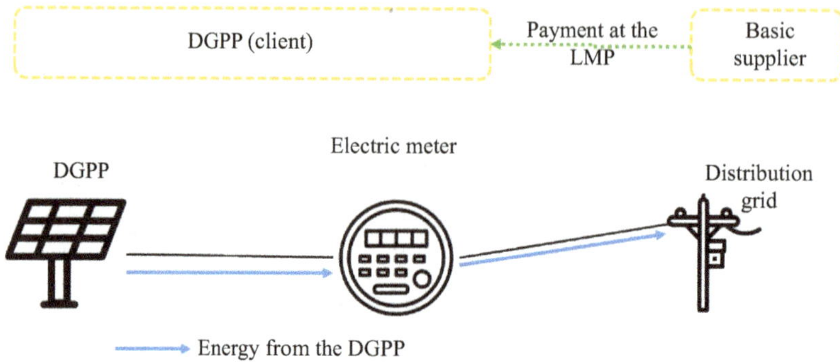

Figure 5.9 The DG total sale scheme

high enough difference between its capacity of generation—affected by its available area for solar PV—and its energy use (e.g., a low consuming warehouse).

The net-metering is more suitable for the residential and small-scale commercial sector since it requires less knowledge of the market. Additionally, the end user tariff uses are higher than the LMP so it results better to discount energy—that is, the essence of net-metering—rather than payments—the essence of net-billing. There are around 40.5 million potential users for this scheme. For the large-scale commercial sector, industry, and in general any user in medium voltage, the definition of the most suitable mechanism is more complex. This is due to the billing methodology applied, which increases uncertainty and uses parameters not controllable by the user. These users pay transmission and distribution costs based on the relationship between their demand curve and hourly generation, as opposed to the maximum demands of the system and the region in which it is located. This methodology causes the financing rates of this type of project to increase and limits the implementation of these mechanisms. As an example, since the reform of the regulatory framework in July 2009, 83,104 interconnection contracts were signed, of which 82,948 were under the net-metering scheme.

If the DGPP is based on a clean energy resource, such as solar, wind, hydro, etc., then it is known as Clean Distributed Generation (CDG) and can obtain CEL from the CRE; these certificates can be traded and therefore represent an additional potential income stream for the project. All load-serving entities (LSE), i.e., both Basic and Qualified Suppliers, have to comply with certain clean energy obligations, creating a competitive market for such certificates.[‡]

[‡]The current federal administration has been trying, as of recently, to overturn or water out some of features of the electricity sector reform. In particular, an attempt was made to grant CELs to all pre-existing renewable energy plants owned and operated by the incumbent utility (CFE)—mostly hydro—which would have flooded the market with CELs, setting their value essentially to zero. Evidently, crediting pre-reform plants with CELs is against the spirit of the reform which tried to incentivize the creation of *new* clean energy plants. This attempt is currently on hold, since a lawsuit was won by one of the private sector companies.

In addition to the potential benefits of the schemes of economic consideration, Mexico has implemented tax incentives for CDG:

- Income tax: the total investment of the CDG system is deductible for income tax purposes.
- Real estate tax: some city governments have implemented a reduction on this tax if a PV system is installed.

5.5 The evolution of DG in Mexico

Regulations explained earlier pretend to promote CDG throughout the country. Furthermore, the law instructs CRE and CENACE to develop the necessary projects of distribution infrastructure for DG. However, there are no rules financing the necessary infrastructure upgrades, so it is up to both potential users and investors to find the best way.

The two most common financing models are direct ownership and third-party ownership. In the first model, the user is the owner of the DG system which is paid through a credit or loan. As a result, the user is also the owner of the energy and CELs produced by the DGPP. In the other model, the user does not own the system, and consequently neither the energy nor the CELs currently held by the investment party. The user pays a fixed price to the system owner—in a scheme that can be seen as a leasing contract—or an energy price (i.e., through a power purchase agreement).

According to CRE [25], by July 2019, there were 817.85 MW of DG spread over 112,660 interconnection contracts. This capacity is expected to increase threefold by 2023, reaching nearly 2,900 MW. Solar PV is the most selected technology for DG so far, accounting for 99.3% of the total capacity. In fact, 29,556 contracts with a total capacity of 247.65 MW are under the scheme previous to the new regulation. This means that in just 2.5 years the capacity increased 321.69 MW on 83,104 new contracts: an increase of 73.7%.

Talking about the contracts under the new regulation, just 0.1% are subscribed to the net-billing or total sales scheme, while the remaining 99.9% use net-metering. This large group consists of companies and domestic users who do not wish to become engaged in activities of the energy market but look for a simple way to reduce their energy costs. Installing rooftop PV is clearly the easiest and most cost-effective model.

5.5.1 DG by the numbers

Since 2007, the DG capacity installed in Mexico has grown exponentially. From 3 kW in 2007, it only took 4 years to reach 1 MW of installed capacity, and another 4 years more to cross the 100 MW barrier in 2015. For 2009, the capacity expected was to settle at 943 MW; by July, it was 818 MW [16]. The number of contracts accounting for that capacity were 1 in 2007, 670 in 2011, 16,979 in 2015, and 2,660 in July 2009 [25]. Figure 5.10 exhibits the growth of the installed capacity

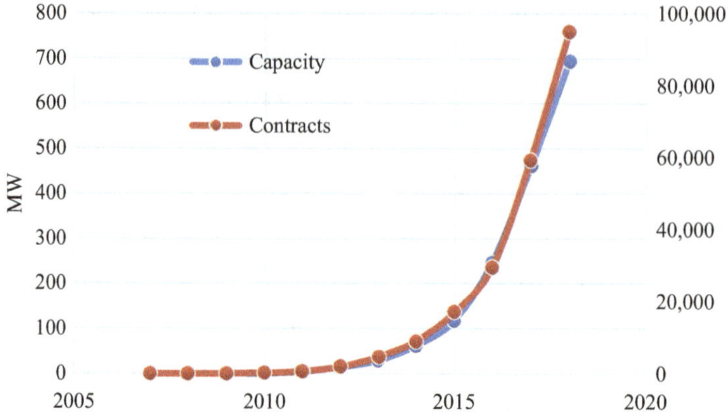

Figure 5.10 The evolution of the existing capacity and the number of contracts for DG [25]

and the number of contracts for the period 2007–2018, showing that the capacity closely mimics the number of contracts.

As a DG system may be installed by almost any kind of user with the main objective of satisfying his or her energy needs, the capacity of these systems is highly variable, ranging from about 1 kW for residential uses to the limit of 500 kW for commercial and industrial purposes. Here, to simplify the analysis, the systems are classified into three categories according to their capacity.

The first category is the small scale where the majority of residential users are included. It corresponds to systems with <10 kW of installed capacity. In 2017, this category accounted for 49,918 users—corresponding to 92% of all DG contracts in the country—and 168 MW, 43% of the total capacity. Of note, 10 of the 32 states in Mexico accumulate two-thirds of the total capacity and the first is *Jalisco* with 28.4 MW, followed by *Nuevo León* with 18.6 MW and *Mexico City* with 11.1 MW. The three major urban zones of Mexico are located precisely in these states, indicating that grid-connected small-scale PV DG is mostly a phenomenon of the big cities (Figure 5.11).

The second category is for the medium scale ranging from 10 to 100 kW with 3,597 contracts and an installed capacity of 98 MW, corresponding to 6.6% of the users and 25% of the total capacity installed in the country. Of note, 69% of the installed capacity in this range (10–100 kW) is located in only nine states where currently Nuevo León is the first with 19.7 MW, followed by Jalisco in second place with 12.3 MW. Mexico City with its 3.2 MW is only in 11th place, with the third going to the sunny Northern state of *Chihuahua* with 7.3 MW (Figure 5.12).

The last category corresponds to relatively large-scale installations, in the 100–500 kW range, where a total of 412 contracts—less than 1% of the total—had been awarded. The total installed capacity of this segment was 123.5 MW, corresponding to 31.7% of the total installed DG capacity in Mexico. Only six states accumulate two-thirds of the installed capacity, but *Estado de Mexico* ("Mexico State") stands out with 48.2 MW of 39% by itself. *Michoacán* and *Jalisco* follow

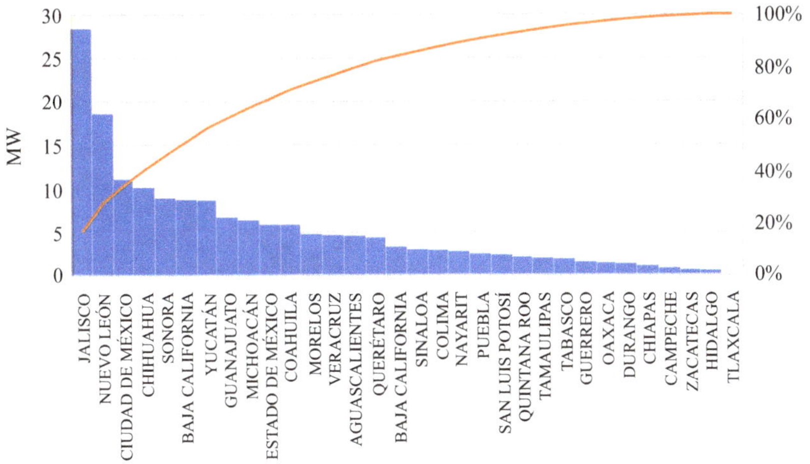

Figure 5.11 Installed capacity of the small-scale DG in different federal states of Mexico and its cumulated share of the total [25]

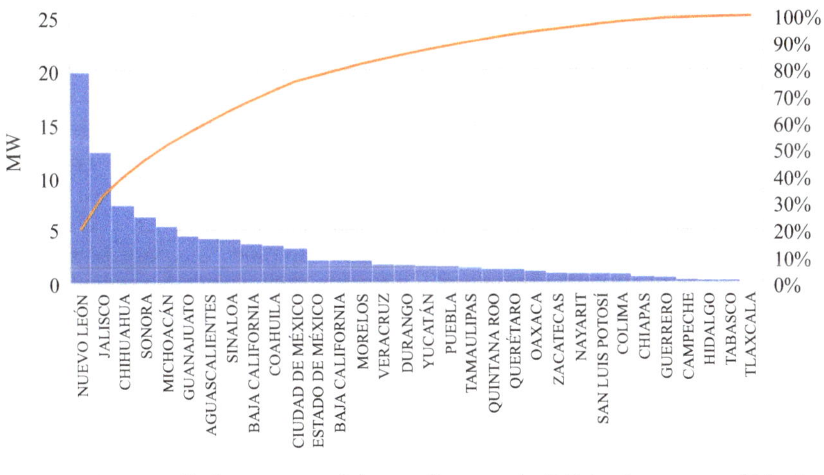

Figure 5.12 Installed capacity of the medium scale DG in the states of Mexico and its cumulated share of the total [25]

with 8.8 and 8.3 MW, respectively. Mexico State is located on the outskirts of Mexico City and features a strong industrial base, which may be the main driver for installing relatively large-scale PV facilities (Figure 5.13).

5.5.2 Barriers for financing and installing DG systems

Financing a DG system in Mexico is not an easy task. Although the projects are profitable, the traditionally high capital cost is often a deterrent for residential and

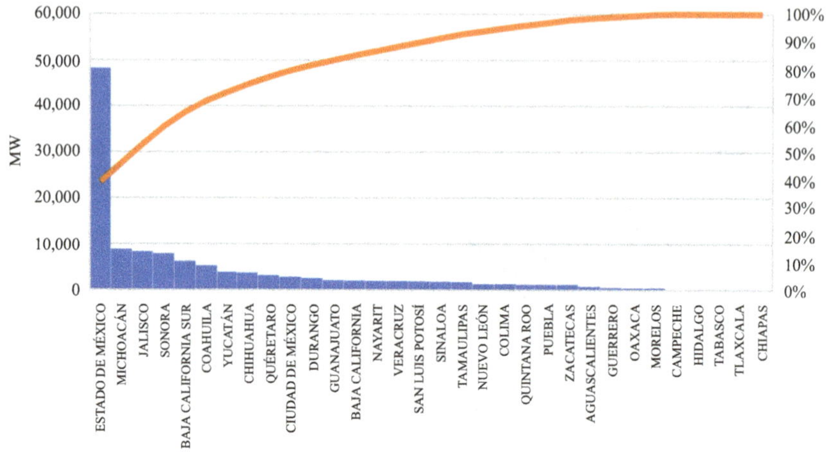

Figure 5.13 Installed capacity of the great scale DG in the states of Mexico and its cumulated share of the total [25]

small commercial users. An additional barrier is the subsidy granted to residential and farming users which is defined every year by the Finance and Public Credit Ministry (SHCP, for its Spanish acronym) under the premise of helping the economy of those sectors. Both sectors sum 90% of the basic supply users and represent 33.7% of the total energy use [26]. Hence, the invoice users pay to CFE-Basic Supply does not reflect the real energy costs. As a result, many users would not find it economically attractive to pay for a PV system since they already have a discount on their consumption that is paid by the government. Consumers with unsubsidized tariffs, particularly those located in the "high consumption" category are, however, potential candidates for residential PV projects.

The financing schemes for a DG system may vary substantially depending on the type of customer. An industrial company is likely to be able to access credit financing at competitive interest rates to develop a DG project in the same way it would develop any other project, for instance, the construction of new facilities or the R&D process for a new product. In a very different group are customers in the residential and commercial sectors. A small- or medium-scale commercial user will often face high interest rates when asking for a loan. The residential customer is more likely to pay its system using his own credit card, at still higher interest rates. However, there are also options like green mortgage schemes although they are focused on new homes.

It is clear then that for small and medium users—residential and commercial—it is harder to finance a DG system since there is a lack of appropriate mechanisms. For higher economic strata users, financing might not be a problem as they may have the purchasing power and the investment is worth enough taking the huge-energy cost savings they can perceive in the medium term. PV systems are currently less expensive than ever, save costs of energy consumption—and can even have earnings, depending on the economic consideration scheme—and return the investment in a very short time. So it is clear now that one barrier for DG is the lack of financial tools that allow the customer to take advantage of the benefits and make the finance more procurable. Nonetheless, this barrier may be overcome as projects like the one proposed in the next section show the virtues of DG.

The most complex issue are the so-called nontechnical losses that consist of electricity theft from people who are illegally connected to the distribution grid or simply refuse to pay. This problem has two faces. First, it is obvious that a consumer who never pays for its energy has no economic gain from using DG. And as easy it may sound, disconnecting the illegal connections or making him pay is a difficult choice because of its political implications and has never been successfully addressed by the authorities. On the other hand, those clandestine connections often do not comply with the standards of a safe electrical installation, causing problems with frequency, harmonics, and peaks, among others, apart from adding a non-quantified extra count load to the distribution grid which may further complicate the installation of DG systems.

The last barrier is limits to the amount of DG that can be installed in a given part of the system. For instance, in South Baja California (BCS) the Energy Regulatory Commission (CRE), upon a request by the incumbent utility CFE, decreed a limit of 10 MW for DG systems under 500 kW, with subsequent admissible annual increase of 1 MW. Although the incorporation of DG into the distribution grid obviously represents challenges, the technical basis for such decision is questionable at best, given that no auditable information was made available to the public which would allow to confirm the rationale of this decision. In BCS the 10 MW was reached in 2017, and ever since the demand for new systems has outpaced the annual installation limit.

5.5.3 Mechanisms for the implementation of DG

As already discussed, the net-billing scheme allows the DG system owner to sell its energy surplus at the LMP whose average is rising with time. This behavior becomes an advantage when it is compared with the behavior of the tariffs. In other words, the LMP at which the energy is sold is higher than the tariff at which the energy is bought. Hence, users need to be informed about this possible benefit and chose the net-billing mechanism over the net metering.

5.5.3.1 Solar Homes

Solar Homes is a program that seeks to help the energy transition and the climate change policies implementation for the accomplishment of the National Intended Contributions (NDC) as it democratizes the access to energy for the residential users. It is a transversal plan that will provide residential users with CDG systems reducing both their energy expenses and the national demand for electricity since each system will generate all the user electricity needs. As a consequence, the power system emissions will be reduced as the generation will come from PV panels installed on the roofs. Finally, the program will also reduce the economic load that the electric subsidy means for the government.

The whole residential sector electric consumption sums up to 28.2% of the national total, with nearly 61.5 TWh from which 95.9% comes from subsidized users which in 2018 was 38.5 million. A residential user loss his subsidy when, in a month, he consumes more than a certain amount of energy depending on the climate region where he lives. In 2018, there were 463,075 non-subsidized residential users, and their energy consumption was 4% of the total residential use.

The project works as follows. First, the government identifies its expenses on the electric subsidy for residential users and transfers it to a fund. Then, that fund is used to finance CDG systems for a certain number of users. Each user is the owner of his system but he will have to pay only 30% or 40% of the cost, and the remaining will be paid from the governmental fund. The user will pay the system through his electric bill, but since his energy uses are already covered by the PV roof he pays only for the CDG system but not for the energy. It is important that this payment should be around 20% lower than the previous energy charge so that the user may perceive savings that make this program attractive for him.

The government will have savings for the reduction of the subsidy applied to the users under Solar Homes. Those savings are meant to be put into the fund so the program may be spread until at least 25 million users got benefited. This is the number of beneficiaries expected by the end of the project, in a time lapse of 15 years. Later, the avoided subsidy is available to be used in other government fundamental issues such as health or education. For the first step of the program, the goal is to install 1.5 million solar in some representative cities of Mexico. It is also expected that the program encourages the rest of the residential users making them more participating in their energy consumption and installing PV roofs.

Solar Homes not only benefits the users and government, but also addresses Climate Change since the electricity generated by the CDG systems is free from greenhouse gases (GHG). For the 1.5 million systems, the expected reduction is 1.2 million tons of CO_2eq per year.

As a starting point, even before the execution of the first step, a pilot project is going to be implemented in Mexico City where 600 systems will be given to residential users from the same neighborhood. The objectives are as follows:

- To quantify the costs and benefits for the residential user, the government, the Basic Supplier and the PV roofs company.
- To measure the impacts that the high density of CDG systems may have on the distribution circuits.
- To benefit the residential users with clean energy and the reduction of their electricity costs.

This pilot project is implemented by the Ministry of Environment of Mexico City, The Climate Initiative of Mexico (ICM), the German cooperation for the Sustainable Development (GIZ), and UK Pact.

As Solar Homes program is implemented, there will be learnings to be obtained through technical and financial studies that will help to impulse the potential of DG for the residential sector. From the technical concern, the analysis of the performance of the PVDG systems and its impact on the distribution circuits is relevant to some entities. The most interested one may be the distribution company—CFE-Distribution. Until at present, the distribution circuits have had to drive energy only in one direction, from the electric substation to the houses, but with the expected growth of DG the direction of the energy will change, not only from the substation to some houses but from generating homes to any others in the circuits. As most of the PVDG systems will not have energy storage, intermittency is also a matter of concern given that at this scale the variation of the electric parameters may be a security issue. The studies that will be made as Solar Homes is implemented will help to understand how much DG would affect the distribution circuits and to give information about the best way both DG and the circuits have to work together to seize all the benefits of DG has to offer.

On the financial aspect, the cost of installing the PVDG systems will be contrasted against the savings for the involved ones. For the beneficiaries, the savings come as it is expected that their energy consumption will be fully satisfied by the PVDG system and that they will only pay a certain amount of the cost of the system that will be lesser than their former energy expense. For the Treasury Ministry—and the government itself—it is expected that although they will pay for the majority of each solar rooftop, there will be savings on the subsidy that is given to residential users every year since this expense will not be needed anymore. It is also worth to say that as the subsidy disappears, DG will become more attractive to other residential consumers as the return of the investment given by the savings on energy expenses will take less time. For the Basic Supplier, economics is not so simple. It will be vital to evaluate the possible reduction of the incomes the supplier gets for energy sales due to the increasing users who are getting PVDG systems. Nevertheless, there may be savings for the company as its total demand decreases, letting it obtain cheaper energy in the market. In addition, the supplier will get the CELs that the PVDG systems will produce, reducing its urgency to get these instruments in the market to fulfill its obligations on clean energy.

5.5.3.2 CSolar

In 2015, the Fund for the Energy Transition and the Sustainable Energy Use (FOTEASE) of SENER approved a program latter known as CSolar, developed jointly by the National Electricity and Clean Energies Institute (INEEL), the United Nations Environmental Programme (UNEP), and ICM and directed by a committee where the National Financial Institution (NAFIN) and the Mexican Banks Association (ABM) were also integrated. ICM acts as project coordinator and manager.

The objective of this program [27] is to eliminate the barriers for the financing of the Photovoltaic Distributed Generation (PVDG) by channeling financial institutions toward the acquisition of such systems. CSolar is a program of public fund guarantees for commercial banks providing financing at preferential rates to the micro, small and medium enterprises (SMEs).

The objective companies—SMEs—are such due to their availability to use PVDG and their difficulty to afford it, explained by the scarcity of credits or financial tools available for the non-big companies which, in many cases, do not even have a credit history. SMEs are under low-voltage low or high demand or medium-voltage high demand tariffs which are not subsidized. There are 4.44 million potential users, most of them (91.8%) on the low-voltage low demand tariff, the one for the less consumption of all, but with only 13.4% of the total energy use (99,843 GWh/year). Meanwhile, the medium-voltage high demand with hourly prices tariff has a share of 70.2% of that total consumption and of 2% of the potential users. The total capacity estimated for the potential users is 58.8 GW although technical and financial issues reduce them to 10 GW [28]

The program works as follows. First, a potential user, i.e., a SME, who is interested in acquiring a PVDG system or only in reducing his energy costs, has to ask one of the CSolar-certified suppliers about a budget for the PVDG system. Then, he has to contact NAFIN or one of the banks affiliated to the program to deal with a credit loan of up to 15 million pesos, i.e., around US$789,000. After the system is installed, the user has to pay the credit like any other bank loan. It may be for 7 years with an interest rate no more than 14.5%. The bank has the guarantee of the payment by the public fund.

One important achievement of CSolar program has come as it has impulse on the professionalization of the Mexican PV industry by certifying any installation company as long as it demonstrates to have the characteristics of the best standards in the world industry. To get this due diligence scheme that is currently recognized by the financial sector NAFIN, banks, and commercial sector, CSolar evaluates the relevant areas of the installation company. The impact of this scheme generates improvements not only for the beneficiaries of the program but also for all the small-scale PV industries, facilities, companies, and the workforce.

The first area to be assessed is the technical capabilities of the workers of the company which are divided into four main points: the capabilities of the installation work, the skills for working on highs, the knowledge to size and sell adequate systems for the client, and supervising the quality of the final installation. For the capabilities of the installation and the working on highs, CSolar helped in the development of national certification of the workforce which include a set of points to be evaluated to verify if a person has sufficient skills, behaviors, and knowledge to perform an activity.

Another area that CSolar evaluates is the quality of the products used in the PV systems. For the electric components as the PV panel, the inverter or microinverter, the interconnection devices, and the cabling, there are currently many standards of electric quality and security all around the world. Something similar is applicable for the mounting parts as its quality is also assessed to assure that they are secure and durable. The program certifies that the company uses components and materials which satisfy the world and national standards.

The final assessing point is the capability and performance of the company. CSolar verifies that it has the capacity to deal with its clients, following the best practices in topics as the administration of the projects or the contact and accompaniment of the clients. For example, one requisite is that the company has a record of all the projects done in which every single one has a folder with data about the client, the location of the system and its features, the type and number of the installed components, etc. Then, CSolar selects some of the facilities and visits them to evaluate that the work done effectively accomplish with the desired standards.

5.6 Prospects for future growth DG in Mexico

5.6.1 Forecast for DG development

For decades, the most common way to generate energy has been with large-scale PPs that use to be far away to the consumption centers; this way is known as centralized generation. Then the energy is transmitted from the generation point to the distribution center through transmission lines which later on sent to the consumption centers by distribution lines. On the other hand, DG is called in that way due to the fact that the PPs are focalized in, or very near, to the consumption centers. As DG is deeply adopted, the transmission lines and the large-scale PPs will be less required.

For the Mexican case, where the national transmission grid is already saturated and yet there are no firm plans to expand it, DG rises as a great solution to the demand of energy that is still growing in the whole country, as it also contributes to the decarbonization of the power sector. The growth of the installed capacity of DG will allow Mexico to reduce the stress in its transmission grid as the time it reduces the transmission component of the LMP.

On the first semester of 2019, CRE forecasted that the exponential growth of the DG installed capacity in Mexico will continue at least until 2023, when it is going to reach 2,898 MW. This means that in the following 4 years, there will be almost 2 GW of new DG systems (Figure 5.14).

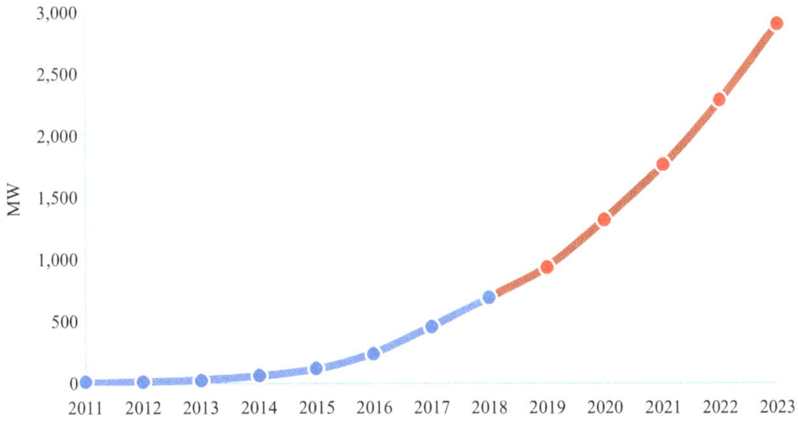

Figure 5.14 Evolution of the installed capacity for DG in Mexico. The blue line shows the existing capacity and the red one the forecast

Only for the SME sector, there is a potential of 51 MW to be installed considering the number of companies under the corresponding tariffs and their average energy consumption. The tariffs at which a SME may be receiving electric service are small demand on low voltage (SDLV), great demand on low voltage (GDLV), great demand on medium voltage with ordinary prices (GDMVO), and great demand on medium voltage with hourly prices (GDMVH) (Figure 5.15).

The tariff with the biggest potential of all is the GDMVH with an estimated capacity of 41.3 GW (70% of the total) whose main states are Nuevo Leon with 4.5 GW, Estado de Mexico with 4.1 GW, Ciudad de Mexico with 3.8 GW, and Jalisco with 2.6 GW (Figure 5.16).

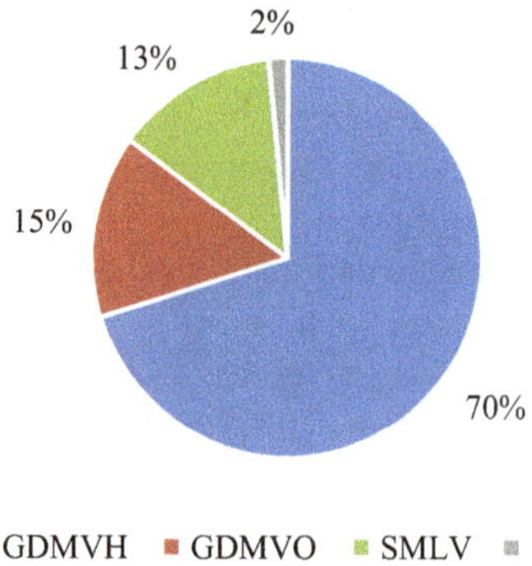

Figure 5.15 Potential for the SMEs depending on the electric tariff

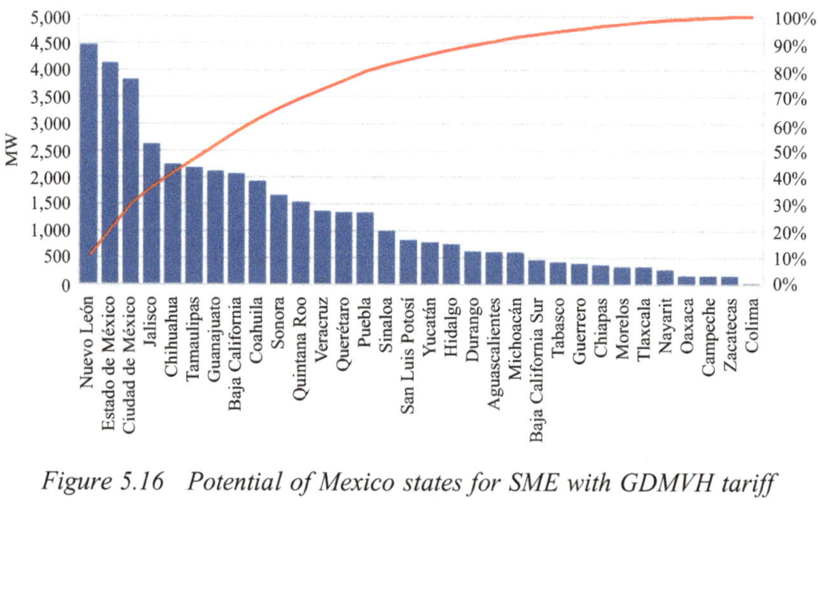

Figure 5.16 Potential of Mexico states for SME with GDMVH tariff

Next, there is the GDMVO with a potential of 8.7 GW for all the country, and the main states are Nuevo Leon with 917 MW, Jalisco with 661 MW, and Veracruz, Baja California, and Estado de Mexico each with nearly 517 MW (Figure 5.17).

The SMLV tariff has a potential of 7.9 GW with Ciudad de Mexico being the highest one, accounting for 1 GW itself. Estado de Mexico and Veracruz follow it with 686 and 665 MW (Figure 5.18).

At last, the GDLV has only 917 MW of potential in the whole country with Ciudad de Mexico being greatly the most important with 582 MW of that total (Figure 5.19).

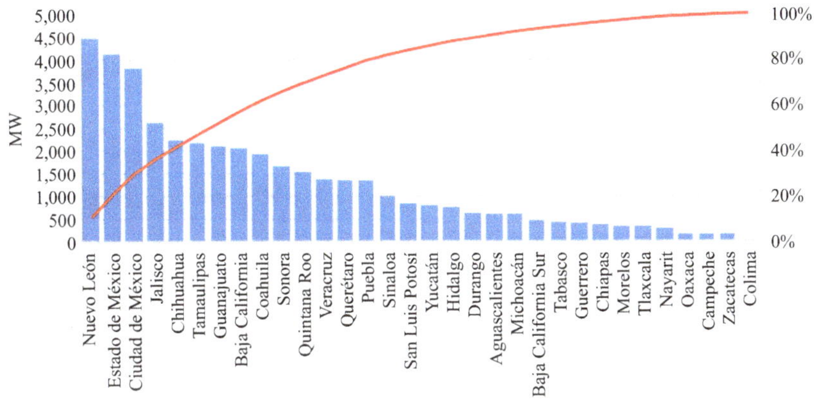

Figure 5.17 Potential of Mexico states for SME with GDMVO tariff

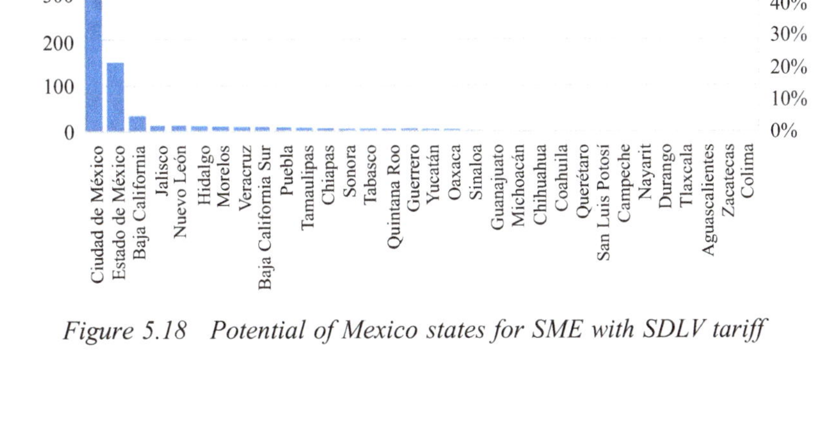

Figure 5.18 Potential of Mexico states for SME with SDLV tariff

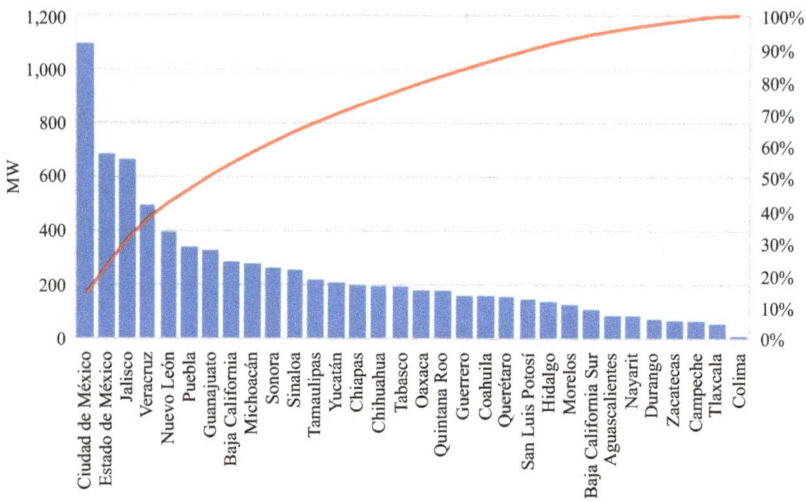

Figure 5.19 Potential of Mexico states for SME with GDLV tariff

As discussed in this chapter, DG is a flexible option for both users and generators to generate clean energy that helps to mitigate the effects of climate change. It is also a technology that contributes to the solution of problems such as the availability of fuels and transmission or distribution lines in certain regions. It also reduces energy costs for users connected to the grid. In the case of users without access to the grid, DG provides them electricity for a better quality life. The exponential spread on the contracts for DG shows that the enormous potential has to be fast adopted by the population.

Yet, there are still many issues that must be addressed to assure the existence of the regulations and mechanisms to fulfill the requirements on energy needs, security, and investments. For instance, there is a process to pass a regulation for a collective DG that would allow the existence of a single figure to represent many DG systems. This figure would have a contract with CFE-Distribution under the net-metering or net-billing schema and would provide energy to many consumers as well. This would be a powerful tool for programs like Solar Homes or similar where a provider may own DG systems installed in the facilities of different users that would receive part of the generated energy.

In the same way, DG is a good option for conflicting power systems such as the peninsulas. For example, the Yucatan peninsula is facing a problem with the availability of the natural gas used for power generation. This has led to the saturation of the transmission lines and some power shortages. If the main consumers, such as the hotels of the touristic places, install PV systems with storage under a DG scheme, the demand could be highly reduced easing the load on the transmission lines without the emissions of the gas turbines or other fast-response technologies.

5.6.2 A new proposal

At the same time, DG is adopted by some Mexican economic sectors, and there are still some opportunity areas where specific circumstances make its adoption harsh. To make DG a desirable option for everyone, it is important to understand such circumstances to find the most appropriate solutions as this technology brings benefits in many ways.

One of the main barriers for many people in residential, commercial, or industrial sectors is the property of the land or buildings since they cannot install a DG system without the permission of the landlord. Without an incentive, it is very probable that the leaser will not make any expense in DG as he will not perceive a direct benefit from the saves on energy costs.

A solution for this circumstance may arise by implementing a mechanism to condone part of the land value tax, or perhaps the rental income tax, for every leased property where a DG system is connected. Thus, not only the tenants will perceive a benefit when adopting DG as savings in energy expenses, but also the owners will have savings in their tax payments.

The energy company could also have economic benefits, especially in the case of departmental buildings, where there is a higher density of energy users. The DG systems could reduce the energy demand in such consumption centers, easing the distribution issues.

Evidently, not all the consumption sectors will see DG as a tempting option for their energy uses as a consequence of their particular circumstances. Yet, understanding those conditions will allow to find applicable solutions for each case; for instance, providing incentives to landlords and making them more willing to install DG so their tenants may take advantage of this kind of clean energy.

5.7 Conclusions

DG, mainly CDG, has a great potential in Mexico which could be highly beneficial for the society, the state, and the climate.

Mexican regulation over DG is quite recent, originated after a huge reform from 2014 that opened the market for the electricity industry, formerly monopolized by the state. At present, the basic regulations are established, but for DG there are still opportunity areas to foster its adoption. Additionally, it is vital that the end users—residential, commercial, or industrial—involve themselves in their own electricity consumption; hence they know their necessities better that they can adopt the better mechanism for them. Every currently available mechanism in Mexico—net-metering, net-billing, and total sales—provides benefits depending on the sector that uses it.

There are financing systems such as green mortgages, credits for DG, or even schemes in which the system is rented instead of sold. These may promote DG adoption as they are deeply adopted. However, they also need to be improved and specifically directed for each sector, so the benefits may be boosted.

Even though the impacts of DG in the distribution grids, the economy, or even the consumption habits are not well known, the outlook for DG is encouraging. Efforts by all the actors in society and private and state institutions are required.

In Mexico, DG is showing benefits for end users mainly by reducing their energy costs; for the state by potentially reducing the expenses in subsidies for important sectors; for the currently congested transmission lines by reducing the energy demand of the cities; and for the environment, since the majority of DG is based on PV, by avoiding the GHG emission for electricity.

References

[1] Pepermans, G., Driesen, J., Haeseldonckx, D., Belmans, R., and D'haeseleer, W. (2003). Distributed generation: Definition, benefits and issues. *Energy Policy*, 787–798.

[2] Department of Energy, *The Potential Benefits of Distributed Generation and Rate-Related Issues That May Impede Their Expansion—A Study Pursuant to Section 1817 of the Energy Policy Act of 2005*. 2007.

[3] Mitra, I., Degner, T., and Braun, M. (2008). Distributed generation and microgrids for small island electrification in developing countries: A review. *SESI Journal*, 18, 1, 6–20.

[4] Walling, R. A., and Miller, N. W. (2002). Distributed generation islanding-implications on power system dynamic performance. *IEEE Power Engineering Society Summer Meeting*. doi:10.1109/pess.2002.1043183

[5] IEA. (2017). *World Energy Outlook 2017*, IEA, Paris [web page: https://www.iea.org/reports/world-energy-outlook-2017

[6] Micangeli, *et al.*, 2017. *Energy Production Analysis and Optimization of Mini-Grid in Remote Areas: The Case Study of Habaswein, Kenya. Energies*. 2017.

[7] REN21. (2018). *Distributed Renewables for Energy Access* [webpage http://www.ren21.net/gsr-2018/chapters/chapter_04/chapter_04/].

[8] Narula, K., Nagai, Y., and Pachauri, S. (2012). The role of decentralized distributed generation in achieving universal rural electrification in South Asia by 2030. *Energy Policy*, 47, 345–357. doi:10.1016/j.enpol.2012.04.075

[9] *Pico Solar PV Systems for Remote Homes. A new generation of small PV systems for lighting and communication*. IEA, 2012.

[10] Movellan J. (2014). Plug-and-play solar systems: Automating the permitting, inspection and interconnection processes. *Renewable Energy World* [webpage: https://www.renewableenergyworld.com/2014/06/19/plug-and-play-solar-systems-automating-the-permitting-inspection-and-interconnection-processes/#gref].

[11] Jamal, T., and Ongsakul, W. (2012). Smart grid in Bangladesh power distribution system: Progress & prospects. *2012 Students Conference on Engineering and Systems*. doi:10.1109/sces.2012.6199096

[12] Twaha, S., Ramli, M. A. M., Murphy, P. M., Mukhtiar, M. U., and Nsamba, H. K. (2016). Renewable based distributed generation in Uganda: Resource

potential and status of exploitation. *Renewable and Sustainable Energy Reviews*, 57, 786–798.

[13] Murphy, P. M., Twaha, S., and Murphy, I. S. (2014). Analysis of the cost of reliable electricity: A new method for analyzing grid connected solar, diesel and hybrid distributed electricity systems considering an unreliable electric grid, with examples in Uganda. *Energy*, 66, 523–534.

[14] Márcia Andréa Rosas Luna, Felipe Barroco Fontes Cunha, Maria Cândida Arrais de Miranda Mousinho, Ednildo Andrade Torres (2019). Solar photovoltaic distributed generation in Brazil: The case of resolution 482/2012. *Energy Procedia*, 159, 484–490.

[15] Renewable Energy World. (2018). *Distributed Solar Makes Up 18.5 Percent of All New Power Generation* [webpage: https://www.renewableenergyworld.com/2018/08/13/of-new-power-generation-how-much-is-on-the-roof-quarterly-update-2018-q1/#gref].

[16] Farrell, J. (2019). Of new power generation, how much is on the roof? Quarterly Update – 2019 Q1, *Renewable Energy World*, 1 July 2019. Available from: https://www.renewableenergyworld.com/2019/07/01/of-new-power-generation-how-much-is-on-the-roof-quarterly-update-2019-q1/#gref [Accessed June 2020]

[17] California Distributed Generation Programs. (2019). *California Distributed Generation Statistics* [webpage: https://www.californiadgstats.ca.gov/programs/].

[18] Joshi, K. A., and Pindoriya, N. M. (2017). Case-specificity and its implications in distribution network analysis with increasing penetration of photovoltaic generation. *CSEE Journal of Power and Energy Systems*, 3(1), 101–113. doi:10.17775/cseejpes.2017.0013

[19] Quiroga, G. A., Kagan, H., Amasifen, J. C. C., Almeida, C. F. M., Kagan, N., and Vicentini, E. (2016). Study of the distributed generation impact on distributed networks, focused on quality of power. *2016 17th International Conference on Harmonics and Quality of Power (ICHQP)*. doi:10.1109/ichqp.2016.7783376

[20] Ismael, S. M., Abdel Aleem, S. H. E., Abdelaziz, A. Y., and Zobaa, A. F. (2019). State-of-the-art of hosting capacity in modern power systems with distributed generation. *Renewable Energy*, 130, 1002–1020. doi:10.1016/j.renene.2018.07.008

[21] Satchwell, A., Mills, A., and Barbose, G. (2015). Quantifying the financial impacts of net-metered PV on utilities and ratepayers. *Energy Policy*, 80, 133–144.

[22] Mountain, B., and Harris, R. (2017). Grid versus distributed solar: What does Australia's experience say about the competitiveness of distributed energy? Innovation and disruption at the grid's edge, 83–99. doi:10.1016/b978-0-12-811758-3.00005-x

[23] Hinz, F., Schmidt, M., and Möst, D. (2018). Regional distribution effects of different electricity network tariff designs with a distributed generation structure: The case of Germany. *Energy Policy*, 113, 97–111. doi:10.1016/j.enpol.2017.10.055

[24] Cambini, C., and Soroush, G. (2019). Designing grid tariffs in the presence of distributed generation. *Utilities Policy*, 61, 100979. doi:10.1016/j.jup.2019.100979
[25] CRE, *Evolución de Contratos de Pequeña y Mediana Escala/Generación Distribuida*, 2019, México.
[26] CFE, *Usuarios y consumo de electricidad por municipio (2010–2017)*, 2017, Mexico.
[27] CSOLAR [webpage https://csolarmexico.com/].
[28] ABM, *Mercado de energía fotovoltaica de baja escala. Generación Distribuida*, 2017, México.

Chapter 6

The role of smart grids for the renewable energy transition

Alexandra von Meier[1], *Mohini Bariya*[2], *Jesus E. Valdez-Resendiz*[3] and *Jonathan C. Mayo-Maldonado*[3]

6.1 Introduction

This chapter focuses on technological innovations in electric power infrastructure and their enabling potential for the integration of variable generation resources, electric vehicles (EVs), and microgrids. Many of the innovations described in this chapter were driven by evolving technical and economic constraints that long predated renewable grids: factors other than *clean energy* have dominated problem-solving and investment in new electric power hardware, software, and design strategies over the past several decades. In the context of the renewable energy transition, however, some of these rather esoteric niche technologies take on a new relevance and interest for a broader community of researchers and practitioners.

The term *smart grid* has been interpreted by different people to mean different things, and we therefore use it with some hesitation—but it serves to evoke the general idea of an electric power system with more information and intelligence to it. We define a *smart grid* broadly here as one that affords an increased capability to observe and control its components, at greater resolution in both time and space. There are many different dimensions of *smartness* that may be realized to varying degrees and for various purposes.

Legacy electric power systems—those with high-voltage networks spanning entire countries—everywhere in the world have been built essentially on the same design concept from the nineteenth century, implemented and expanded throughout

[1]California Institute for Energy and Environment, University of California, Berkeley, CA, USA
[2]Department of Electrical Engineering and Computer Science, University of California, Berkeley, CA, USA
[3]School of Engineering and Sciences, Tecnológico de Monterrey, Monterrey, Mexico

the twentieth century [1]. A crucial design constraint since the inception of electric grids was that detailed information about the actual operating state of an alternating-current (a.c.) network was prohibitively expensive, or simply impossible to obtain. Planners and operators have long worked with approximate load forecasts, unreliable specifications of the physical network itself, limited real-time communications, and much residual uncertainty. The uncertainties concern not only what any given component is doing at the moment or might do next, but also the true operating limits that, if exceeded, would result in a blackout. Consequently, electric grids had to be robust and, in some sense, forgiving: components were oversized and redundant; large and responsive generators had to be ready to take up the slack in assuring real-time power balance, with stored energy in the form of prime mover fuel and rotational inertia; wherever possible, distribution systems, which carry electricity to individual customers, were built as radial structures with one-way power flow and straightforward design rules; and human operators retained discretion for making whatever adjustments necessary to keep the lights on.

Two unrelated but equally profound changes arose in the 1980s and 1990s. First, legacy grids reached the limits of economies of scale [2], while also encountering new technical problems associated with regional interconnectedness and large power transfers over long distances, such as oscillations [3]. It was no longer easy to reduce costs by growing and expanding the network. With new economic pressures came new incentives to cut costs, along with restructuring and deregulation of the industry in various countries. Second, semiconductors and the revolution of information technology presented an enormous array of new possibilities for application to the electric power domain. Simply put, the cost of information fell sharply relative to the cost of energy and infrastructure hardware. This is the central motivation of *smart grids*.

While the idea of *smart grids* did not have its origins in renewable energy or climate protection efforts, there are important areas of intersection where *smarter* technologies in the electric grid can directly or indirectly enable more effective use and higher penetration levels of renewable energy. That intersection, where more refined and granular information and control can address challenges associated with the integration of renewable resources, is the focus of this chapter. We organize these challenges into five areas: power balance, transmission constraints, voltage management, protection, and integration.

6.2 Power balance

It is often stated that in an electric grid, the power produced at any instant must equal the power consumed. Solar and wind resources with highly variable power output pose an obvious difficulty here. There are important subtleties, however, which depend on the timescale under consideration.

First, note that production and consumption are never *exactly* equal. To satisfy energy conservation, whatever discrepancy exists between power injection to and extraction from the electric transmission and distribution infrastructure must be absorbed by some form of energy storage. The relevant storage medium varies with timescale.

6.2.1 Timescales

Most commonly, we think about power and energy transfers on the order of minutes, hours, days, months, or years. On these timescales, three basic types of balancing tools are reserve generation, energy storage, and controllable demand. Traditional reserve generation, or units ready to respond to a signal requesting additional power, implies energy storage in the form of chemical fuel, nuclear fuel, or an elevated water reservoir at the power plant. By contrast, explicit energy storage technologies, such as hydroelectric pumped storage, batteries, flywheels, etc., imply a two-way energy conversion in a charge–discharge cycle. As discussed in Chapter 8, a variety of storage technologies is technically proven and advancing on a spectrum of economic competitiveness, where different storage media and conversion technologies favor different timescales of response and duration of storage. The third balancing tool is controllable loads. Historically, the only way to control electricity demand was to shed load at the crude level of an entire distribution circuit—an option that grid operators naturally prefer to avoid. Electric demand response programs offer a more refined and voluntary approach, with much untapped potential remaining [4].

Power balance, most simply, is observed through the a.c. system frequency: if generation exceeds load, the frequency increases; if generation is less than demand, the frequency decreases. On timescales of minutes and more, frequency is by definition equal across an entire *synchronous* grid. The fact that the system-wide frequency can be observed locally, and generator governors respond to system needs by way of *droop control*—a simple feedback loop where turbine power is increased if the rotation rate slows down, and vice versa—has been essential for allowing power systems to be balanced at all, even with nineteenth-century technology.

As illustrated in Figure 6.1, power balance extends down to the finer timescales of seconds, cycles, and sub-cycle energy exchange between devices. In these regimes, it is not always meaningful to think of a uniform system frequency, since imbalances can be observed as local or system-wide oscillations of frequency or phase angle (i.e., generator rotational velocity or position). At these shorter timescales, where a.c. grid behavior was not historically measurable, we rely on electromechanical feedback and rotational inertia within large, synchronous machines. Electromechanical forces, acting through magnetic fields in relation to generator windings stabilize the rotational *position* of multiple synchronous machines relative to each other, keeping them in lockstep by exerting an increased torque on a generator that attempts to pull forward, and a reduced torque on a generator that falls behind. These restoring forces give rise to the damped harmonic oscillations observed when generators experience step changes in load. At present, a crucial quantity monitored in large grids is the amount of damping that the entire system (including the rotating machines as well as network impedances) exerts to subdue such oscillations [5].

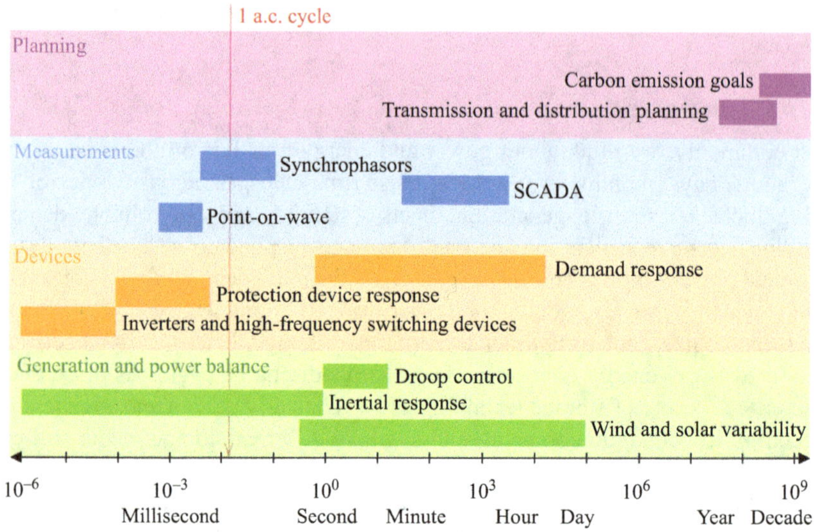

Figure 6.1 Timescales in electric grid operation. This figure situates problems and solutions on a logarithmic timescale. Energy is stored, information is gathered, decisions are made, and control actions are taken with very different time characteristics, for which different technologies will be suitable

Rotational inertia, or the tendency of large spinning objects to maintain their rate of rotation, tends to stabilize rotational speed and thus electrical frequency. The inertia supplied by the sum total of rotating machines in an a.c. system counteracts sudden changes in frequency due to loss of generation or load and buys precious seconds to engage control systems (such as governor actuation for primary frequency response). Meanwhile, megajoules of rotational kinetic energy are gently dissipated, as the system slows down gradually.

Another subtlety of grid power balance is that it includes both real and reactive power. Instantaneous power transfer always equals instantaneous voltage × instantaneous current. But since both current and voltage vary (we assume sinusoidally) and may be shifted in time relative to each other due to physical properties of the loads (specifically, inductive and capacitive reactance), instantaneous power will reverse direction during the cycle. Real or active power is the average power transferred into or out of a circuit over one or more cycles; multiplied by time, real power represents the physical work done by or on a device in terms of movement or heat. Reactive power represents the oscillating component of time-varying instantaneous power, which is alternately stored in electric and magnetic fields. An imbalance of real power in an a.c. system is observed as a change in frequency, whereas an imbalance of reactive power is observed as a change in voltage. Transmission lines, transformers, and a majority of loads tend to act as inductive reactances. The associated reactive power needs of the system are met traditionally by adjusting the excitation voltage of synchronous generators (which effectively shifts the timing of current relative to voltage) and additionally by installing capacitors throughout the transmission and distribution system.

6.2.2 The role of inverters

The vast majority of new electricity generation capacity based on renewable resources, specifically solar and wind, uses inverters to produce a.c. at the grid frequency (50 or 60 Hz) from the direct current that photovoltaics or batteries naturally supply or, in the case of modern wind turbines, that has been rectified from "wild" a.c. associated with variable mechanical speed.* Inverters are thus the key interface component. Crucially, modern inverters are capable of addressing power balance at each of the timescales described earlier and are therefore a cornerstone of any high-renewable *smart grids*. At the device level, the technology for controlling today's power electronics to meet the needs of the electric grid is almost trivial—in the sense that off-the-shelf components are available to do the job. The main challenge lies in the intelligent recruitment and coordination of large numbers of inverters throughout the grid.

Inverters can adjust both real and reactive power output and thus support both dimensions of power balance. Like conventional rotating machines, inverters have a reactive capability curve that bounds the vector sum of their real plus reactive power transfer (which is thermally limited by total current, irrespective of the current's timing relative to voltage). This implies some trade-off between providing real and reactive power when the inverter is operating at its rated limit; that is, real power output has to be modestly curtailed in favor of reactive power, for which some appropriate financial incentive must be provided. However, when the solar or wind resource is not at its maximum, inverters typically have headroom available. The optimal provision of reactive power in the steady state is, in that sense, an economic issue rather than a technical one. Managing time variations is a distinct problem, discussed later in the context of voltage regulation.

Regarding time variations in real power, grid operators are increasingly concerned about the loss of both frequency response and inertia, as large synchronous generators are gradually replaced with inverter-based resources. Load frequency regulation includes both primary and secondary frequency response. This is traditionally achieved via droop control of rotating machines, where generators autonomously adjust power output in response to locally measured frequency to stabilize any rapid changes in frequency (primary response) and subsequently respond to commands for adjusting their frequency/power set point to reconcile discrepancies between actual and nominal frequency, as well as economically desired generation schedules and power transfers (secondary response) [6].

Until recently, it has been very rare for solar and wind resources to be recruited for frequency regulation services in large grids. However, such opportunities and

*A noteworthy exception is solar thermal power that uses a conventional steam generation cycle, with the attendant controls and conveniences, but its contribution at present is small.

requirements are beginning to emerge, e.g., in Central America, with the economic implications for renewable generators not yet entirely clear. A common concern in the industry is that when intermittent renewables replace conventional generators but do not participate in frequency regulation, they introduce more variability for the remaining resources to compensate and can thus erode frequency regulation reserves from both directions [7].

However, there is no fundamental technical reason why advanced inverter control, paired with a modicum of energy storage in battery or other form, cannot be assigned to perform frequency regulation services much like a conventional machine. The problem is about appropriate tariffs, incentives, and priorities, but not hardware or software capabilities. It is worth emphasizing that the amount of stored energy required for fast-responding frequency regulation on the order of seconds is small compared with the more familiar applications of energy storage for arbitrage or reliability which span multiple hours.

Inverters can also be programmed to emulate the inertia of rotating machines [8,9]. Again, the energy storage required along with suitable control algorithms to enable this functionality amounts to only seconds worth of peak power generation. Thus, inertia has been proposed as an ancillary service to be included in tariffs [10], and there is no fundamental physical constraint that would preclude solar and wind resources from participating in such a market. Controllers on wind turbines can also be programmed to simulate the inertial response of generators, by leveraging the rotational kinetic energy of the turbine [11].

Arguably, inverters could do even better than emulating synchronous generators. The vital importance of inertia to power balance and therefore grid stability can itself be understood as a consequence of the limitations on the control of large rotating machines. The delay in primary governor response is precisely what necessitates inertia in the system, so as to reduce the rate of change in frequency relative to a power imbalance and prevent a frequency excursion large enough to cause problems (e.g., trip under frequency relays that will disconnect loads or generators, with potentially cascading effects). In sum, inverter-based renewables—either augmented with energy storage or oversized to allow for some headroom of upward control—can be leveraged as a uniquely flexible and agile resource for the grid, precisely because their control is so fast, and decoupled from unwieldy physical masses. Optimal strategies for recruitment and coordination remain a subject of ongoing research.

6.3 Transmission constraints

At present, a significant portion of solar photovoltaic installations are on or near buildings, co-located with load [12]. However, the most attractive renewable resources tend to be located farther away from load centers, partly because the sunniest deserts and windiest areas tend to be less hospitable environments for people to live. Conversely, wind resource benefits from smooth, unobstructed terrain. For example, the Gansu Wind Farm in China, one of the world's largest wind resources, is located in the Gobi Desert, thousands of kilometers from China's coastal population centers [13]. Offshore wind farms, such as the Netherlands' Gemini 600 MW wind farm 55 km off the Dutch coast, capitalize on consistent strong winds with low turbulence, at the expense of undersea transmission cables to the shore [14].

Hydroelectric power—distant from loads—was a key historical driver of long-distance transmission development, with one of the first high-voltage a.c. transmission lines built in 1895 from Niagara Falls to New York [1]. More recent examples include Egypt's Aswan high dam connected to Cairo by a.c. transmission, and the gargantuan Itaipu dam on the border of Brazil and Paraguay, over 1,000 km from the load center of Sao Paulo [15,16]. Transmission networks further evolved in the twentieth century based on the location of fossil fuel resources (especially coal) and cooling water for thermal generation (including nuclear). This legacy infrastructure happens not to be ideally matched to the geography of solar and wind resources.

Strategic investment in new transmission lines to access distant renewable resources, especially wind, is being undertaken in various countries including China and Northern Europe [17,18]. However, creating new transmission capacity is an expensive process that can take many years, especially when permitting new rights-of-way is involved. This provides incentive for smart grid technologies to leverage better utilization of existing transmission assets, while observing the inherent physical constraints.

In electricity markets, limitations imposed on power transfers due to transmission loading concerns are known as *congestion*. Transmission congestion compels the use of higher cost generators to meet power demand, when lower cost generators are inaccessible due to insufficient transmission capacity linking them to demand locations. Transmission congestion therefore increases the overall cost of electricity generation. Studies suggest that renewable growth will increase transmission congestion, especially when renewable generation is situated far from load [19]. In the worst case, transmission congestion can lead to the curtailment of renewable output, making expansion of renewable generation infrastructure less attractive. Therefore, transmission infrastructure expansion is a critical part of harnessing renewable energy from distant regions.

6.3.1 Thermal and stability limits

Multiple factors limit the power that can be transferred through individual lines in the transmission network. The current that can be carried by an individual conductor, also known as the ampacity, is constrained by a thermal limit. Resistive (I^2R) heating causes a transmission or distribution line to thermally expand and sag, which can bring it dangerously close to trees or other objects below and increase the risk of a fault. Since heat dissipation to the environment is important, ambient conditions including temperature, wind speed, rainfall, and solar irradiance have a significant impact on ampacity [20]. Thermal limits tend to be the dominant constraint on the majority of transmission lines, especially shorter ones. On longer lines, voltage and angle stability can dominate—and with solar and wind farms located in remote areas, these stability limits will likely gain significance.

Voltage stability relates to theoretical limits on a.c. power that can be transferred across a transmission line, given its impedance and the nominal operating voltage. Voltage stability can only be fully understood in the multidimensional space of nodal voltage phasors and real and reactive power transfer, which are linked through the complex line impedance. For an intuitive simplification, consider a line with a fixed sending end voltage and a variable receiving end voltage, V_R. To increase the power delivered, we must increase the current flow on the line; however, this leads to a greater voltage drop along the line, reducing V_R, which acts to decrease power delivered. At the point of maximum power transfer, the system becomes unstable, and an attempt to drive current beyond this point would cause voltage collapse. For mainly inductive transmission lines, the problem can be mitigated by injecting more reactive power, but this approach, too, is ultimately constrained by reactive power losses or unacceptably high voltages at the sending end [21].

Angle stability concerns the synchronization of rotating machines. Normally, in steady-state conditions, interconnected synchronous generators rotate at precisely the same speed. The relative angles—which can be seen either as rotor position in space or voltage waveform in time—are directly associated with generator power output. Mostly for inductive transmission lines, the real power transfer is approximately proportional to the sine of the angle difference between the two ends, with a theoretical maximum at a voltage phase angle difference of one-quarter cycle. As power flow (and thus the angle difference) increases, the stabilizing electromechanical interaction between machines at either end is weakened, and an attempt to stretch the steady-state angle separation beyond 90° would lose synchronicity.

6.3.2 Oscillations

Even before the point of losing synchronicity, large angle separations reduce an a.c. system's ability to withstand disturbances such as a transmission line fault or the sudden loss of a generator. Such a disturbance will cause generator angles to swing as harmonic oscillators (described by the *swing equation*) about the operating point. Electrically distant generators can swing against each other across the transmission network in non-obvious ways, exchanging power back and forth in *inter-area oscillations* that threaten overall system reliability. For example, the Pacific Intertie in the western United States, a long transmission corridor delivering hydropower from the northwest to Southern California, was unexpectedly disconnected by an automatic relay operation in 1996 due to large but previously unnoticed oscillations, causing a widespread blackout. This came as a surprise to operators, since the system had been operating well within its thermal limits [3].

Inter-area oscillations are one example of sub-synchronous oscillations, i.e., power oscillations occurring at lower frequencies than the a.c. grid frequency (50 or 60 Hz). Distant renewable integration via long transmission can produce sub-synchronous oscillations in additional ways. The U.S. State of Texas has substantial wind generation in remote areas, connected by long transmission lines to distant load centers. In 2009, interaction between the control system of some wind plants and the series compensation capacitors of the transmission lines led to large oscillations that damaged the wind turbines [22]. This was not an isolated incident and has been observed again elsewhere in the United States and in China [23].

6.3.3 HVDC

High-voltage direct-current (HVDC) lines are a promising technology for addressing the transmission challenges of distant renewable resources. An HVDC link connects two a.c. systems through a d.c. line via a.c.–d.c. converters at each end. HVDC is advantageous for transmitting power over long distances for several reasons. It allows for greater power transfer than a comparable a.c. system and does not suffer from an angle stability constraint, which can be especially restrictive in long a.c. lines with higher inductance. Since currents are d.c., the link has no inductance or capacitance. HVDC is therefore the only viable option for long undersea cables, which would have prohibitive capacitance for 50 or 60-Hz a.c. Similarly, HVDC links do not require expensive capacitive compensation, as long a.c. links do. HVDC can also be used to connect a.c. systems that are not synchronized or are operating at different frequencies. Finally, power flow along an HVDC link is more controllable, potentially enabling methods to stabilize the entire system by controlling HVDC power flows [24].

For these reasons, numerous HVDC links are already operational across the world, with more planned for the near future. Several HVDC lines connect offshore wind farms in the North Sea to mainland Europe. The world's two largest hydroelectric power stations—the Three Gorges Dam in China and the Itaipu Dam in Brazil—are connected by HVDC links to distant load centers. The longest HVDC link, nearly 2,500 km long, is the Rio Madeira system that delivers hydropower from the Amazon Basin to large load centers in Southeastern Brazil [25].

Conventional HVDC transmission systems use line commutated converters (LCC) in their a.c.–d.c. conversion process. The LCC is based on thyristor valves that require a synchronous and stable voltage to operate. Thyristors are semiconductor devices that perform a switching action each half cycle but are not fully and independently controllable; rather, thyristors rely on an existing a.c. signal that causes the current direction to reverse, which in turn triggers the switch. For this reason, LCCs are also known as current source converters (CSCs). LCC technology is still the most common because it can handle very high voltage and power levels for HVDC systems. Also, there has not historically been a need to make these converters responsible for controlling voltage and frequency.

The alternative to LCCs are voltage source converters (VSCs), introduced to the power sector in the late 1990s. These converters use insulated gate bipolar transistors (IGBTs) which are fully controllable semiconductors. This means that the current can be switched on and off at any time, independent of an external a.c. signal. Consequently, these devices can independently control power quality and also perform a *black start* in case of a power outage. Drawbacks of VSCs include capacity limitations as well as higher electrical losses and expense for the dielectric of the insulation components. Based on their particular electrical characteristics, LCCs and VSCs are suited for different HVDC applications.

One particularly important configuration that draws on VSC technology is the Modular Multilevel Converter (MMC), introduced in 2002 [26]. As the name suggests, the MMC can generate voltages at multiple levels; it does this by rapidly switching in and out circuit branches within the converter, which effectively add and subtract voltages in series. The modular character of these branches means that they can be stacked in greater numbers to increase the overall voltage capability of the converter. An MMC can thus synthesize a very precisely controlled a.c. signal, which resembles a sinusoidal waveform with many small steps: the more levels, the smaller the steps. This translates into a low harmonic content of the resulting waveform, which in turn means less filtering and less waste heat. Unlike familiar electronic devices that can create such waveforms, the key innovation of the MMC was the ability to operate at high voltages and transfer large amounts of current. The first commercial use of MMCs in power systems was in California, with the 400-MW HVDC Trans Bay Cable that went into operation in 2010 to provide transmission security for the San Francisco Peninsula [27].

6.3.4 Flexible a.c. transmission systems

Flexible a.c. transmission systems (FACTS) are a versatile enabling technology, formed by the collection of controller devices. These can be applied in coordination for one or more purposes including power flow control, load sharing, voltage regulation, reactive power compensation, enhancement of transient stability, or mitigation of system oscillations. The principal objective of FACTS is to enhance the controllability and power transfer capability of specific transmission lines or paths. Of course, power flow is governed strictly by physics of Ohm's and Kirchhoff's laws, based on the amount of power injected or demanded at any given node in the network, which determines the voltage at that node. The current on any given transmission line is then the ratio of the difference in voltage between the two ends and the line's impedance. Conventionally, we assume that this impedance has a constant value, determined by the conductor material, size, and shape (neglecting minor variations due to temperature). The crucial innovation of FACTS is that by throwing many semiconductor switches at high speed, it can alter the appearance of the line impedance, as presented to the a.c. network. In essence, a FACTS controller is a power converter that just reshapes power in subtle ways. Consequently, FACTS can control power system parameters including series impedance, shunt impedance, current, voltage, and phase angle.

Since the manipulation of the waveform by a FACTS device may be small, it does not need to internally process the entire amount of power conveyed, as in a complete a.c.–d.c. conversion. The cost of FACTS controllers is therefore considerably less (about 10%–30%) than HVDC converters for similar power ratings. Thus, FACTS controllers are an attractive choice for stabilizing a.c. transmission when frequency variations are not an issue (since FACTS cannot change the fundamental frequency) and the distance is not too long (in which case the loss reduction benefits of d.c. transmission dominate).

At present, FACTS controllers provide new options where the construction of new transmission lines or introduction of HVDC technology is prohibitive.

A sample application in the context of renewable resource integration would be to enable the operation of a highly economical solar or wind farm some distance away from a load center, which is found in power flow simulation to increase the vulnerability of the transmission network to overloads or contingencies. Clever use of FACTS in such a case could limit or redirect power flow on specific lines to ensure that all transmission security constraints are observed.

6.4 Voltage management

Modern solar and wind farms at the transmission level are treated similarly to other large generators in terms of managing reactive power injection and voltage magnitude at their interconnection point. In the distribution system context, where much of the renewable generation capacity involves small solar PV inverters connected behind customer meters, their role in voltage control is less obvious. Historically, when PV generation represented at most a small fraction of load, it was typically just considered *negative load* and treated essentially as noise on the system. In this situation, inverters are expected to maintain a unity power factor (i.e., refrain from exchanging reactive power) and simply inject whatever solar power they can. As the amount of PV generation relative to load increases, however, the legacy voltage regulation equipment on utility distribution circuits can become overtaxed. Two related but distinct issues of concern are overvoltage and voltage volatility.

On a traditional, radial distribution circuit, the voltage is assumed to be the highest at the substation (feeder head) and drop according to Ohm's law as a function of distance (line impedance) and load (current) along the circuit. Overvoltage can result when the net load and thus the voltage drop on the circuit is much smaller than the design expectation. With large clusters of PV generation, the power flow and voltage profile may even be inverted, increasing rather than decreasing toward the end of a distribution feeder. Typical legacy equipment includes tap-changing transformers and line voltage regulators which change the effective transformer turns ratio in steps to elevate voltage, or capacitor banks with simple in/out switching logic, which raise voltage by compensating for reactive power demand. These devices may not readily adapt, their range may be exceeded, or their control logic may be unfit for the new situation. For example, a voltage regulator can be fooled by the combination of reverse power flow and overvoltage to step the voltage up instead of down.

Voltage volatility can result from large and correlated variations in sunshine, as caused by passing clouds of a size comparable with the distribution circuit. In the extreme case, the associated voltage variation might exceed the standard service range (e.g., nominal voltage ±5% or 10%). More likely, such volatility would prompt legacy voltage regulators to actuate and change their tap settings much more frequently than they were designed to do over the course of a routine day, *hunting* the voltage as solar generation changes. Since these legacy electromechanical controls are vulnerable to wear, while also expensive to repair or replace, this has been cited as a concern by utilities [28].

As power electronic devices, *smart* inverters are readily capable of contributing to voltage management by adjusting their real and/or reactive power output—very quickly if needed—in response to a measurement or external control signal. The challenge is to codify suitable rules for inverter behavior which, when followed by a large number of devices, will result in the desired stable aggregate behavior—especially considering interactions with the legacy equipment.

The definitive standard is IEEE 1547-2018, revised in 2018 to articulate a suite of advanced inverter capabilities that are required by many jurisdictions. Note that compliance with this standard requires hardware capability to be present, but the local interconnection authority governs which inverter functionalities will be activated. Guidance for inverters performing voltage regulation services is given in terms of droop curves, where real or reactive power injection is adjusted based on measured local voltage magnitude. The *Q–V* or *P–V* droop curve is piecewise linear, flat within a central voltage band, and increasing/decreasing proportionally with voltage outside this band, up to the inverter's maximum. Inverter behavior is thus defined by the four inflection points [29,30].

Although there is a theoretical possibility of adverse interaction among such multiple controllers to cause an oscillating behavior in aggregate, available evidence suggests that this is an unlikely problem in practice, owing to the amount of network impedance between the devices [31]. A very different issue is that any curtailment of real power injection impacts payment or billing credit received, and thus it needs to be addressed in the interconnection tariff. Reactive power is somewhat but not completely independent of real power injection, as it counts against an inverter's thermal budget and associated kVA capacity and thus limits the amount of real power that can simultaneously be injected. At peak sunshine, this is an opportunity cost that similarly needs to be addressed.

One alternative approach to recruiting solar inverters for voltage regulation is to equip distribution circuits with dedicated power electronic devices for this purpose [32]. Like inverters, these can respond very quickly to adjust reactive power injection to help maintain a desired voltage level. The advantage is that they can be coherently controlled, without conflicting objectives or potential adverse control interactions, while the obvious disadvantage is the cost of a separate set of devices.

In sum, managing voltage in power systems even under very high penetrations of variable and distributed renewable resources is an entirely solvable problem from a technical standpoint. However, it may not be obvious how to solve it most efficiently and cost-effectively in any given setting, or how to allocate the costs among participants in the grid.

6.5 Protection

Protection engineering is concerned with maintaining safe operation in the face of faults on electrical circuits, while minimizing customer interruptions. Faults—i.e., any type of unintended contact between different parts of the network, or between the network and surrounding objects—can lead to dangerously large current flows, high voltages that humans might come into contact with, or arcs that can spark fires. Overhead distribution systems, which are more exposed and vulnerable than underground systems or high-voltage transmission networks, experience faults quite often—for example, when tree branches or animals touch lines.

Protective devices vary in complexity. The simplest and most common in lower-voltage systems is a *fuse*. When a large enough current flows through the fuse, the material melts, opening the circuit path. Fuses are inexpensive and reliable, but they must be manually replaced after use, which adds to customer interruption time. *Circuit breakers* mechanically open the path on detecting an overcurrent, e.g., with a spring-loaded mechanism, and can be reset. These devices consist of a sensing component such as a current transformer (CT), an electromechanical or digital relay whose setting determines whether the measured current exceeds the threshold to *trip*, and the actual breaker mechanism for physically opening the electrical contact that will isolate the faulted circuit section and de-energize the fault. A more sophisticated version of a circuit breaker is a *recloser*, which automatically reconnects the opened contacts after a preprogrammed delay (e.g., several seconds), sometimes repeatedly, to determine if the fault has cleared by itself. The use of reclosers greatly reduces customer interruption time, although it comes with some risk of re-energizing a faulted circuit.

A key challenge in protection settings, and an inherent limitation of overcurrent protection, is to distinguish between a fault current and a high load current. For example, it would be a nuisance if protective devices tripped under a large motor starting current. Conversely, if the fault impedance is high (e.g., a downed line on a dry asphalt surface), the current through it may not exceed the set protection threshold, and the fault would go undetected by legacy technology.

Multiple protection devices on the same circuit must be coordinated so that the circuit is opened and de-energized as close as possible to a fault, minimizing customer interruption. This is accomplished by comparing time–current coordination curves for each device on a circuit and estimating the expected fault current for likely scenarios. The analysis then shows which device would trip first, at a given fault current. The general design intent is for the device closest to a fault to open first, which minimizes customer interruption.

To facilitate protection coordination, legacy distribution systems are mostly radial, with the substation as the single source from which power flows unidirectionally outward. When it can be assumed that any fault will be fed by a single source along a single path, coordination of protective devices is more straightforward. By contrast, distributed generation (DG) leads to multiple power sources and multidirectional power flows in the network, potentially bedeviling existing protection schemes. When DG is introduced, it will contribute fault current in addition to the substation. This may change the direction of power flow and can either increase or decrease the fault current observed at various points on the circuit. Consequently, existing fuses and protective relays may not operate as intended in some fault scenarios because their sensed current fails to meet the expected threshold, in a phenomenon known as *relay desensitization*.

Another impact of DG on protection is inverter tripping. Inverter behavior is governed by digitally programmed control schemes. According to present standards, under any unusual system conditions—namely, abnormal network voltages or system frequency—inverters must quickly trip offline, disconnecting DG from the circuit [33]. The objective is to prevent unintentional *islanding*, where one or more inverters continue to energize an isolated section of a distribution circuit. Such an island may pose a safety risk to utility workers who assume it to be de-energized. Depending on the load-generation balance, a power island may also have unstable frequency and voltage, potentially damaging equipment [34]. Anti-islanding protection is therefore mandatory in grid-connected inverters. However, just as it can be difficult to diagnose a fault current, it is not obvious what amount of deviation from normal frequency and voltage should be taken as a threshold value for detecting an island.

Even brief, transient system disturbances can lead to the disconnection of all inverters and DG on a network. This is a highly defensive control strategy, which diminishes system reliability as DG penetrations increase. Inverter response to transient voltage disturbances is a canonical example. Under normal operating conditions, distribution network voltages are regulated within a tight band around the nominal system voltage. When a fault occurs, however, the extra current drawn can cause a *voltage sag*, a drop in voltage outside the permissible band. Voltage sags are generally very short (in the order of seconds), as many faults are transient and protection equipment quickly operates to clear the fault and return conditions to normal without interrupting service. However, even a brief voltage sag can cause inverters to disconnect. This can cause inconvenience and frustrate customers whose DG output, as well as the utility's, is curtailed. On long, rural feeders, the DG may provide voltage support, and its unintentional disconnection may lead to voltage issues [35]. DG may also mask a significant amount of feeder load. When a large amount of DG simultaneously trips offline, the substation load abruptly increases, affecting equipment and protection and aggravating voltage issues.

6.5.1 Smart protection

The previous paragraphs highlighted the emergent challenges of protection in distribution networks with DG. Fortunately, new solutions are gradually making their way into the field. Most long-term solutions to protection design in the presence of DG rely on improved sensing, communications, or both. This highlights a general feature of future distribution network operations: with the growth of DG, distribution networks can no longer be managed as predictable yet opaque systems. Real-time network information and dynamic operation enabled by novel sensing and control will become essential. A short-term approach to successful protection in the presence of DG is the installation of rapid trip relays at the DG point of interconnection which quickly disconnect the DG on detection of a fault. While this approach addresses the concern of DG sustaining faults, rapid tripping may become a reliability issue as DG penetrations increase. If a large amount of DG trips offline due to transient faults or disturbances originating elsewhere in the network, the sudden loss of generation can cause cascading system failures [36]. Another short-term approach suggests limiting the output of DG to levels that avoid protection coordination issues. This is undesirable as it curtails the DG resource and discourages further installations [37].

A more sophisticated tack is taken by adaptive protection schemes. Adaptive protection proposes to leverage increased communication and sensing for dynamic protection coordination. Real-time network data, such as DG outputs, nodal voltages, etc., would be analyzed to detect and localize faults. Messages would be sent over communication channels to networked relays to open or remain closed, isolating the fault while minimizing disruption and avoiding unnecessary DG trips. Unlike fuses, which operate based on immutable overcurrent settings, adaptive protection allows protection settings to be chosen based on real-time system conditions [38]. An important component of adaptive protection is fault detection and localization, for which a plethora of algorithms have been proposed. Approaches include estimating the fault distance using measurements of nodal voltages and currents [39], analyzing frequency decompositions of high-resolution measurements of voltage transients produced by the fault [40,41] or using multiple phasor measurements [42]. All these methods in general require a level of sensing and communication that is currently absent in distribution networks. It seems likely that concerns about DG integration impacts will drive utilities to install the sensing and communication needed to make adaptive protection a reality.

Improved communications and sensing also underlie solutions for unintentional islanding. Transfer trip schemes use sensors to detect the opening of breakers that isolate a section of the circuit from the rest of the feeder. On detection, messages are transmitted to DG on the isolated section to trip offline, preventing the formation of an unintentional power island. Of course, this requires all circuit breakers and DG inverters to be linked through communication channels.

In the future, the formation of power islands may be an asset rather than a liability to grid operations. *Intentional* islanding could be used to improve system reliability. On detection of a disturbance, rather than tripping all DG offline, circuit sections containing DG and load would now intentionally island from the rest of the system, isolating themselves from the disturbance while continuing to provide power. Controlled intentional islanding could also be used during planned system maintenance, allowing customers to continue to receive power from local DG, even when crews are working on circuit sections between them and the substation. In fact, some researchers have gone further with the concept of intentional islanding, proposing to operate the entire future DG as an accumulation of several microgrids that have both local generation and load and can be operated in both *grid-connected* or *islanded* mode [43]. Intentional islanding could also enable faster post-disturbance restoration of the system. On resolution of a system disturbance, islands could seamlessly reconnect to the main network, re-forming the connected DG without interrupting service [43].

6.6 Integration and coordination

The key challenges of integrating variable and distributed resources at high penetration levels can be described as a problem of coordination in both time and space.

In essence, renewable and distributed resources introduce new constraints on resource availability: we cannot always have the resources when we want them, where we want them. Although the same is true to some extent for every energy resource, the constraints associated with renewables can be more stringent and simply different from the constraints that legacy power systems were historically designed and built around.

The earlier sections of this chapter addressed temporal coordination in terms of power balance, and spatial coordination in terms of transmission constraints, while voltage management and protection include both temporal and spatial characteristics. As stated in the introduction, a *smart grid* is one that provides for such coordination at increasingly fine resolution in both time and space, allowing us to more actively and precisely manage the many old and new components of a synchronous grid. This means many more frequent decisions need to be made and communicated, based on larger volumes of information. An overarching problem thus concerns the infrastructure to support decision-making and execution of all this smart management.

6.6.1 Communications

Perhaps the most obvious challenge is the need to establish and maintain communication channels with larger numbers of smaller, distributed devices, including generation, energy storage, and controllable loads. Ethernet and wireless connectivity have obviously revolutionized this field, so that the mere availability or extension of a physical communication layer to devices on the electric grid is no longer a major technical hurdle, particularly in buildings and urban environments. Still, field devices on the transmission and distribution system in areas with poor wireless coverage may require adding dedicated radio or hard-wired communications at substantial cost.

The traditional options for physical communication layers in Supervisory Control and Data Acquisition (SCADA) include telephone lines, radio, and power line carrier (PLC). The cost of providing communication has often been the limiting factor in decisions about whether to install remote terminal units (RTUs) on T&D equipment that can be operated from a control center by humans or automated systems. These investments were made in most countries at the transmission level in the latter part of the twentieth century, but often not on distribution circuits, where comparatively small amount of load served has not economically justified remote control or automation. For example, it is not unusual that field switches require a person to go on location to operate them or verify their open/closed status.

To achieve fast, reliable communications, especially in the distribution network, utilities are turning to new technologies. Communicating economically with remote devices, which cannot piggyback on existing telecommunications infrastructure, is a challenge. In PLC, the communication signal is transmitted via the power lines themselves, by modulating and injecting the signal into the grid voltage waveform. PLC communication is popular because it avoids installation of a dedicated communication link, instead of using the grid directly. However, this makes PLC communication vulnerable to grid failures like downed lines, which could sever communication links precisely when they are most needed. Alternatively, optical fiber communication, like copper line communication, is isolated from the grid itself. While copper communication links are generally not owned by the utility and support slower speeds, optical links are utility owned and enable rapid communication. However, rolling out optical fiber communication throughout the entire distribution network, especially to remote systems, is extremely expensive [44].

6.6.2 Internet of Things

The Nest thermostat, which appeared on the market in 2011, has exemplified a new era where the Internet of Things (IoT) meets the electric grid. At present, it seems commonplace to control electronic devices from a smartphone. But what stands out about thermostats is (1) their functional coupling to the grid infrastructure that was previously well isolated from external information systems, and (2) their association with a very substantial electric load category, heating and air conditioning. In their potential aggregate significance to the grid, actively controlled heating and cooling loads will only be matched by EV charging [45]. EV charging stations are generally equipped with communications and sophisticated software to manage charging (e.g., kW charge rates, queuing, time delays, and intermittent charging) based on numerous technical and economic criteria. These criteria could easily include advanced electric grid support services, if the right information is provided to the EV chargers [46,47].

Timing the behavior of heating, cooling, EV charging, and other loads to provide services to the grid requires connecting intelligence beyond the edges of the grid, leading to the development of *smart buildings*. Smart buildings can adjust their behavior in response to grid conditions through building energy management systems. The most sophisticated energy management systems allow interfacing with every controllable device in a building, from lights to thermostats to other appliances. A diversity of data from within the building and from the grid, including time-series measurements, shared through the energy management system, inform the building's behavior as it optimizes for internal and grid-support objectives. Commands are then sent to controllable devices. Creating energy management systems that elegantly handle the diversity of data and devices in a building is a challenging task and remains an active area of research [48]. As these systems are deployed, they will greatly increase the coupling of customers and the grid, upending the traditional view of electric load from the grid's perspective as immutable.

The coupling of IoT with the electric grid has introduced new concerns about cybersecurity, although IoT devices are by no means the only possible attack vector. Vulnerabilities have been highlighted by the attack on Ukraine [49] and a U.S. power plant [50], where hackers gained access to functional control systems through unrelated software on utility employees' personal computers [51]. The traditional approach for cybersecurity in grid operations has been air gapping, or strict physical isolation between computer systems involved in grid operations and anything on the outside, but such gaps are becoming more difficult to maintain. Some experts have argued for a shift in focus on attack detection and mitigation [52]. Leveraging existing external communications networks as well as computing resources for smart grids certainly demands a new emphasis on cybersecurity, whose cost and risks (whether actual or perceived) may dominate over the physical infrastructure investment.

6.6.3 Smart aggregation

Beyond simply communicating with large numbers of small and distributed resources on the grid, there is the problem of how to efficiently structure transactions with them. Different approaches have been proposed [53,54] for how to aggregate devices, information, and decision-making—at what level and what location—so as not to overwhelm central grid operations and appropriately delegate responsibility. For example, a building management system or a microgrid may aggregate individual devices at the scale of, say, an office building or a university campus and present a single, controllable resource to a grid operator. While some forms of control such as frequency regulation services may be aimed at the bulk grid level and transacted with a balancing authority, there is also a growing need for locationally specific control, such as managing voltage on a distribution circuit or preventing overloads on a particular transformer. At this higher level of resolution, distribution system operators (DSOs) may be a more logical entity tasked with the coordination of resources [55].

6.6.4 Situational awareness

At any level of aggregation, it is important to use information effectively to support system visibility, situational awareness, and functional understanding by operators. The reality of distributed devices introducing greater variance and uncertainty in power flows—along with the possibility of harnessing them creatively for grid support services—calls for a new level of knowledge and insight: knowledge of real-time status and conditions, and insight into the mechanisms of interaction that create risks for grid stability and reliability.

The 2003 Blackout in the Northeastern United States and Canada revealed some of the challenges in this area. This event predated high penetration levels of solar and wind generation, but it can be understood as an outcome of a growing complexity of the grid, with larger power flows over longer distances and thus stronger, less transparent interdependencies. Underlying causes identified by the postmortem analysis included a lack of situational awareness and a lack of system understanding by grid operators [56].

To make correct decisions (e.g., whether to pre-emptively curtail load to prevent a larger cascading failure), operators need awareness of the real-time status of grid components (e.g., transmission line loading), along with an understanding of the possible consequences—all in an environment where effects can be highly nonlinear, and the number of interactions grows combinatorially with the number of devices considered. For instance, a standard metric for grid operational security is the N-1 criterion (or sometimes N-2), where a simulation determines whether the system can withstand the unexpected failure of any one component (or two). It is far from obvious which contingencies must be included among these simulations. This was illustrated, for example, by the 2011 Blackout in the Southwestern United States, where a transmission line failure in Arizona caused a major outage in San Diego—to the considerable surprise of grid operators, who had believed their system to be secure with respect to any one contingency [57].

6.6.5 Grid data

One strategic response to the growing complexity and need for situational awareness of the grid has been the deployment of advanced sensor data collection, with precisely time-synchronized Phasor Measurement Units (PMUs) that stream detailed synchrophasor data. Synchrophasors typically include voltage and current root-mean-square (rms) magnitudes and phase angles in degrees (i.e., the time offset of the sinusoidal waveform), along with the computed a.c. frequency and rate of change of frequency (ROCOF). These data frames are typically reported once or twice per cycle, or every other cycle. This quickly accumulates a fairly large data volume compared with traditional utility operations. Over the past two decades, PMUs have been increasingly deployed in transmission systems around the world [58,59]. Synchrophasor data provide unique insight into aggregate frequency response of the grid, as well as wide-area sub-synchronous oscillations that can threaten large interconnected a.c. systems.

The ultimate high-resolution grid data comes from continuous point-on-wave (CPOW) measurements. Unlike PMUs, which abstract synchrophasors from many analog samples per cycle through a lossy compression algorithm, CPOW sensors retain and communicate the detailed waveform data. This type of explicit information promises to be especially valuable for observing transient disturbances, harmonic distortion, rapid dynamic response, or any electrical condition not well characterized by an ideal sinusoid [60]. Traditionally, such measurements have only been made intermittently and locally in the context of power quality, without time alignment or cross-reference to distant nodes in the grid.

Naturally, the prospect of continuously streaming large volumes of grid measurement data for ready consumption in grid operations and engineering analysis hinges on recent advances in information technology. Powerful databases such as the Berkeley Tree Database (BTrDB), developed specifically for managing time-series sensor data [61], or kdb+ from high-frequency financial trading [62], are an important area of innovation or novel application to the grid. Beyond turning to new platforms for hosting a barrage of new sensor data, the industry will also need to learn how best to integrate heterogeneous data streams, old and new, with analytic and decision-making functionality across various applications, not just isolated silos. Rather than keeping data from dedicated sensors separate for their respective purposes—e.g., protection, voltage regulation, and customer billing—the ensemble of measurements can support comprehensive algorithms such as state estimation, for which no individual data set would suffice. Such integrative platforms will enable an experimental approach to grid science [63], including the application of rapidly evolving machine learning techniques to the electric grid space [64].

To date, most electrical sensing and data collection efforts for grid operational purposes have been limited to the transmission level. Traditionally, distribution systems comprised only passive loads and were largely one-directional. Planning and operating decisions could be based on an envelope of maximum loads and fault currents, without the need to track real-time status. With the proliferation of active distributed devices, of course, this situation is changing dramatically. Beyond simply monitoring, a more interventionist approach to distribution system operations may be desired, which requires visibility into quantities such as voltage magnitudes and power flows at various points along a distribution feeder. At the same time, transmission level operations have a growing interest in real-time information from behind the substation, since the aggregate behavior of distributed resources can now have a significant impact on the bulk grid.

For one, the variability of solar PV generation implies a need for adequate generation and ramping reserves, unless local storage resources and/or controllable loads can guarantee smoothing the net load. Perhaps most important, the bulk grid is exposed to the possible simultaneous loss of a large number of inverters in response to a grid disturbance, as mandated by interconnection standards. One recent example of this phenomenon was the 2016 Blue Cut Fire incident [65], where a step change in the voltage phase angle due to a transmission line outage was interpreted by countless inverters on the Western U.S. grid as a frequency deviation, requiring them to disconnect. Note this behavior was consistent with the letter, but not the intent, of the IEEE 1547 standard governing inverters. The Western grid consequently experienced a sudden loss of ~1.2 GW [65], where the additional loss of behind-the-meter rooftop PV was difficult to estimate.

The Blue Cut Fire incident also illustrated the need for accurate dynamic models of grid components like inverters to predict how they will respond individually and in aggregate to various disturbances [66,67]. Models of power systems have always been simplified to make their analysis mathematically tractable, especially since they predate the age of computing that today allows for numerical solution of large systems of equations. For example, standard methods assume balanced three-phase operation with sinusoidal a.c. waveforms and analyze steady-state operation separately from dynamic or transient conditions. While we have sophisticated models of rotating machines to describe their individual electromechanical behavior, their collective response, which depends on the electrical network between them and possible interactions through the logic of their control loops, is much less rigorously understood. Consequently, emergent properties of the interconnected system—such as the response to forced oscillations, or unintended unit trips—can surprise engineers and operators alike. Inverter-based resources, though eminently capable and controllable as individual components, introduce another major uncertainty to this mix. Reliable smart grids will need to build a solid foundation of system understanding with realistic models for both offline analysis and real-time situational awareness.

A key challenge for building and validating good models is access to empirical data of appropriate quality. The same is true for exercising control. Data quality can mean temporal and spatial resolution, but may also include accuracy, time synchronization, latency, and continuity [68]. Good grid data require not only physical sensors at suitable locations in the field, but also an information infrastructure that may include streaming communications, voluminous data archives, rapid search capabilities, flexible analytics, Application Programming Interfaces (APIs), and helpful user interfaces for humans. At the time of this writing, researchers and industry developers are actively exploring a spectrum of new approaches for gathering and synthesizing intelligence for the electric grid to help this venerable infrastructure perform its vital role in the twenty-first century.

6.7 Economic considerations

The economics of smart grid technologies are challenging to quantify because of the complexity of the electric grid, and the fact that many smart grid technologies aim to address impending future problems rather than acute pain points.

Creating appropriate incentives for desirable technologies and behaviors is generally a work in progress. Demand response programs, for example, incentivize customers to adjust their load with time varying prices or other signals from utilities. Studies have shown that demand response can enable peak shaving, reduced losses, and improved voltage control [69]. Yet these programs are still limited, with most customers continuing to see traditional, fixed electricity prices. As they expand, demand response programs will increase the economic incentives for smart appliances and building energy management systems that can respond to time-varying price signals. Indeed, comprehensive demand response is only feasible with the improved communications and measurement infrastructure of a smart grid.

More broadly, there are many research efforts to develop new electricity pricing models that reflect the increasingly dynamic and distributed nature of the electric grid. Under existing models, environmentally beneficial technologies may actually become financially undesirable at high penetration levels, as is the case for rooftop solar in Australia [70]. New approaches aim to better incentivize DG and storage and even to enable more decentralized transactions such as peer-to-peer electricity trading [71]. In light of these changes and uncertainties, the cost-saving potential of the smart grid technologies in this chapter can only be discussed qualitatively and somewhat speculatively.

Smart grid technologies associated with maintaining power balance span communications infrastructure, measurement technologies, and novel devices such as inverters and energy management systems. At present, grid frequency—a proxy for power balance—is carefully controlled by adjusting the output of large centralized generators. System inertia enables the grid to ride through transitions and periods of imbalance, but the growth of highly variable renewable generation and reduced system inertia may reduce the effectiveness of these traditional methods. Smart grid technologies will enable the recruitment of more diverse and distributed resources for ensuring power balance. Besides maintaining system balance, these technologies may reduce the need for expensive *peaker* power plants, by shifting load to flatten the load curve. However, economic models are still evolving to capture these value streams and compensate distributed resources for participating in frequency support, peak shaving, and other *ancillary services*. There are proposals to create or adapt ancillary services markets to better enable this. Overall, we expect smart grid technologies to help maintain power balance with less overcapacity.

From the transmission perspective, smart grid technologies have several potential economic benefits. By enabling more DG and therefore increased *local* balancing of generation and consumption, as well as dynamic load shifting by customers in response to real-time signals, smart grid technologies can reduce transmission congestion. Transmission congestion is already a significant issue, costing one transmission system operator US$2 billion in the eastern United States in 2005 [72]. With reduced congestion, expensive transmission capacity expansion projects can be delayed or even avoided altogether. Smart grid technologies will also enable better diagnosis of transmission outages, faster post-outage restoration, and even preemption of some outages. At present, owing to overbuilt capacity and conservative operational limits, transmission system outages are rare in most countries, but when they do occur, they can be immensely costly. The cost of the August 2003 Northeastern blackout was estimated between US$4.5 and US$8.2 billion [73].

Outage risk is likely to increase with the growth of renewables and more extreme weather events. Northern California, for example, has recently experienced pre-emptive outages intended to reduce the risk of arcs, faults, or other grid malfunctions sparking wildfires. Such power shutoffs are likely to become increasingly frequent as drought conditions in the state continue. One outage in October 2019 is estimated to have cost the California economy over US$2 billion [74]. Smart grid technologies can hypothetically reduce this cost by mitigating the need for preemptive outages altogether. For example, techniques have been developed which use novel measurements to detect and de-energize falling lines before they can hit the ground and spark a fire [75]. Such smart grid technologies would enable more focused wildfire mitigation strategies rather than blanket blackouts.

At the distribution level, smart grid technologies can potentially reduce operation and maintenance costs by enabling more efficient operation and allowing issues to be quickly diagnosed and even preempted. Grid outages predominantly arise in the distribution network, the most sprawling and exposed portion of the entire electric infrastructure. In many places, these outages must be localized manually by inspection crews—a time-consuming and therefore expensive task.

Smart grid technologies, in particular novel measurements, communications, and algorithms, can automatically localize the source of outages, greatly reducing cost and outage durations. For example, the deployment of such technology by Pacific Gas & Electric in 2017 saved an estimated US$2.5 million in the first year by reducing the number of person-hours needed to localize outages [76]. These technologies also reduce the frequency and duration of outages experienced by customers: an important benefit, although difficult to economically quantify. Using one model for the value of reliability, PG&E estimates that smart grid upgrades for the automatic localization and isolation of faults and eventual service restoration, termed FLISR, saved US$808.5 million by avoiding 303.9 million customer minutes (minutes of outages × number of customers experiencing the outages) [76]. The annual savings due to FLISR technology are growing as more extreme weather events increase the number of outages and therefore the benefit of automated localization and restoration.

Another pressing incentive for smart grid technology deployments on distribution systems is the growth of rooftop solar PV generation and EVs. Numerous countries have ambitious targets for EV use in the coming decades, and EV charging will place significant new load on distribution networks. For example, in the United States, electricity consumption may be 38% higher in 2050 with aggressive adoption of EVs compared with reference forecasts [45]. Without controlled charging, EVs may greatly increase peak load, which is the most expensive to satisfy. Smart grid technologies that enable and incentivize optimized charging schedules for peak reduction, and altogether flatten the load profile, will lead to great cost savings. These technologies can transform PV and EV integration from a liability and strain on grid infrastructure, into an asset. Fundamentally, they will be *required* to enable the levels of PV generation and EV charging mandated by many national policies.

Overall, smart grid technologies tend to have several economic benefits across the system, rather than a single incentive. For sensing and measurement technologies in particular, it can be difficult for a utility to justify an investment based on an individual use case. Instead, greater benefits emerge when considering an entire portfolio of applications that are enabled by the technology. These investment decisions will become more clear-cut as the industry and stakeholders define specific expectations or capabilities that the electric grid should deliver as a matter of strategic policy. The authors certainly hope that the operation of power systems with 100% carbon neutral resources will be a goal on which all countries in the world can agree in the near future.

References

[1] Hughes T. P. *Networks of Power: Electrification in Western Society, 1880–1930*. Baltimore: Johns Hopkins University Press; 1983.

[2] Hirsh R. F. *Technology and Transformation in the American Electric Utility Industry*. New York: Cambridge University Press; 1989.

[3] Venkatasubramanian V., and Li Y. 'Analysis of 1996 Western American electric blackouts'. Presented at *Bulk Power System Dynamics and Control Conference*; Cortina d'Ampezzo, Italy, 2004,

[4] Gils H. C. 'Assessment of the theoretical demand response potential in Europe'. *Energy*. 2014; **67**:1–18.

[5] Huang Z., Zhou N., Tuffner F. K., *et al. Mango-Modal Analysis for Grid Operation: A Method for Damping Improvement through Operating Point Adjustment*. PNNL Report 19890, 2010.

[6] Eto J. H., Undrill J., Roberts C., Mackin P., and Ellis J. *Frequency Control Requirements for Reliable Interconnection Frequency Response*. LBNL Report 2001103, 2018.

[7] Huang S. H., Maggio D., McIntyre K., Betanabhatla V., Dumas J., and Adams J. (eds.). 'Impact of wind generation on system operations in the deregulated environment: ERCOT experience'. *Proceedings of the IEEE Power & Energy Society General Meeting;* Calgary, Canada, July 2009. New Jersey: IEEE; 2009. pp. 1–8.

[8] Driesen J., and Visscher K. (eds.). 'Virtual synchronous generators'. *Proceedings of the IEEE Power and Energy Society General Meeting—Conversion and Delivery of Electrical Energy in the 21st Century*; Pittsburgh, USA, July 2008. New Jersey: IEEE; 2008. pp. 1–3.

[9] Tamrakar U., and Shrestha D., Maharjan M., Bhattarai B. P., Hansen T. M., and Tonkoski R. 'Virtual inertia: Current trends and future directions'. *Applied Sciences*. 2017; **7**(654):1–29.

[10] Ela E., Gevorgian V., Tuohy A., Kirby B., Milligan M., and O'Malley M. 'Market designs for the primary frequency response ancillary services—Part I: Motivation and design'. *IEEE Transactions on Power Systems*. 2013; **29**(1):421–431.

[11] Gonzalez-Longatt F., Chikuni E., and Rashayi E. (eds.). 'Effects of the Synthetic Inertia from wind power on the total system inertia after a frequency disturbance'. *Proceedings of the IEEE International Conference on Industrial Technology*; Cape Town, South Africa, February 2013. New Jersey: IEEE; 2013. pp. 826–832.

[12] Bolinger M., and Seel J. *Utility-Scale Solar: Empirical Trends in Project Technology, Cost, Performance, and PPA Pricing in the United States—2018 Edition*. LBNL Report, 2018.

[13] Fairley P. 'China's potent wind potential', *MIT Technology Review*, 2009. Available from https://www.technologyreview.com/2009/09/14/209991/chinas-potent-wind-potential/

[14] Agence France-Presse. *Full Tilt: Giant Offshore Wind Farm Opens in North Sea* [online]. London: The Guardian; 2017. Available from https://www.theguardian.com/environment/2017/may/09/full-tilt-giant-offshore-wind-farm-opens-in-north-sea [Accessed December 2019].

[15] Salama A. Y., Shalaby M., Aziz A. A., and Refai R. A. 'Aswan High Dam/Cairo proposed HVDC transmission system'. *Beni-Suef University Journal of Basic and Applied Sciences*. 2014; **3**(1):37–44.

[16] Rudervall R., Charpentier J. P., and Sharma R. 'High voltage direct current (HVDC) transmission systems technology review paper'. Presented at *Energy Week*; Washington, DC, USA, 2000.

[17] Temple J. 'China's giant transmission grid could be the key to cutting climate emissions'. *MIT Technology Review*, 2018.

[18] entso-e. *Executive Report Connecting Europe: Electricity 2025–2030–2040 Version for Public Consultation*. entso-e TYNPD Report, 2018.

[19] Hitaj C. 'Location matters: The impact of renewable power on transmission congestion and emissions'. *Energy Policy*. 2015; **86**:1–16.

[20] Karimi S., Musilek P., and Knight A. M. 'Dynamic thermal rating of transmission lines: A review'. *Renewable and Sustainable Energy Review*. 2018; **91**:600–612.

[21] Van Cutsem T., and Vournas C. *Voltage Stability of Electric Power Systems*. New York: Springer; 2007.

[22] Adams J., Carter C., and Huang S. H. (eds.). 'ERCOT experience with sub-synchronous control interaction and proposed remediation'. *Proceedings of PES T&D;* Orlando, USA, May 2012. New Jersey: IEEE; 2012. pp. 1–5.

[23] Shair J., Xie X., Wang L., Liu W., He J., and Liu H. 'Overview of emerging subsynchronous oscillations in practical wind power systems'. *Renewable and Sustainable Energy Reviews*. 2019; **99**:159–168.

[24] Setreus J., and Bertling L. (eds.). 'Introduction to HVDC technology for reliable electrical power systems'. *Proceedings of the 10th International Conference on Probabilistic Methods Applied to Power Systems*; Rincon, Puerto Rico, May 2008. New Jersey: IEEE; 2009. pp. 1–8.

[25] Saksvik O. (eds.). 'HVDC technology and smart grid'. *Proceedings of the 9th IET International Conference on Advances in Power System Control, Operation and Management*; Hong Kong, China, November 2012. London: IET; 2012. pp. 168.

[26] Marquardt R. 'A new modular voltage source inverter topology.' *Conf. Rec. ECE*, 2003.

[27] Westerweller T., Friedrich K., Armonies U., Orini A., Parquet D., and When S. (eds.). 'Trans bay cable—World's first HVDC system using multilevel voltage-sourced converter'. *Proceedings of CIGRE*; Paris, France, 2010. Paris: CIGRE; 2010.

[28] Yan R., Marais B., and Saha T. K. 'Impacts of residential photovoltaic power fluctuations on on-load tap changer operation and a solution using DSTATCOM'. *Electric Power Systems Research*. 2014;**111**:185–193.

[29] Yan R., Marais B., and Saha T. K. 'IEEE standard for the interconnection and interoperability of distributed energy resources with associated electric power system interfaces'. *IEEE Std 1547*. 2018.

[30] Ropp M. *Guide to the IEEE 1547-2018 Standard and Its Impact on Cooperatives*. NRECA Report, 2019.

[31] Chakraborty A., Hoke A., and Lundstrom B. (eds.). 'Evaluation of multiple inverter volt-VAR control interactions with realistic grid impedances'. *Proceedings of IEEE Power & Energy Society General Meeting*; Denver, USA, July 2015. New Jersey: IEEE; 2015. pp. 1–5.

[32] Bloemink J. M., and Green T. C. 'Benefits of distribution-level power electronics for supporting distributed generation growth'. *IEEE Transactions on Power Delivery*. 2013; **28**(2):911–919.

[33] Xue Y., Divya K. C., Griepentrog G., Liviu M., Suresh S., and Manjrekar M. (eds.). 'Toward next generation photovoltaic inverters'. *Proceedings of IEEE Energy Conversion Congress and Exposition*; Phoenix, USA, September 2011. New Jersey: IEEE; 2011. pp. 2467–2474.

[34] Seguin R., Woyak J., Costyk D., Hambrick J., and Mather B. *High-Penetration PV Integration Handbook for Distribution Engineers.* NREL Report TP-5D00-63114, 2016.

[35] Dugan R. C., and Mcdermott T. E. 'Distributed generation'. *IEEE Industry Applications Magazine*. 2002; **8**(2):19–25.

[36] Lopes J. P., Hatziargyriou N., Mutale J., Djapic P., and Jenkins N. 'Integrating distributed generation into electric power systems: A review of drivers, challenges and opportunities'. *Electric Power Systems Research.* 2007; **77**(9):1189–1203.

[37] Chaitusaney S., and Yokoyama A. 'Prevention of reliability degradation from recloser-fuse miscoordination due to distributed generation'. *IEEE Transactions on Power Delivery*. 2008; **23**(4):2545–2554.

[38] Ackermann T., and Knyazkin V. (eds.). 'Interaction between distributed generation and the distribution network: Operation aspects'. *Proceedings of IEEE/PES Transmission and Distribution Conference and Exhibition*; Yokohama, Japan, October 2002. New Jersey: IEEE; 2003. pp. 1357–1362.

[39] Zhu J., Lubkeman D. L., and Girgis A. A. 'Automated fault location and diagnosis on electric power distribution feeders'. *IEEE Transactions on Power Delivery*. 1997; **12**(2):801–809.

[40] Borghetti A., Corsi S., Nucci C. A., Paolone M., Peretto L., and Tinarelli R. 'On the use of continuous-wavelet transform for fault location in distribution power systems'. *International Journal of Electric Power & Energy Systems*. 2006; **28**(9):608–617.

[41] Magnago F. H., and Abur A. 'Fault location using wavelets'. *IEEE Transactions on Power Delivery*. 1998; **13**(4):1457–1480.

[42] Ren J., Venkata S. S., and Sortomme E. 'An accurate synchrophasor based fault location method for emerging distribution systems'. *IEEE Transactions on Power Delivery*. 2013; **29**(1):297–298.

[43] Lasseter R. H. 'Smart distribution: Coupled microgrids'. *Proceedings of the IEEE*. 2011; **99**(6):1074–1082.

[44] Kuzlu M., and Pipattanasomporn M. (eds.). 'Assessment of communication technologies and network requirements for different smart grid applications'. *Proceedings of IEEE PES Innovative Smart Grid Technologies Conference*; Washington, DC, USA, February 2013. New Jersey: IEEE; 2013. pp. 1–6.

[45] Mai T. T., Jadun P., Logan J. S., *et al. Electrification Futures Study: Scenarios of Electric Technology Adoption and Power Consumption for the United States.* NREL Report TP-6A20-71500, 2018.

[46] Liu M., Phanivong P., Shi Y., and Callaway D. 'Decentralized charging control of electric vehicles in residential distribution networks'. *IEEE Transactions on Control Systems Technology*. 2017; **27**(1):266–281.

[47] Kempton W., Tomic J., Letendre S., Brooks A., and Lipman T. *Vehicle-to-Grid Power: Battery, Hybrid, and Fuel Cell Vehicles as Resources for*

Distributed Electric Power in California. UC Davis Report UCD-ITS-RR-01-03, 2001.

[48] Fierro G., and Culler D. E. 'XBOS: An extensible building operating systems'. Presented at *2nd ACM International Conference on Embedded Systems for Energy-Efficient Built Environments*; 2015.

[49] Lee R. M., Assante M. J., and Conway T. *Analysis of the Cyber Attack on the Ukrainian Power Grid*. E-ISAC Report, 2016.

[50] Greenberg, A. *The Highly Dangerous 'Triton' Hackers Have Probed the US Grid* [online]. 2019. Available from https://www.wired.com/story/triton-hackers-scan-us-power-grid/ [Accessed December 2019].

[51] Kshetri N., and Voas J. 'Hacking power grids: A current problem'. *Computer*. 2017; **50**(12):91–95.

[52] Liu C. C., Stefanov A., Hong J., and Panciatici P. 'Intruders in the grid'. *IEEE Power and Energy Magazine*. 2011; **10**(1):58–66.

[53] Horta J., Kofman D., and Menga D. 'Novel paradigms for advanced distribution grid energy management' 2016. Available from https://arxiv.org/abs/1712.05841.

[54] Halu A., Sala A., Khiyami A., and Gonzalez M. C. 'Data-driven modeling of solar-powered urban microgrids'. *Science Advances*. 2016; **2**(1):1–9.

[55] Kristov L., De Martini P., and Taft J. D. 'A tale of two visions: Designing a decentralized transactive electric system'. *IEEE Power and Energy Magazine*. 2016; **14**(3):63–69.

[56] U.S.–Canada Power system Outage Task Force. Final Report on the *August 12, 2003 Blackout in the United States and Canada: Causes and Recommendations*. EPA Report, 2004.

[57] FERC & NERC Staff. *Arizona–Southern California Outages on September 8, 2011: Causes and Recommendations*. FERC & NERC Report, 2012.

[58] Lu C., Shi B., Wu X., and Sun H. 'Advancing China's smart grid: Phasor measurement units in a wide-area management system'. *IEEE Power and Energy Magazine*. 2015; **13**(5):60–71.

[59] Data and Network Management Task Team. *NASPI 2014 Survey of Synchrophasor System Networks—Results and Findings*. NASPI Report NASPI-2015-TR-015, 2015.

[60] Silverstein A., Follum J. *High-Resolution, Time-Synchronized Grid Monitoring Devices*. North American Synchrophasor Initiative Technical Report (NASPI-2020-TR-004), March 2020. Available from https://www.naspi.org/sites/default/files/reference_documents/pnnl_29770_naspi_hires_synch_grid_devices_20200320.pdf

[61] Andersen M. P., and Culler D. E. (eds.). 'BTrDB: Optimizing storage system design for timeseries processing'. *Proceedings of 14th USENIX Conference on File and Storage Technologies*; Santa Clara, USA, February 2017. ACM SIGOPS, pp. 39–52.

[62] Gutierrez D. *Kdb+ and the Internet of Things/Big Data* [online]. 2015. Available from https://insidebigdata.com/2015/01/19/kdb-internet-things-big-data/ [Accessed December 2019].

[63] Murphy S. P., Andersen M., Jones K. D., Bariya M., and Schuman J. (eds.). 'A universal sensor analytics and artificial intelligence platform for the grid'. *Proceedings of CIGRE Grid of the Future Conference;* Virginia, USA, October 2018. Paris: CIGRE; 2018.

[64] Stewart E. M., Hendrix V., Chertkov M., and Deka D. *Integrated Multi-Scale Data Analytics and Machine Learning for the Distribution Grid and Building-To-Grid Interface.* LLNL Report LLNL-TR-727125, 2017.

[65] Stewart E. M., Hendrix V., Chertkov M., and Deka D. *1,200 MW Fault Induced Solar Photovoltaic Resource Interruption Disturbance Report.* NERC Report, 2017.

[66] Mather B., Aworo O., Bravo R., and Piper D. (eds.). 'Laboratory Testing of a Utility-Scale PV Inverter's Operational Response to Grid Disturbances'. *Proceedings of IEEE Power & Energy Society General Meeting;* Portland, USA, August 2018. New Jersey: IEEE; 2018. pp. 1–5.

[67] Mather B., Shah S., Norris B. L., *et al. NREL/SCE High Penetration pV Integration Project: FY13 Annual Report.* NREL Report TP-5D00-61269, 2014.

[68] PMU Applications Requirements Task Force. *PMU Data Quality: A Framework for the Attributes of PMU Data Quality and a Methodology for Examining Data Quality Impacts to Synchrophasor Applications.* NASPI Report 2017-TR-002, 2017.

[69] Vardakas J. S., Zorba N., and Verikoukis C. V. 'A survey on demand response programs in smart grids: Pricing methods and optimization algorithms'. *IEEE Communications Surveys & Tutorials.* 2014; **17**(1):152–178.

[70] Gilmore J., Vanderwaal B., Rose I., and Riesz J. 'Integration of solar generation into electricity markets: An Australian National Electricity Market case study'. *IET Renewable Power Generation.* 2014; **9**(1):46–56.

[71] Sorin E., Bobo L., and Pinson P. 'Consensus-based approach to peer-to-peer electricity markets with product differentiation'. *IEEE Transactions on Power Systems.* 2018; **34**(2):994–1004.

[72] Market Monitoring Unit. *2007 State of the Market Report.* PJM Report, 2008.

[73] Market Monitoring Unit. *The Economic Impacts of the August 2003 Blackout.* ELCON Report, 2004.

[74] Stevens, P. *PG&E Power Outage Could Cost the California Economy More Than $2 billion* [online]. 2019. Available from https://www.cnbc.com/2019/10/10/pge-power-outage-could-cost-the-california-economy-more-than-2-billion.html [Accessed November 2019].

[75] O'Brien W., Udren E., Garg K., Haes D., and Sridharan B. (eds.). 'Catching falling conductors in midair—Detecting and tripping broken distribution circuit conductors at protection speeds'. *Proceedings of 69th Annual Conference for Protective Relay Engineers;* College Station, USA, April 2016. New Jersey: IEEE; 2017. pp. 1–11.

[76] O'Brien W., Udren E., Garg K., Haes D., and Sridharan B. (eds.). *Pacific Gas and Electric Company Smart Grid Annual Report—2017.* PG&E Report, 2017.

Chapter 7

Demand response technologies in buildings for curbing and shifting electric loads

Therese E. Peffer[1]

7.1 Introduction

The mismatch between electrical supply and demand, both in space and time, presents a growing problem found worldwide as customers seek reliable energy. Large natural gas-, nuclear-, or coal-fired power plants supply electricity from a central location, with distribution lines stretching across long distances to reach end uses such as commercial and residential buildings. Electrical demand changes throughout the day and seasons. Following the activities and schedule of people, demand is generally higher during the day and lower at night. Seasonally, electrical demand can peak in the summer afternoons from residential and commercial buildings (as is the pattern in California) or in winter mornings (such as in Korea). Figure 7.1 shows that the small percentage of time electricity demand was quite high in California. These short durations of peak demand can cause electricity shortages and power outages, or contribute to air pollution by utilities bringing on older "peaker" plants.

The burgeoning efforts to reduce greenhouse gases have added new challenges and present new opportunities for coordinating electrical supply and demand. Various companies, states, and countries are setting goals to increase the percentage of electricity from renewable power, reduce carbon emissions, increase distributed generation, and add zero emission vehicles [2]. More and more renewable generation is decentralized and colocated with demand, such as rooftop solar. However, electricity generated by solar, wind, and hydro is affected by weather and climate change, providing intermittent and not fully predictable power. Solar generation installation in California has caused concerns when the sun goes down leaving utilities scrambling to find generation resources quickly to contend with steep ramp rates, as in the neck of California's "duck curve" (Figure 7.2). Electric Vehicle (EV) charging also creates substantial loads on the grid as yet another

[1]California Institute for Energy and Environment, University of California at Berkeley, Berkeley, CA, USA

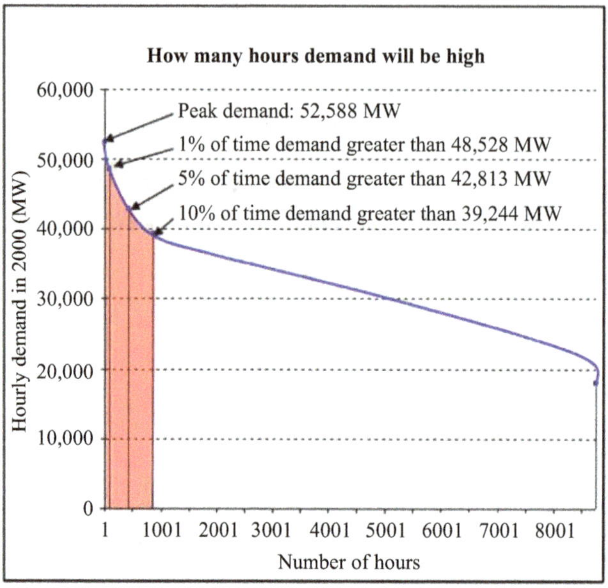

Figure 7.1 Hourly demand in megawatts (MW) for 1 year in California (reproduced from Figure II-1-12 in California Energy Commission [1])

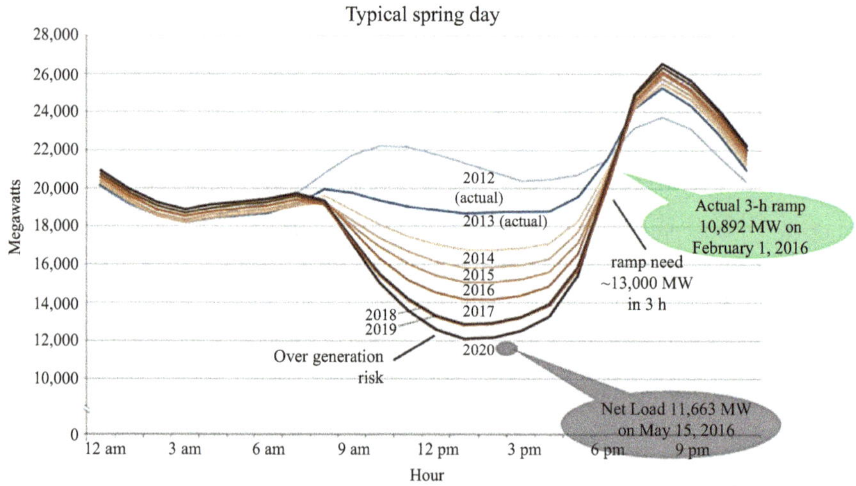

Figure 7.2 The net electricity load curve in California for 1 day (from California Independent System Operator [2])

stochastic load. Recent advances in data science are finding patterns in these seemingly random patterns that may help with scheduling or regulation.

Residential and commercial buildings are a natural focus for matching or balancing supply and demand as they consume 72% of electricity in the United States [3] and 40% of global energy [4]. Buildings are where we spend up to 90% of our time [5] working, playing, eating, sleeping, and so on. At present, the connected technologies in buildings can lead to cost-effective solutions for electric grid imbalances as well as for greenhouse mitigation [4].

This chapter describes various technologies for electricity customers (demand) to curtail or shift electrical loads (response) to balance electricity demand with supply.

7.1.1 Background

The term "demand response" (DR) refers to a means of motivating changes in energy use by non-generation resources—typically reducing or modifying demand by consumers as opposed to increasing supply—and includes policy, technology, and social–behavioral factors. Electrical utilities have used DR techniques and programs for decades. As early as 1966, customers were asked to curtail electricity in the United States, when the nation experienced blackouts due to air conditioning loads in residential and commercial buildings in the summer [6]. Demand side management (DSM) describes the influence of the demand side or customer use of energy to produce the desired changes in the utility's load shape. In the vernacular, DSM is another term for energy efficiency,* although when first introduced, it included load management (reducing or shaving peaks and shifting load in time) through *direct load control* [7] (see inset), for the utility to directly turns off or cycles loads such as air conditioning [8,9]. Another technique is the customer such as a business reduces load to an agreed-upon level [10]. The last two decades have seen a trend away from direct load control to price-sensitive programs, such as dynamic pricing and demand bidding. In the last decade, the emphasis has evolved toward distributed energy resources (DER) or less commonly integrated DER (IDER).

Direct load control

An early DR technology was *direct load control* [7]. One study in 2006 found that 80% of utility DR programs across the United States used direct load controls for air conditioning [9]. Typically, an electric utility offered a program such as a discounted electrical rate (e.g., monthly bill reduced by US$5) or money rebate (e.g., one time incentive of US$30) to install a radio-controlled switch directly on an electric water heater or air conditioner compressor. When

*Energy efficiency refers to providing the same service (e.g., heating and lighting) with less energy; energy conservation describes reducing energy consumption, which can result in reduced service (e.g., less heat or light).

electrical demand was high, the utility would directly control these end loads by turning off or periodically cycling on and off these devices for a period of time. For example, the utility might turn off the unit for 15 min and on for 15 min (called a 50% cycling rate) for up to 4 h for a maximum of 60 h over the season [7]. Communication occurred in one direction, and the customers did not have much choice or notice.

Two distinguishing features describe DR approaches (Table 7.1). One key distinction is whether the DR program is *dispatchable* or *non-dispatchable*. Dispatchable or interruptible programs, including direct control, interruptible tariffs, and demand bidding, are not triggered by the end customer. Dispatchable programs can respond to events, such as hot weather. A recent bifurcation in dispatchable programs is *supply side* versus *load modifying* resources [11] depending on which called the event. The Independent Service Operator can call for an event (supply-side event). Utilities can alert customers periodically (e.g., 10–15 days/year) to reduce or shift loads (load-modifying event) during certain hours (e.g., 14:00–18:00 or 15:00–20:00), such as Peak Day Pricing or Critical Peak Pricing events. Non-dispatchable programs include Time of Use (TOU) tariffs where the price per kilowatt-hour of electrical energy changes based on the time of day and is designed to encourage customers to consume energy when it is abundant and reduce consumption during peak periods.

Case study: a brief history of DR in California

In the late 1970s, regulators at the state level encouraged utilities to promote energy efficiency from their customers. But utilities had little incentive to push

Table 7.1 DR approaches

***Dispatchable* (event-based, interruptible, not triggered by end customer)**		***Non-dispatchable* (not event-based, continuous, not interruptible)**
Supply side (triggered by Independent Service Operator)	*Load modifying* (triggered by load servicing entity, such as a utility)	
Example: Ancillary service program where customers bid to a certain load reduction and agree to perform this curtailment if the bid is accepted, and receive payment if called by Independent System Operator [12]	Example: Utility periodically calls for event day where price of electricity is high during certain hours	Example: TOU tariff (price of electricity changes based on time of day)

conservation, because they lost revenue; moreover, regulators wanted to see effort, but not necessarily results. Early programs were based on little consumer research or cost–benefit analysis; the utilities portrayed energy conservation as cutting back and depriving oneself, and saw lukewarm response [8]. By the late 1980s, several forces led to high supply; the regulators again changed the incentives for the investor-owned utilities (IOUs), so they could make a profit from energy efficiency. The utilities were rewarded to spend money but were not held accountable for the programs saving energy [8]. In the early 1990s, utilities were spending US$2 billion per year on DSM up from zero in the 1970s [13].

Insufficient oversight of deregulation allowed market manipulations that created an energy crisis in California in 2000–01; this energy shortage led to a renewed look at DR policies. In 1996, California passed an energy deregulation law to increase market competition; the large IOUs sold a large portion of their power generation to private companies, which then sold electricity to the utilities in wholesale. The California Independent System Operator (ISO) emerged in 1998 as a not-for-profit public benefit corporation to manage the transmission of electrical power and to operate competitive markets for electrical power [14]. However, lack of oversight led to the manipulation of prices in the electricity market. In addition, deregulation reduced incentive to maintain infrastructure of the electricity grid. In 2000, the capacity margin was <10%, whereas before 1990 there was a 15%–20% margin [15]. In addition, the transmission network grew by <15% between 1988 and 1998; the generating capacity grew by only 0.1% during 1995–2001 [16].

The California Public Utility Commission (CPUC) and California Energy Commission (CEC) created the first Energy Action Plan in 2003, which emphasized DR as part of a "loading order" in improving grid reliability. The order begins with (1) decreasing demand by increasing energy efficiency and continues with (2) implementing DR, (3) meeting new needs for generation with renewables, (4) with distributed generation resources (e.g., rooftop solar, co-generation, storage), before (5) clean fossil-based generation [10].

The IOUs were required to install networked interval meters—11.7 million meters across California. This technology vital to implementing this DR policy was called *smart, advanced, or interval meters*—kilowatt-hour meters that could record electrical usage at set intervals (average of continuous (e.g., 10–30 s) measurements of electrical power consumption over 15 min or 1 h) and transmit data to the utility automatically (and typically once a day). Old mechanical meters provided the accumulation of energy consumed between two manual reading of the meters, typically once per month in the United States. These new meters record when the electricity is used. Rather than charging a static rate months or days after energy is used, finally the utilities would have a tool to charge a dynamic price based on the actual value of energy. In California, these meters were first installed to large customers (mid-2000s), and then to residential. These meters are replacing electromechanical meters across the United States and internationally. According to the

Edison Foundation, in 2016, electric utilities had deployed 72 million smart meters; estimated installations should reach 90 million in 2020 in the United States (Edison Foundation, 2017) and over 200 million in the European Union and Norway [17].

Smart, advanced, or interval meter

A key enabler of modern DR is the interval electricity meter to measure energy between the power source and the circuit breaker panel leading to the loads. The meter is often referred to as a smart meter or part of the Advanced Metering Infrastructure (AMI). The primary features are the ability to collect measurements at sub-hour intervals (e.g., 15 min, 10 s) and two-way radio frequency communication between the utility and the meter. Some interval meters have two low-powered radios: one that communicates within the building and one to communicate to the utility data center via a local AMI aggregation point.

7.2 Coordination of DR with energy efficiency

DR can coordinate with energy efficiency, especially within utility programs. Researchers at the Lawrence Berkeley National Lab depicted the relationship as part of a continuum (Figure 7.3) depending on the timescale of delivery. Energy efficiency strategies—using less energy to perform the same task—can be applied on a continual or daily basis all year. A TOU tariff is a price per kilowatt rate that changes during the day and often between seasons; in California, electricity during the summer is more expensive than in winter and during late afternoons/evening than during night/morning. A DR event with day-ahead notification allows the customer to plan and adjust curtailment. Real-time DR is best suited to loads that can instantly change. Often DR strategies that are implemented for buildings on event days (e.g., raised temperatures) can act as a driving force for deeper energy efficiency. Strategies are tested, and occupants and facilities managers can experience the changes for a brief period before implementing them more permanently.

A more recent representation of DR strategies over grid dispatch frequency is the "shape–shift–shed–shimmy" moniker, shown in Figure 7.4. *Shape* refers to overall energy efficiency and reduction of load. *Shift* relates to moving energy consumption away from peak demand periods into peak generation periods. *Shed* consists of reduced demand during DR events designed to reduce peak load periods. *Shimmy* describes quick responses to smooth out net load.

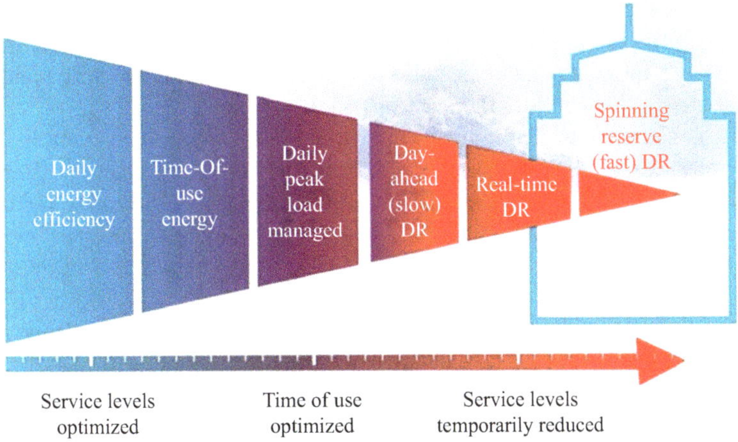

Figure 7.3 Energy efficiency and DR as a time continuum [18]

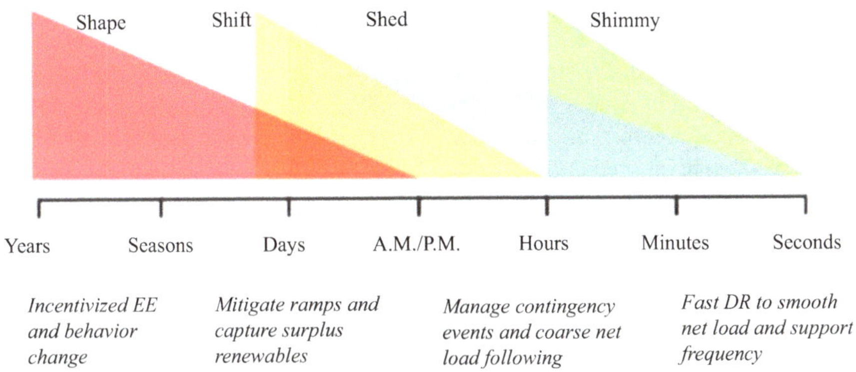

Figure 7.4 DR service types over time continuum of grid dispatch frequency or response [19]

7.3 Building characteristics

Buildings vary in their capability and potential for energy efficiency and DR based on four elements: the physical structure (building materials, design, and orientation), the fixed equipment (lighting, and heating, ventilation, and air conditioning (HVAC)), the appliances (vary with building type), and customer behavior [20].

7.3.1 Residential buildings

Even though residential customers provide the vast majority of utility service accounts, residential buildings have much smaller loads than commercial customers. Nonetheless, DR for residential buildings is growing globally and may provide faster response times than commercial buildings [21]. Three trends are driving this growth [22]. First, technology development is rapidly expanding in advanced metering, data analytics, and smart controls. Second, as more renewable energy sources come online, there is an increase in load shifting to move peak electricity consumption (e.g., in the evenings) to match renewable generation (e.g., midday solar). Finally, DR is integrating with other DSM programs to widen service offerings, and integrating more DERs, so that electricity consumers are turning into electricity "prosumers"; both consuming and producing electricity. One electric services company, Green Mountain Power, has a program to draw power from home batteries [22].

Figure 7.6 shows an example of the average daily load shape for all fuel types and energy usage levels for residential customers in a single forecast zone and building type in California in 2014 [24]. The peak is typically at 19:00 PM, driven by space heating or cooling, cooking, dishwasher, lighting, and television loads. The load shape shows the evening peak of the duck curve from Figure 7.2.

Thermal comfort and DR

Thermal comfort [21] is defined as "the condition of mind that expresses satisfaction with the thermal environment and is assessed by subjective evaluation." Early studies focused on the physics and physiology of thermal sensation and comfort, based on steady-state laboratory conditions, resulting in a mathematical prediction of comfort given six factors (see below).

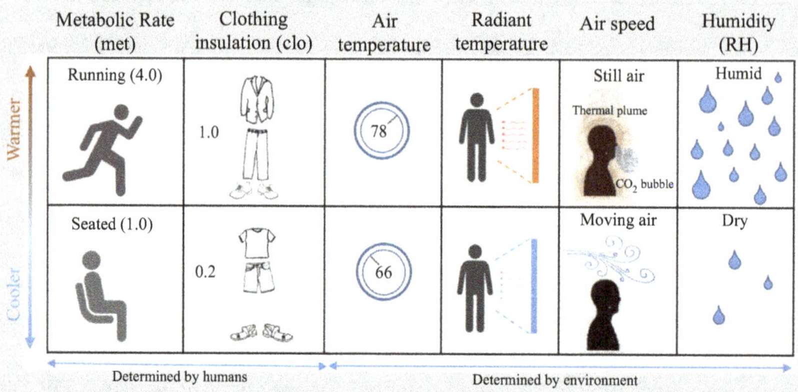

Based on surveys from people in commercial buildings, a modified model called the Adaptive Comfort Standard (ACS) introduced the influence of cultural expectations, social settings, and conditioning. This model includes the physiological and psychological adaptation and behavioral adjustments people make to find a much wider range of temperatures comfortable. In naturally ventilated buildings, people were comfortable at a wider range of temperatures that correlated to outside air temperature than people in air-conditioned buildings. Prevalent air conditioning use leads to physiological "addiction" that causes people to become less tolerant of temperature excursions outside a narrow temperature range. Residential buildings typically have operable windows and can use natural ventilation or air conditioning for cooling. In homes, people have more control over their environment (e.g., open a window, change clothing) than in a work setting, and they often pay for their energy. Thus, they accept a wider temperature range at home than at work. During the summer, the ACS allows a higher temperature comfort range than in the winter, which bodes well for the raised temperatures during DR events. Finally, people are accepting more of conditions outside normal operation if they are informed of them.

US residential buildings are dominated by air conditioning (in 90% of homes), water heating, and space heating (Figure 7.5). Typical US residential DR programs use smart or connected advanced thermostats to reduce air conditioning during peak summer days. These thermostats can provide two-way communication: receive signals and transmit data. Thermostats are often offered in utility programs, or in Bring Your Own Device programs, and participants can purchase any thermostat from a preapproved list [22]. Networked thermostats receive a DR event signal from the utility or aggregator; the thermostat can then increase the cooling set point or target temperature a few degrees (e.g., 4 °F or 2.2 °C) to reduce the cycling of the air conditioning system during the event period and thus reduce load.

This increase in temperature introduces the issue of thermal comfort [25] and DR (see inset). Air movement, such as through ceiling fans, can cool a person

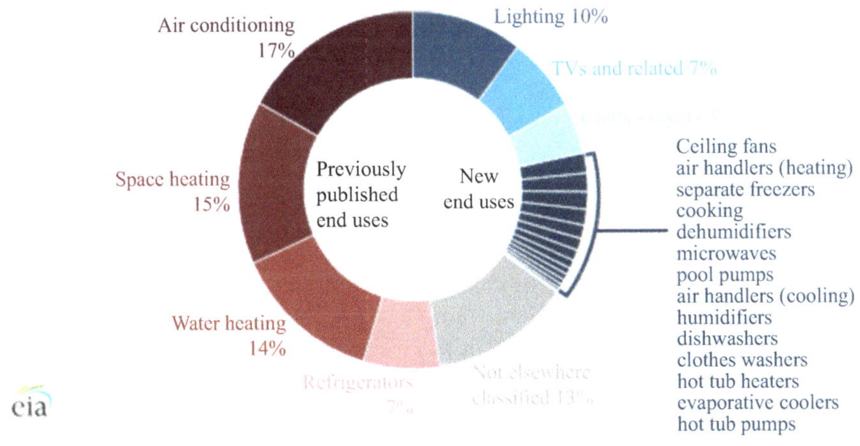

Figure 7.5 US residential electricity consumption by end use in 2015 [23]

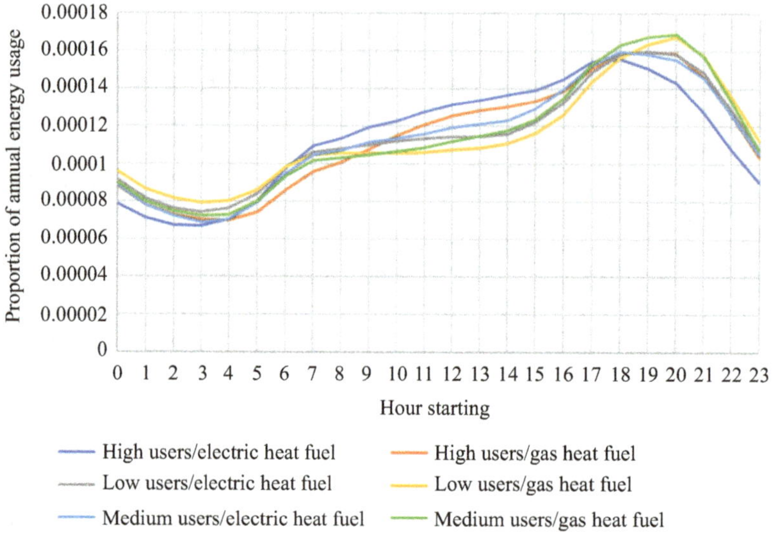

Figure 7.6 Average daily load shape for California residential customers [24]

indoors in a similar manner as lowering the temperature, but uses only a tiny fraction of the energy required by HVAC systems. New fans, using only 1–8 W (compared with 2,000–3,500 W for the typical 1.5–3 ton air conditioning system) can offset a 3 °C (6 °F) increase in indoor air temperature. This improves the occupant's comfort and perceived air quality while decreasing energy consumption.

Allowing higher indoor temperatures reduces a building's total HVAC energy an average of 5% per 0.5 °C (1 °F), and even greater in climate zones where natural ventilation or evaporative cooling systems are used instead of compressor-based cooling, or where there are a large number of airside economizer hours (such as in California). Thus ceiling fans could be used as the first stage of cooling, for year-round energy efficiency or for DR or resiliency events (Figure 7.7). A recent study [26] found 60% energy savings and improved comfort when ceiling fans were used when the air conditioner failed, and showed slightly improved comfort (89% satisfied compared with 82% with air conditioning only).

Home electrical energy storage, such as batteries, is gaining traction. However, buildings and appliances with thermal mass can provide DR capability for less cost. Thermostatically controlled loads besides air conditioning include refrigerators and tank electric water heaters, both providing thermal energy storage. These thermostatically controlled loads have shown promise especially in providing fast time-scale reserves [28].

In places where air conditioning is less prevalent, controls target other loads. Besides water heaters, smart controls may be found in EV chargers, pool pumps, and embedded controls on appliances, such as dishwasher and clothes washers.

Home energy management systems (HEMS) are another way of controlling residential loads from a central place and also can provide feedback and

Figure 7.7 *Move air, then cool it: a two-stage low-cost means of cooling, especially during DR events [27]*

Figure 7.8 *Nielsen's model of system acceptability [33]*

information for people to understand their energy consumption. For DR, ideally the HEMS would (1) alert the consumer to price changes, (2) replicate thermostatic control functions and add additional control functions appropriate to DR, and (3) provide information, advice, and tools to help the consumer manage electrical energy usage [29]. The consequent strategy of direct feedback with some form of advice has shown to be effective in reducing energy consumption [30]. In addition, providing information to help the user develop insight leads to better comfort and satisfaction [31]. A list of product offerings may be found [32].

Acceptance of technology, for example, thermostats, greatly impacts energy efficiency as well as DR. Figure 7.8 shows Nielsen's diagram of system acceptance [33], including practical acceptability, which includes usefulness—the balance of utility and usability. Usability refers to the extent to which a product can be used by specified users to achieve specified goals with effectiveness, efficiency, and satisfaction in a specified context of use [34]. People may withstand a device with poor usability if it has high utility (such as a mobile phone). Studies from 10 years ago found that programmable thermostats are not installed properly, are generally misunderstood, and have poor usability [35,36]. In general, the new advanced thermostats are an improvement, using touchscreens, color, graphics, and advanced algorithms to reduce the drudgery of programming and setting up the thermostat. However, many thermostats remain difficult to learn or remember how to program; many people simply do not understand how their thermostats work. Sectors of the population unfamiliar with smart phones or Internet of Things devices may find advanced technologies unnecessarily complicated. A recent project found that an advanced thermostat had five methods of controlling the blower fan, which most people rarely use; occupants accidentally scheduled the fan on so that it consumed more energy and needlessly increased the energy cost [26].

Decision-making and motivation remain other parameters that affect the success of residential DR, arguably more than technology. The study of behavioral decision-making applied to energy programs has greatly evolved in the past 15 years [37–44], leading to a better understanding of social norms, feedback [45,46], and choice architecture: how decisions are influenced by how the choices are presented [38]. Energy consuming behavior is not always based on economic rationale and can benefit by "nudges," such as feedback, comparisons with neighbor's energy use, appropriate defaults for settings, and so forth. The rollout of TOU rates for residential customers in California, planned for the end of 2020, assumes that the high afternoon prices (US$0.32/kWh) from June to September will motivate a shed or shift to reduce the peak load. Fifteen years ago, the California Statewide Pricing Pilot (SPP) showed positive results using price to reduce peak electrical demand [47]. Yet price alone may not be the most effective motivator, nor be persistent over time; nudges can help prompt curtailment. A lab study showed no effect on customer behavior in temperature setback during DR events when the subjects were shown either energy or price information. However, *how* the question was framed showed significant results. The subjects who thought a nonprofit entity was asking them to curtail energy were more likely to choose higher temperature offsets than those who thought the utility was asking them to curtail energy [29].

Another factor concerning residential DR is customer choice. Historically, utilities designed mandatory programs, however, with the advent of the Internet of Things connected devices, utilities have seen improved results when they allowed their customers to "opt-out" of DR events [48,49].

California's greenhouse gas reduction goals are driving fuel-switching and electrification of both new and existing homes. In highly insulated Net Zero Energy homes with energy efficient appliances and equipment (all new homes built in California must be zero net energy beginning in 2020), heating and cooling loads become less prominent. Miscellaneous loads are expected to dominate [50]. Improved controls and machine learning can play a role in shaping energy loads; simulation studies show that even net zero residential buildings can substantially

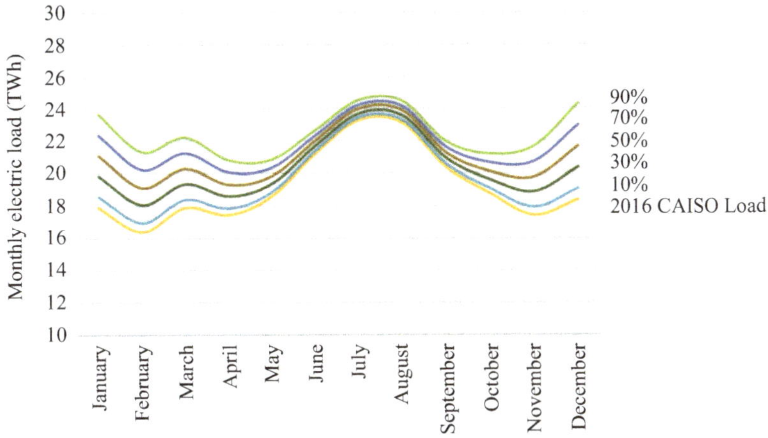

Figure 7.9 Projections in California's annual energy consumption under different percentages of space and water heating electrification [55]

benefit by machine learning [51]. As retrofits of existing homes continue to electrify traditional fossil-fueled loads (space and water heating, cooking loads) by replacing with efficient heat pumps and induction cooktops, we expect that left unchecked, electrical space and water heating loads could further exacerbate electric load balancing. In California, the peak electricity demand occurs during summer afternoons driven by air conditioning, and the net load has shifted toward evening due to solar. Peak loads are used to plan system-wide capacity and investments. Thus electrification of space heating could contribute to winter heating peaks. Some think these peaks are expected to remain lower than summer peaks [52]; some have shared concern that electrification of water heating plus space heating might shift to a winter peaking system (see Figure 7.9) [53]. Currently, the vast majority (nearly 90%) of water heating in California is fueled by natural gas, using a 40–50 gallon storage tank [54]. One study found that if all of these water heaters were replaced with efficient electric water heaters, such as heat pumps, California's electricity consumption would increase by 19% [55]; another study found that California would need almost 50% more electricity if all passenger vehicles were electric [56].

7.3.2 Commercial buildings

Commercial buildings are quite heterogeneous; a myriad of businesses and organizations are housed by commercial buildings: churches, retail stores, restaurants, auto dealerships, education, and so on. In 2012, in the United States there were 5.6 million commercial buildings; most of the commercial buildings are relatively small: less than half are (465 m^2), and nearly three-fourths are smaller than (929 m^2) [57]. In 2012, in California commercial buildings consumed the largest portion of the electricity consumption at 37% and contributed 37% of the peak load [58].

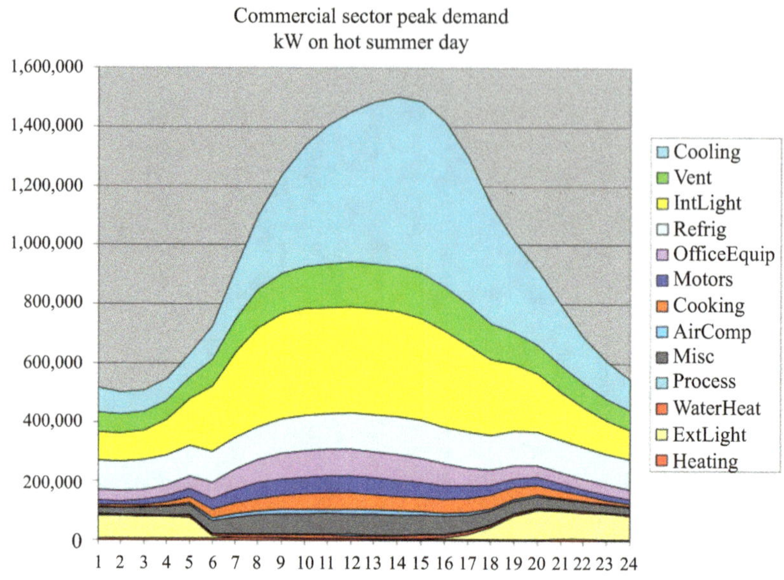

Figure 7.10 Electrical loads in commercial buildings on a hot summer day [57,58]

Figure 7.10 shows the end use profile of the peak demand on a hot summer day in California. Most of the peak load is due to cooling followed by interior lighting.

In a recent behavioral meta-study on commercial building energy efficiency potential, three types of buildings—office, retail, and education—showed the greatest energy efficiency potential (16%–33%), followed by health care and lodging [59]. Lighting, HVAC, and computers showed the most promise as targets of energy efficiency measures [59]. One study (Figure 7.11) showed that HVAC-related savings potential was the highest (HVAC savings by end use: large buildings: 5%–15%, medium 7.5%–17%, small 10%–23%), followed by equipment (5%–15%) and lighting (2%–3%) [60]. The large savings in small commercial buildings are primarily driven by controls of HVAC systems (e.g., networked thermostats), since these buildings do not benefit from the Building Automation Systems that are found in nearly all large commercial buildings [61]. Hot water was another end use which showed potential for energy savings. In addition, the size of the building made a difference; the saving estimates were larger for smaller buildings.

Most services—heating, lighting, security—in a building cannot communicate with others because they do not share the same communications protocol; vendor variants are incompatible; or no interfaces exist. The market is already transitioning to open protocols, since they allow more manufacturers to sell and service the market, and Internet Protocol (IP)-enabled devices can ease this transition.

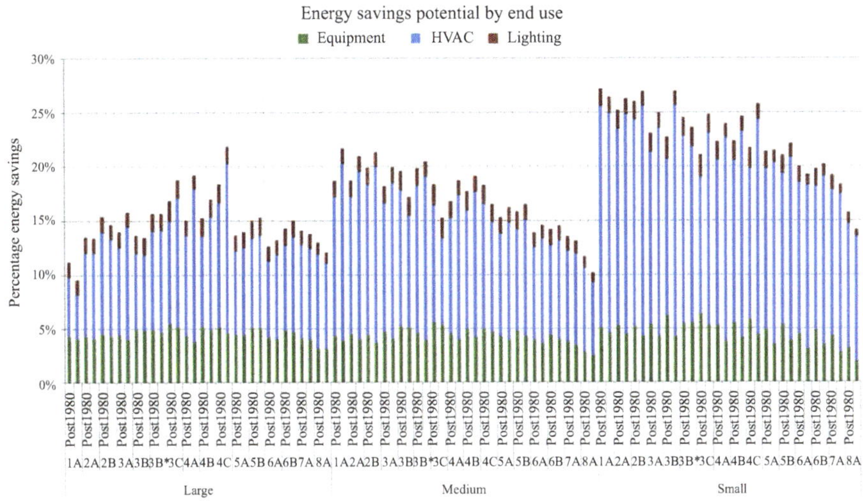

Figure 7.11 Simulation results for energy savings potential by end use and commercial building size [60]

Interoperability between services can enable improved energy savings and augment existing services. A small commercial building on a campus in California uses the occupancy sensors from the advanced lighting system to control the set points of the thermostats in the classrooms; when no occupancy is detected, the lights automatically dim and the temperature set points are expanded to reduce heating and cooling.

For large commercial buildings, Auto-DR, where a building management system automatically initiates energy saving strategies upon receipt of a signal [62], has been proven successful. At present, the OpenADR communication specification developed by LBNL is used worldwide, has been adopted by many vendors, used in utility programs, and implemented in hundreds of commercial buildings [63,64]. Typical peak reductions are approximately 10%, with some buildings achieving as much as 20% [62]. The following section describes a study on deep DR in a large commercial building.

Case study: deep DR in a large commercial building

The goal of the Distributed Intelligent Automated DR (DIADR) [65] project from 2010 to 2012 at University of California, Berkeley, was to reduce peak electricity of the Sutardja Dai Hall building by 30% while maintaining a healthy, comfortable, and productive environment for the occupants. Working with Siemens and LBNL, we sought to provide "deep" DR at the appliance level as well as for typical lighting and HVAC applications. We focused on the office portion of the building. Annual energy consumption is approximately 8,053 MWh; the office portion is estimated as 1,924 MWh. The maximum peak load during the study period was 1,175 kW.

The extensive submetering and nimble access to this data provided great insight into the details of the building operation as well as quick diagnostics and analyses of tests. For example, students discovered a short-cycling chiller, a stuck damper, and a leaking cooling coil in the first field tests.

Most of the reduction in peak demand came from the HVAC system by increasing supply air and zone temperatures, and reducing ventilation rates. We conducted a study of demand-controlled ventilation using CO_2 sensors in the conference rooms. Other reduction involved dimming and turning off lights. The web interface for the distributed load controller or gateway allowed the prioritization of curtailable appliances (such as lamps, laptops, fans, heaters, and printers) and visualization of energy consumption. We found challenges in achieving this goal in the office portion of the building, given an overcooled building not fully commissioned, grossly oversized chillers, continually increasing demand, and difficulties isolating the office portion of the building from the nanofabrication laboratory. We discovered that the chiller load decreased 30–60 min after the DR HVAC curtailment control measures. Nonetheless, we reduced peak electricity by 14%–24% and cooling by 25–78 tons using the absorption chiller, the deeper reduction for hotter weather conditions. We also applied many of the DR strategies (such as reducing minimum ventilation rates and expanding the zone temperature range) toward daily energy efficient strategies for overall energy savings.

Case study: DR in small commercial buildings

While large commercial buildings can use Building Automation Systems or Energy Management systems to control HVAC and lighting to respond to DR signals, small commercial buildings tend not to have these centralized systems and must rely on thermostats and other energy management systems. The high cost of technological upgrades—such as replacing equipment with energy efficient products—is a major barrier to energy efficiency goals; a software solution that relies on simple hardware and can leverage existing systems is a much less expensive and nimble approach to achieve energy goals.

We developed a platform, the eXtensible Building Operating System (XBOS) shown in Figure 7.12, which provided a secure data bus, data storage, metadata management, and drivers to various devices (networked thermostats, interval meter data, environmental sensors, and lighting gateways). Working with Siemens, Carnegie Mellon University—Silicon Valley, and Quantum Energy Services & Technology (QuEST), the research team developed a user interface, developed the system architecture, recruited about 20 small commercial buildings, installed the platform, and tested cost-saving strategies in 16 buildings in PG&E and SCE territory in summer 2018 and 2019.

The platform itself successfully communicated price signals, provided aggregated data (weather, multiple thermostats, whole building data), and allowed different control algorithms to act on the systems. Installing the networked thermostats alone saved up to 29%. Expanding temperature set points during price event days reduced demand 5%–25% and reduced costs. The team also explored forcing the two-stage HVAC systems to run on the first stage (single compressor);

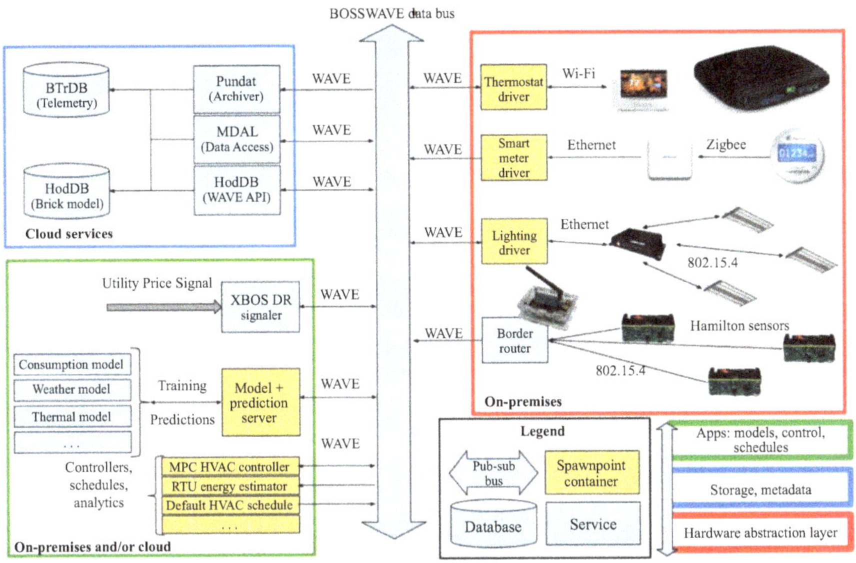

Figure 7.12 Schematic of the XBOS platform

the thermostat interface did not allow this intervention, so it required reverse engineering to conduct this test. The team was able to integrate lighting reduction in one building and EV charging and appliance loads in another building. We concluded that there is an incredible value of incorporating real-time building energy data with thermostat data: one can achieve building system identification, conduct diagnostics, and improve control.

7.4 Behavioral DR

Behavioral DR refers to information-only interventions, such as public appeals broadcast on media to reduce energy, personalized communications, and data analysis as compared with direct control technologies and price structures. A recent Randomized Control Trial by Opower (at present, part of Oracle), a major developer of behavioral DR, showed a 3.4% reduction in peak demand just based on a public appeal [66]; the number went up to 5% on a hot day [67]. Interval meters are key to this type of analysis: Opower has developed a platform that can pull in customer data, analyze it, and send personalized communication to customers before and after a peak event. The customer is informed the time of the event, and later how much they saved. While the platform works without a variable price rate, the performance improves with a paired price structure [67].

Another company that uses behavioral DR but on the supply side is OhmConnect, which pays residential customers for reducing electrical energy during peak periods. OhmConnect offers these aggregated megawatts of demand reduction capacity into California's Proxy DR program run by CAISO as well as participating in the DR Auction program. OhmConnect developed a gamification approach to engaging customers by providing contests and ways to build community with a graphic smartphone interface. The app seems to appeal to environmentalists and hard-to-reach customers such as low-income customers by reminding customers that they are reducing energy that would have been provided by peaker plants. Customers can buy or use credits for smart thermostats or smart plugs. This Energy-Savings-as-a-Service company has saved 500,000 Californians over 2 million kWh of energy in October 2019 [68].

WattTime is a nonprofit startup that provides local information for consumers on how carbon-intensive their source of electrical energy (power plant) is in real time. This information allows any Internet-connected device to run at times when energy is clean. WattTime coined the practice Automated Emissions Reduction [69].

7.5 DR and renewable energy

This chapter describes DR as a consumer-based source of electrical power system flexibility to maintain the balance between supply and demand. As sustainability goals drive increasing amounts of renewable energy generation on the grid, how can DR enable this transition, reduce solar curtailment, and mitigate the duck curve?

The DR solution includes policy, technology, and social–behavioral methods working together. To scale rapidly, the most cost-effective solutions must be identified. As more EVs enter the market, policies and smart charging technologies can encourage charging vehicles during the day and at workplaces to coincide with solar generation. Machine learning techniques can optimize charging on demand to reduce range anxiety. Residential TOU rates combined with gamification, nudges, and smart technology can reduce evening loads; precooling can shift the air conditioning loads and smart switches can schedule dishwashers, clothes washers, and dryers to avoid evening hours. Electrification of water and space heating needs to consider the impact on energy costs when transitioning to TOU rates; there is potential for storage technologies to mitigate the costs (e.g., small point-of-use under-sink water heaters to heat water during the day and possibly phase-change materials to store heat or cool mechanically). Education and feedback can help all customers, whether commercial or residential. Bundling DR services with other services (e.g., maintenance and security) will improve adoption. Advanced thermostats can enable many small commercial buildings not only to improve energy efficiency but also to reduce load from late afternoon.

In short, buildings can become efficient and grid-interactive on a local level. Both residential and commercial customers can play a role in enabling a tighter alignment between supply and demand.

7.6 Conclusion

DR is helping to transform the electricity system from a centralized, one-way power network where customers passively receive electricity to a much more dynamic and responsive two-way flow of information and where customers and loads can contribute to system operations [11]. This chapter described DR, a brief history, residential and commercial strategies for DR, and behavioral DR. Challenges remain on usability of devices, education and outreach, and developing the right motivation for the right audience. With the Internet of Things and machine learning/ Artificial Intelligence, opportunities abound to truly make a modern and smart grid that enables and engages its customers to coordinate a better match between supply and demand, and enables greater integration of renewable generation.

References

[1] California Energy Commission (CEC). *2002–2012 Electricity Outlook Report*. Sacramento; 2002.

[2] California Independent System Operator (ISO). *What the Duck Curve Tells Us about Managing a Green Grid* [Internet]. 2016 [cited November 1, 2019]. Available from: https://www.caiso.com/Documents/FlexibleResourcesHelp Renewables_FastFacts.pdf

[3] US Department of Energy (DOE). *Energy Efficiency Trends in Residential and Commercial Buildings* [Internet]. 2008 [cited November 1, 2019]. Available from: https://www1.eere.energy.gov/buildings/publications/pdfs/corporate/bt_stateindustry.pdf

[4] Global Alliance for Building and Construction (GABC). *2018 Global Status Report* [Internet]. 2018. [cited November 1, 2019]. Available from: https://globalabc.org/

[5] US Environmental Protection Agency (EPA). *Report to Congress on Indoor Air Quality: Volume 2*. Washington, DC; 1989.

[6] Cooper G. *Air-Conditioning America: Engineers and the Controlled Environment, 1900–1960*. Baltimore: Johns Hopkins University Press; 1998. 228 p.

[7] Cook JD. Residential Air conditioner direct load control "Energy Partners Program." In: *Proceedings of the Ninth Symposium on Improving Building Systems in Hot and Humid Climates* [Internet]. Arlington; 1994 [cited February 28, 2020]. Available from: https://oaktrust.library.tamu.edu/bitstream/handle/1969.1/6651/ESL-HH-94-05-31.pdf?sequence=4&isAllowed=y

[8] Hirsh RF. *Power Loss*. Cambridge: MIT Press; 1999.

[9] Rosenstock R. *EEI Member and Non-Member Residential/Commercial/ Industrial Efficiency and Demand Response Programs for 2005/2006* [Internet]. 2006. [cited March 1, 2006]. Available from: http://www.eei.org

[10] Bender S, Doughman P, Hungerford D, *et al. California Energy Commission Implementing California's Loading Order for Electricity Resources Staff Report* [Internet]. Sacramento; 2005 [cited November 1, 2019]. Available from: https://ww2.energy.ca.gov/2005publications/CEC-400-2005-043/CEC-400-2005-043.PDF

[11] Panfil M, and Fine J. *Putting Demand Response to Work for California* [Internet]. 2015. [cited November 1, 2019]. Available from: https://www.edf.org/sites/default/files/demand-response-california.pdf

[12] Peak Load Management Alliance. *Training Course Resource Demand Response Acronyms & Glossary First Edition Demand Response Training Series* [Internet]. 2017 [cited February 28, 2020]. Available from: www.peakload.org

[13] Geller HS, and Walmet GE. Preface. In SM Nadel, MW Reid & DR Wolcott (Eds.), *Regulatory Incentives for Demand-Side Management*. Washington DC: American Council for an Energy Efficient Economy; 1992.

[14] California Independent System Operator. *California ISO—Our Business* [Internet]. [cited October 21, 2019]. Available from: http://www.caiso.com/about/Pages/OurBusiness/Default.aspx

[15] Fleay B. *USA's Triple Energy Whammy in Electric Power, Natural Gas, & Oil*. [Internet]. 2001. [cited November 1, 2019]. Available from: http://www.mnforsustain.org/fleay_bj_usa's_energy_whammy.htm

[16] Electric Power Research Institute. *The Western States Power Crisis: Imperatives and Opportunities*. Palo Alto: EPRI; 2001.

[17] Escan SI Consulting, *USmartConsumer. European Smart Metering Landscape Report "Utilities and Consumers"* [Internet]. Madrid; 2016 [cited October 22, 2019]. Available from: www.usmartconsumer.eu

[18] Demand Response Research Center (DRRC). *Demand Response Best Practices, Design Guidelines and Standards*. Berkeley: Lawrence Berkeley National Laboratory; 2008.

[19] Alstone P, Potter J, Piette MA, *et al. 2025 California Demand Response Potential Study* [Internet]. Berkeley; 2017. [cited October 22, 2019]. Available from: https://www.cpuc.ca.gov/General.aspx?id=10622

[20] National Action Plan for Energy Efficiency. *Coordination of Energy Efficiency and Demand Response* [Internet]. 2010 [cited November 1, 2019]. Available from: www.epa.gov/eeactionplan

[21] Comverge. *Residential_Demand_Response_Infographic_-_Final.jpg (2552× 6504)* [Internet]. 2013 [cited November 3, 2019]. Available from: https://s3.amazonaws.com/dive_static/diveimages/Residential_Demand_Response_Infographic_-_Final.jpg

[22] Mehrhoff J. *Welcoming the Next Generation: Residential Demand Response 3.0/Utility Dive*. April 2019 [cited November 3, 2019]. Available from: https://www.utilitydive.com/news/welcoming-the-next-generation-residential-demand-response-30/551947/

[23] U.S. Energy Information Administration (EIA). *Residential Energy Consumption Survey (RECS)* [Internet]. 2015 [cited November 1, 2019]. Available from: https://www.eia.gov/consumption/residential/

[24] Baroiant S, Barnes J, Chapman D, Keates S, and Phung J. *California Investor-Owned Utility Electricity Load Shapes California Energy Commission* [Internet]. Sacramento; 2019 [cited February 29, 2020]. Available from: https://ww2.energy.ca.gov/2019publications/CEC-500-2019-046/CEC-500-2019-046.pdf

[25] ANSI/ASHRAE. *Standard 55—Thermal Environmental Conditions for Human Occupancy*. Atlanta; 2017.

[26] Raftery P, Miller D, Zhang H, *et al. Integrating Smart Ceiling Fans and Communicating Thermostats to Provide Energy-Efficient Comfort*. California Energy Commission: Sacramento; 2020.

[27] Miller D. *Move Air Then Cool It: Integrating Air Movement for Energy Efficient Comfort*. Berkeley: Cool Buildings Workshop; 2019.

[28] Mathieu JL, Dyson M, and Callaway DS. Using residential electric loads for fast demand response: The potential resource and revenues, the costs, and policy recommendations. In: *Summer Study for Energy Efficiency in Buildings* [Internet]. Washington, DC: American Council for an Energy Efficient Economy (ACEEE); 2012 [cited November 3, 2019]. Available from: https://aceee.org/files/proceedings/2012/data/papers/0193-000009.pdf

[29] Peffer TE. *California DREAMing: The Design of Residential Demand Responsive Technology with People in Mind* [Internet]. University of California, Berkeley; 2009. [cited October 22, 2019]. Available from: https://escholarship.org/uc/item/8rk0g6mh

[30] Darby S. Making it obvious: Designing feedback into energy consumption. In: *Energy Efficiency in Household Appliances and Lighting*. New York: Springer Verlag Berlin Heidelberg; 2001. pp. 685–96.

[31] Wyon DP. Individual control at each workplace for health, comfort, and productivity. In: *Paper presented at the Creating the Productive Workplace Conference*. London; 1997.

[32] Northeast Energy Efficiency Partnerships (NEEP). *Home Energy Management Systems* [Internet]. 2018. [cited October 22, 2019]. Available from: https://neep.org/initiatives/integrated-advanced-efficiency-solutions/home-energy-management-systems#HEMS Product List

[33] Nielsen J. *Usability Engineering*. San Francisco: Morgan Kaufmann; 1993. 358 p.

[34] International Standardization Organization (ISO). *Guidance on Usability ISO 9241-11*. Geneva; 1998.

[35] Peffer T, Perry D, Pritoni M, Aragon C, and Meier A. Facilitating energy savings with programmable thermostats: Evaluation and guidelines for the thermostat user interface. *Ergonomics*. 2013;56(3):463–79.

[36] Peffer T, Pritoni M, Meier A, Aragon C, and Perry D. How people use thermostats in homes: A review. *Build Environ*. 2011;46(12):2529–2541.

[37] Cialdini RB. *Influence: The Psychology of Persuasion*. New York: Harper Collins; 2006.

[38] Thaler RH, and Sunstein CR. *Nudge: Improving Decisions about Health, Wealth, and Happiness* [Internet]. Yale University Press; 2008 [cited February 29, 2020]. 293 p. Available from: https://archive.org/details/nudgeimprovingde00thal

[39] McKenzie-Mohr D. *Fostering Sustainable Behavior: An Introduction to Community-Based Social Marketing* [Internet]. Third Ed. Gabriola Island: New Society Publishers; 2011 [cited February 29, 2020]. Available from: https://www.cbsm.com/book

[40] Heath C, and Heath D. Switch: How to Change Things When Change is Hard. New York: Broadway Books; 2010. 305 p.

[41] Lutzenhiser L. Social and behavioral aspects of energy use. *Annu Rev Energy Environ*. 1993;18(1):247–89.

[42] Wilhite H, Shove E, Lutzenhiser L, and Kempton W. The legacy of twenty years of energy demand management: We know more about individual behaviour but next to nothing about demand. In: *Society, Behaviour, and Climate Change Mitigation*. New York: Kluwer Academic Publishers; 2006. pp. 109–26.

[43] Stern PC. New environmental theories: Toward a coherent theory of environmentally significant behavior. *J Soc Issues* [Internet]. January 1, 2000 [cited February 29, 2020]; 56(3):407–24. Available from: http://doi.wiley.com/10.1111/0022-4537.00175

[44] Allcott H, and Mullainathan S. Behavior and energy policy. *Science*. 2010;327(5970):1204–5.

[45] Darby S. The effectiveness of feedback on energy consumption: A review of the literature on metering, billing and direct displays. Oxford: Environmental Change Institute; 2006.

[46] Ehrhardt-Martinez K, Donnelly KA, "Skip" Laitner JA, York D, Talbot J, and Friedrich, K. Advanced metering initiatives and residential feedback programs: A meta-review for household electricity-saving opportunities. Washington DC: American Council for an Energy Efficient Economy; 2010.

[47] Charles River Associates. *Impact Evaluation of the California Statewide Pricing Pilot Final Report* [Internet]. 2005 [cited November 3, 2019]. Available from: https://www.smartgrid.gov/files/Impact_Evaluation_California_Statewide_Pricing_Pilot_200501.pdf

[48] *Three Ways Opt-Out can Improve Your Demand Response Program* [Internet]. [cited November 3, 2019]. Available from: https://www.energyhub.com/blog/three-ways-opt-out-can-improve-your-demand-response-program

[49] *Opt-In, Opt-Out Options Vie for Top Billing in Efforts to Reduce Electricity Use—Chicago Policy Review* [Internet]. [cited November 3, 2019]. Available from: https://chicagopolicyreview.org/2014/10/28/opt-in-opt-out-options-vie-for-top-billing-in-efforts-to-reduce-electricity-use/

[50] Sparn B, Earle L, and Christensen C. *A Net-Zero Energy Home Grows Up: Lessons and Puzzles from 10 Years of Data* [Internet]. Washington, DC; 2016 [cited November 2, 2019]. Available from: https://aceee.org/files/proceedings/2016/data/papers/10_772.pdf

[51] Javaid N, Mehboob Hussain S, Ullah I, *et al. Demand Side Management in Nearly Zero Energy Buildings Using Heuristic Optimizations. Energies* [Internet]. 2017 [cited November 2, 2019];10:1131. Available from: www.mdpi.com/journal/energies

[52] Mahone A, Li C, Subin Z, Sontag M, Mantegna G, and Karolides A. *Residential Building Electrification in California Consumer Economics, Greenhouse Gases and Grid Impacts* [Internet]. San Francisco; 2019 [cited November 3, 2019]. Available from: www.ethree.com

[53] Wei M, Nelson JH, Greenblatt JB, *et al.* Deep carbon reductions in California require electrification and integration across economic sectors. *Environ Res Lett*. 2013;8(1):014038.

[54] California Energy Commission (CEC). Chapter 5, *2016 Residential Compliance Manual* [Internet]. Sacramento; 2016. [cited October 22, 2019]. Available from: https://ww2.energy.ca.gov/2015publications/CEC-400-2015-032/chapters/chapter_5-Water_Heating_Requirements.pdf

[55] Hopkins AS, Takahashi K, Glick D, and Whited M. *Decarbonization of Heating Energy Use in California Buildings Technology, Markets, Impacts, and Policy Solutions* [Internet]. Cambridge; 2018 [cited November 3, 2019]. Available from: https://www.synapse-energy.com/sites/default/files/Decarbonization-Heating-CA-Buildings-17-092-1.pdf

[56] Davidson FT, Tuttle D, Rhodes JD, and Nagasawa K. *Is America's Power Grid Ready for Electric Cars?* [Internet]. City Lab. 2018 [cited November 3, 2019]. Available from: https://www.citylab.com/transportation/2018/12/americas-power-grid-isnt-ready-electric-cars/577507/

[57] U.S. Energy Information Administration (EIA). *Commercial Buildings Energy Consumption Survey (CBECS)* [Internet]. 2012 [cited November 1, 2019]. Available from: https://www.eia.gov/consumption/commercial/data/2012/bc/cfm/b6.php

[58] Peeking Under the Hood: *The Energy Characteristics of California's Commercial Building Sector Martha Brook, P.E. California Energy Commission*—ppt download [Internet]. [cited November 1, 2019]. Available from: https://slideplayer.com/slide/4482388/

[59] Ehrhardt-Martinez K. Behavior-based Energy Savings Opportunities in Commercial Buildings: Estimates for Four U.S. Cities. In *Proceedings from 2016 Summer Study on Energy Efficiency in Buildings*. Washington, DC: American Council for an Energy Efficient Economy (ACEEE); 2016 [cited November 1, 2019]. Available from: https://aceee.org/files/proceedings/2016/data/papers/8_940.pdf

[60] Azar E, and Menassa CC. A comprehensive framework to quantify energy savings potential from improved operations of commercial building stocks. *Energy Policy*. 2014; 67:459–72.

[61] Katipamula S, Piette MA, Kuruganti T, *et al.* Small- and medium-sized commercial building monitoring and controls needs: A scoping study. PNNL-22169. Richland: Pacific Northwest National Laboratory; 2012.

[62] Piette MA, Watson D, Motegi N, Kiliccote S, and Linkugel E. Automated demand response strategies and commissioning commercial building controls LBNL-61013. In: *National Conference on Building Commissioning*. San Francisco; 2006.

[63] Kiliccote S, Piette MA, Mathieu J, and Parrish K. Findings from 7 years of automated demand response in commercial buildings. In: *Summer Study for Energy Efficiency in Buildings*. Washington, DC: American Council for an Energy Efficient Economy (ACEEE); 2010.

[64] OpenADR Alliance. *OpenADR* [Internet]. [cited November 4, 2019]. Available from: https://www.openadr.org/

[65] Auslander D, Culler D, Wright P, Lu Y, and Piette M. *A Distributed Intelligent Automated Demand Response Building Management System*

[Internet]. Pittsburgh, PA, and Morgantown, WV (United States); March 2013 [cited November 3, 2019]. Available from: http://www.osti.gov/servlets/purl/1172982/

[66] Hopkins AS, and Whited M. *Best Practices in Utility Demand Response Programs* [Internet]. Cambridge; 2017. [cited October 22, 2019]. Available from: https://www.synapse-energy.com/sites/default/files/Utility-DR-17-010.pdf

[67] Walton R. *How Opower is Pushing Behavioral Demand Response into the Mainstream | Utility Dive. Utility Dive* [Internet]. May 27, 2015 [cited November 4, 2019]. Available from: https://www.utilitydive.com/news/how-opower-is-pushing-behavioral-demand-response-into-the-main-stream/399790/

[68] Galford C. *New California Law Requires Development of Smart Appliances—Daily Energy Insider. Daily Energy Insider* [Internet]. October 19, 2019 [cited November 4, 2019]. Available from: https://dailyenergyinsider.com/news/22348-new-california-law-requires-development-of-smart-appliances/

[69] Chaturvedi A. *Californian Non-Profit to Monitor Emissions from All Power Plants Globally. Geospatial World* [Internet]. October 18, 2019 [cited November 4, 2019]. Available from: https://www.geospatialworld.net/blogs/californian-non-profit-emissions-power-plants/

Chapter 8

Storage regulations and technologies

David Fernandes[1], Ricardo de Azevedo[1] and Oliver Probst[2]

It is an often stated claim that the power generated and consumed in an electric grid has to be balanced at all times. Evidently, this statement is more a figure of speech than an actual reality, since unbalances constantly arise on a number of time scales. Long before the advent of massive amounts of variable renewable generation on the electric grid, the main concern of grid operators revolved around securing sufficient reserves for regulation, in order to keep variations in frequency and voltage within certain prescribed bounds. The amounts of energy kept in store, if you will, traditionally correspond to the quantities of extra fuel which can be mobilized if scarcity in power supply arises; conversely, fuel supply to generators can be throttled down if demand decreases.

Even in an electric system with a very high degree of renewable energy penetration, say 80%, this conventional approach of balancing works perfectly well as long as the regulating capabilities of the power plants charged with load balancing are at least as large as the largest variations in the net load (here defined somewhat colloquially as the difference between the actual load and the variable renewable generation). Fossil-fuel power plants continue to have an important role in an electric grid dominated by renewables, but their role has to evolve from satisfying base loads to providing regulation services. For very high degrees of penetration, where fossil fuels are ultimately phased out, other options for regulation have to be made available. A conceptually simple approach is to use existing hydropower facilities for regulation, but this approach may be limited by the availability of water (both seasonally and due to long-lasting droughts) and the transmission infrastructure connecting hydro resources with the parts of the grid where regulation is required.

[1]ON Energy Storage, Miami, FL, USA
[2]Tecnológico de Monterrey, School for Engineering and Sciences, Monterrey, N.L., Mexico

Based on the reasoning above, an electric grid with a generation mix almost exclusively based on variable renewable energy technologies will need a set of storage technologies which can be deployed in a modular way and without major geographic constraints. Battery energy storage systems (BESSs) have been leading the way in this effort, though other technologies exist and may become relevant on a larger scale in the near future. Interestingly, even long before a grid starts to run short on regulation capacity (at very high renewable energy penetration levels), the deployment of modular energy storage systems (ESSs) may make both technical and economic sense. One of the reasons is the fact that the modernization of the electric grid is a challenging task, facing manifold barriers ranging from economic to regulatory and political. Though the grid serves everybody, and everybody will ultimately benefit from a cleaner and more robust grid, the line from beneficiaries to those footing the bill cannot always be traced easily, making it difficult for resources to be allocated in a sufficient and timely manner. Modular storage units, on the other hand, can be developed, financed, installed and operated on a level far below national transmission and distribution planning, providing a pathway to a much more agile transformation of the grid.

However, a requirement for such a transformation to happen is the existence of clear and consistent regulations, allowing the market to identify suitable business models and present convincing narratives to investors and lenders. Regulations are closely related to energy (and power) services that can be provided by storage devices and allow for their remuneration. The first part of this chapter will therefore focus on those services and their contributions to the betterment of the grid. We will then dwell somewhat on the crucial distinction between storage services in the transmission vs. the distribution grid. The latter have traditionally been designed for radial flow (with power being injected at the distribution feeder and withdrawn along the way), leading to a more or less predictable voltage profile as one gets farther away from the feeder. Protections based on such radial profiles have long dominated the design of distribution grids. With distributed generation emerging on a massive scale, particularly residential and commercial solar photovoltaic (PV), much of the traditional design logic no longer applies, calling for a "smarter" grid. The corresponding implications are discussed in far greater detail in Chapter 6 of this book. Here, it may suffice to note that distributed storage may go a long way toward accommodating larger amounts of PV generation by limiting the excursion of critical parameters such as frequency and voltage.

After investigating in some detail, the services provided by storage devices and the regulatory aspects required for an optimal utilization of such services, the technological options available to date and expected to become available in the near future will be visited. Some final thoughts will be dedicated to the synergies that are likely to arise at the interface between traditional grid uses and the mobility sector, fueled by the considerable interest in electrified transport.

8.1 Grid services provided by storage

8.1.1 Overview: energy vs. power services, grid vs. user services

There are different approaches to classifying the ways in which storage can help electric grids and promote the integration of variable renewable electricity generation (VRElec). It should be noted that most services function for conventional generation and VRElec alike, even though its utilization will be especially beneficial in electric sectors committed with a high participation of clean energy.

The first classification refers to the type of application, commonly divided into "energy" and "power" applications. This rather colloquial distinction is based on the characteristic time scale T_c of the service required, defined by $T_c = E_c/P_c$, where E_c is the characteristic amount of energy stored and P_c is a characteristic power scale, often the name-plate capacity of the power converter of the storage device. In *energy* applications, T_c is at least of the order of several hours, allowing to reduce charges for users with time-of-day tariffs and/or demand charges or, conversely, helping generators to obtain the maximum benefit from their generation. A classic example is *arbitrage* in Section 8.1.2.1, where energy is either withheld (in behind-the-meter (BTM) applications) or purchased from the grid (for in front-of-the-meter (FTM) applications) at times of low (market) prices, and sold back during periods of high prices. Similarly, abundant generation during early morning hours could be stored in order to be sold to the grid during peak hours, thereby generating capacity credits.

Power applications, on the other hand, have very short associated time scales ranging from a few seconds to a few minutes. Rather than displacing energy between hourly periods, storage systems configured for power applications are capable of injecting or absorbing large amounts of real or reactive power in order to stabilize the grid. A classical application is *frequency response*, where typically additional *active* power has to be injected into the grid in response to a loss-of-generation event elsewhere on the grid, which would otherwise result in an unacceptable downward excursion of the system frequency, possibly followed by a cascade of load-shedding events, triggered by automatic relays, and eventually even blackouts. Similarly, voltage support can be provided if *reactive* power is injected into or absorbed from the grid, a capability readily provided by modern-day electronic power converters such as those found in most energy storage systems.

8.1.2 Grid services

8.1.2.1 Energy arbitrage

As mentioned earlier, *energy arbitrage* is based on the idea of buying cheap and selling expensive, usually based on different moments in time, although the concept may be readily expanded to include a spatial dimension in markets with *Locational Marginal Prices* (LMPs); see Chapter 10 of this book for a discussion of electricity markets and LMPs.

In order to create some insights into the potential and the limitations of energy arbitrage, let us consider a grid-connected storage facility, in an FTM configuration where the power G (MW) fed into the grid during a diurnal cycle is given by $G = -C$ for $t \leq t_1$, $G = D$ for $t_2 \leq t_1 \leq T$, and an idling period $(G = 0)$ between t_1 and t_2. Here, C is the power absorbed from the grid during the charging half-cycle, and D is the power injected into the grid during the discharging half-cycle. Evidently, C and D are related, since the charge level L of the storage device is expected to be the same at the beginning and at the end of the full diurnal cycle.

Neglecting self-discharge, the charge level L can be written as

$$L = \eta_C C t_1 \theta(t_1 - t) - \frac{D}{\eta_D}\theta(t - t_2)(t_2 - t) + L_0 \tag{8.1}$$

where η_C and η_D are the efficiencies for charging and discharging, respectively. Writing $t_1 = \alpha T$ and omitting the idling period $(t_1 = t_2)$, we can relate the charging and discharging power values by

$$D = \eta_C \eta_D \frac{\alpha}{1 - \alpha} C = \eta_{RT} \frac{\alpha}{1 - \alpha} C \tag{8.2}$$

where η_{RT} is the *round-trip efficiency* of the storage unit, i.e. the combined efficiency for charging and discharging. Note that in (8.2), it has been assumed that charging occurs during the first part of the cycle (during time αT) and discharging during the rest of the cycle. In addition to (8.2), C and D are bound by the nominal capacity for charging and discharging, assumed to be same:

$$C \leq C_0, D \leq C_0 \tag{8.3}$$

After these observations functional relationships can be constructed for $D(\alpha)$ and $C(\alpha)$, i.e. the (dis)charging power as a function of the duty cycle value α. We first define a critical α_c by

$$\alpha_c = 1/(1 + \eta) \tag{8.4}$$

defined by the condition $D = C$.

Assuming first that the storage system will charge at full capacity $(C = C_0)$, corresponding to the first part of the cycle in our example (see (8.1)), we note that (8.2) is obeyed as long as $\alpha \leq \alpha_c$. For larger values, the discharge capacity will be the limiting factor $(D = C_0)$, so the charging power has to be adjusted by solving (8.2) for C. As a result, the expressions for charging and discharging power are the following:

$$\begin{array}{lll} C = C_0 & D = \eta_{RT} \dfrac{\alpha}{1 - \alpha} C_0 & \alpha \leq \alpha_c \\ C = \dfrac{1 - \alpha}{\alpha} \dfrac{1}{\eta_{RT}} C_0 & D = C_0 & \alpha > \alpha_c \end{array} \tag{8.5}$$

After these preparations, it is easy to state an expression for the net revenue per cycle (with period T) and unit charging capacity (C_0):

$$r = \frac{R}{TC_0} = -\alpha \, \mathrm{LMP}_1 \, c + (1 - \alpha) \, \mathrm{LMP}_2 \, d \tag{8.6}$$

where c and d are the values of per-unit charging and discharging power, respectively, i.e. $c = C/C_0$ and $d = D/C_0$. LMP_1 and LMP_2 are the average (Locational) Marginal Prices at which electricity is sold or purchased at the market at a given location during the periods $[0, \alpha T]$ and $[\alpha T, T]$, respectively*. The task of the storage designer now comes down to calculating (from historical data) or predicting the average LMP values during in store-in and store-out periods.

Box 8.1: A simplified arbitrage example

In order to get a feel for the potential of energy price arbitrage, we will first assume a symmetric rectangular price signal with price spread $\Delta\mathrm{LMP} = \mathrm{LMP}_2 - \mathrm{LMP}_1$ and average price $\overline{\mathrm{LMP}} = \frac{1}{2}(\mathrm{LMP}_1 + \mathrm{LMP}_2)$. We will also assume that the storage unit is operated at $\alpha = 0.5$, i.e. the unit is charged and discharged in a symmetric fashion. From (8.6), we have

$$r = \frac{1}{2}(-\mathrm{LMP}_1 \, c + \mathrm{LMP}_2 \, d) \tag{8.7}$$

It is easy to see from (8.4) that $\alpha = 0.5 < \alpha_c$, so the equations on the first row of (8.2) apply. Inserting into (8.7), we obtain $c = 1$ and $d = \eta$, and therefore

$$r = \frac{1}{2}(-\mathrm{LMP}_1 + \mathrm{LMP}_2 \, \eta). \tag{8.8}$$

Introducing the relative price spread $\beta = \Delta\mathrm{LMP}/\overline{\mathrm{LMP}}$, we can rewrite the net revenue to first order in the small quantities $\delta\eta$ and β as

$$r \approx \frac{\overline{\mathrm{LMP}}}{2}(\beta - \delta_\eta) \tag{8.9}$$

where the round-trip efficiency has been written as $\eta = 1 - \delta\eta$. From the second expression, the trade-off between round-trip efficiency and relative price spread becomes apparent. Considering an example of $\eta = 0.8$, it is easy to see that a relative price spread of $\beta = 20\%$ is needed in order to make up for the losses involved in charging and discharging the storage unit.

*Electricity markets with locational pricing are nowadays quite common, particularly in the US (see Chapter 11), while in other regions (e.g. predominantly in Europe) a market-wide common marginal price is used.

The calculations presented above can be easily extended to less simplistic price signals, e.g. with a sinusoidal behavior: $\mathrm{LMP}(t) = \overline{\mathrm{LMP}} - \mathrm{LMP}_0 \sin 2\pi t/T$. The reader is encouraged to work out the corresponding results.

Price variations in electricity markets occur mainly because of the underlying concept of the (Locational) Marginal Price discussed previously. Many wholesale energy markets use what is known as an order of merit system to assign generation to meet demand. Under this system, generators are dispatched in order of the lowest variable cost of generation with technologies such as wind and solar being dispatched first. Fuel cost is generally the primary driver of a plant's variable cost of generation. Given the mechanical nature of thermal generation, via CCGT (combined-cycle gas turbines) or ICE (internal combustion engines), there are additional variable costs of generation for maintenance, lubricants, accruals for long-term overhauls, etc. Variable costs of generation are minimal with renewable technologies. The operational costs associated with a PV generation facility, for example, are nearly the same regardless of whether the sun is out or not. Thus, an independent system operator places them higher in the order of merit to be dispatched before higher cost alternatives. In essence, the order of merit system is a reverse auction that assigns the lowest cost supply bids until demand of electricity can be met.

To illustrate this mechanism, we consider the following example:

Box 8.2: An example of demand vs. generation stack

Imagine a generation supply stack with the following power plants and their respective variable cost of energy:

Generator	Variable cost of energy ($/MWh)	Available capacity (MW)	Cumulative capacity (MW)
PV	0	100	100
Hydro	0	300	400
Wind	5	200	600
CCGT	35	600	1,200
ICE	60	300	1,500

In the example above, we can see how wholesale prices of energy change holding generation the same. An increase in demand can require the introduction of the subsequent unit of generation in the order of merit. By definition, that unit of energy will have the same, or higher, variable cost of generating the next unit of energy demanded by the system. Depending on the generation and consumption characteristics of a given market, these price variations can be small or large and easier or more difficult to predict.

In Figure 8.1, we see that around 8.00 AM demand begins to surpass the generation from zero/low marginal cost renewable units. The system needs to incentivize or compensate this higher cost unit by offering a price that meets its variable cost of generation. We can see that later in the evening, an even higher cost reciprocating engine plant is needed to meet the peak demand around 10.00 PM. Wholesale energy markets take into account additional variables such as start and stop costs and minimum dispatch levels, but the fundamental principal of this system remains unchanged: when demand increases, prices can only increase when supply remains the same.

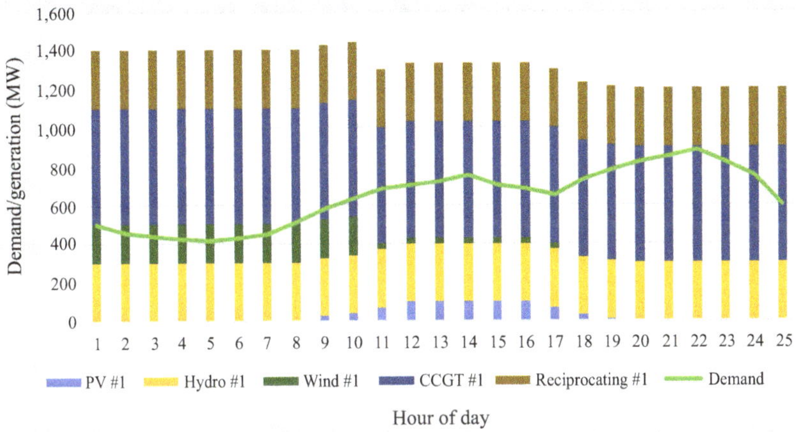

Figure 8.1 Example of the evolution of an order-of-merit generation stack and the corresponding demand time series

BESS systems can charge, store and discharge electricity on nearly instantaneous intervals. As such, any temporal variation in energy prices open commercial possibilities for energy arbitrage. In many power markets, electricity is priced on an hourly basis. Often, there are clear temporal price tendencies which are correlated to total demand. In many instances, but not always, prices begin the day at their lowest. From midnight until about 6.00 AM, total electricity demand is low and stable due to the fact that most of us are sleeping. As we begin our days and switch on our devices and appliances, and as commercial and industrial loads come online, total system demand rises sharply. It generally levels out throughout the work day until the late afternoon with many systems experiencing a daily peak between 5.00 and 10.00 PM.

After these conceptual examples, we will visit a more realistic example, taken from the Mexican wholesale electricity market.†

Box 8.3: A practical case of arbitrage calculations

In Figure 8.2, we see the prices from the Aguascalientes Node in the Mexican wholesale market for February 14, 2020; prices are in Mexican pesos (MX$). It can be seen that there are 2 h of prices that are well above baseline. Considering that a storage time of 2 h is a standard offer from manufacturers, let us assume that storage project developer will propose a 1 MW/2 MWh BESS to his investors. The assumed efficiencies for charging η_C and discharging η_D are assumed to be both equal to 90%. The following charge/discharge strategy is proposed:

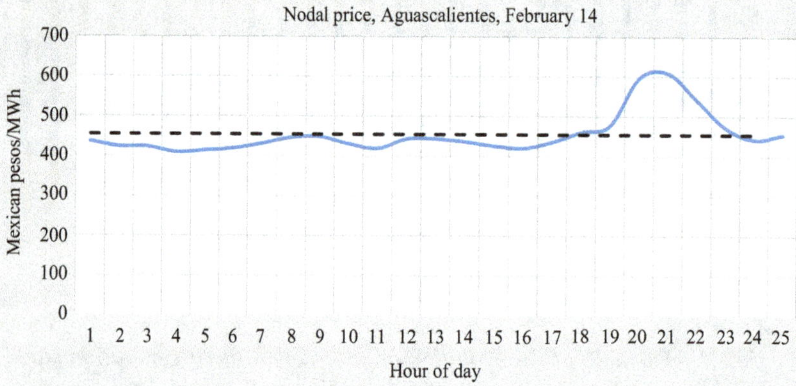

Figure 8.2 Example of an LMP time series in Mexico

†A number of mysteries in wholesale electricity market have emerged since the new federal government took office in late 2018. February 2020 LMPs are now around 25 US$/MWh, down from about 65 US $/MWh in 2019 and nearly 80 US$/MWh in 2018. Though in some regions the large-scale introduction of solar and wind generation has contributed to depressing prices, mainly because of inadequate transmission capacity, a reduction in pricing due to the horizontal shifting of the generation stack (see the example in Box 8.2) was somewhat expected, with the role of the marginal unit shifting from fuel oil-fired to single-cycle and even combined-cycle gas turbine plants. However, even in regions with low availability of low-cost generation such as in the Yucatán peninsula, LMPs have been unusually low. Evidently, in-depth studies are warranted to shed further light on these mechanisms, but initial evidence points to the hypothesis that high-sulfur fuel oil (now precluded from its use in maritime vessels) is being sold at a bargain to the incumbent electric utility company, without regards for environmental concerns. In either case, it seems clear that relying on merchant sales in an emerging and interventionist market is a risky business proposition.

- The BESS will strive to accumulate 2 MWh every morning during the lowest cost hours. Given the finite efficiency for charging ($\eta_C = 0.9$), the somewhat larger amount of 2 MWh/η_C = 2/0.9 MWh = 2.2 MWh has to be charged, at an average price of MX$412/MWh.
- The BESS will discharge every evening between 7.00 and 9.00 PM at a price of MX$602/MWh. Given the finite efficiency for discharging ($\eta_D = 0.9$), only 0.9 × 2 MWh = 1.8 MWh will be delivered to the grid.
- This price curve is assumed to be stationary.
- No secondary effects are created by installing this battery and carrying out this strategy. In particular, the BESS can safely be assumed to be small enough, in order to create any effect on the LMP.

With these assumptions, the total annual revenue generated can be calculated as (1.8 MWh/day × MX$602/MWh – 2.2 MWh/day × MX$412/MWh) × 365 days = MX$64,648/year = US$3,234/year, assuming a typical exchange rate of 20 MX$/US$ for 2019/2020. Further simplifying, let us assume there are no operational costs or taxes to be paid on this revenue and that our developer has an investor who will place capital only if it can be recovered in 5 years or less. Our example developer needs to be able to install this system at a total cost of 5 × US$3,234 = US$16k or less in order to meet the conditions of his capital. Evidently, this is completely unrealistic as current costs for this ESS would range around $300k–$600k, so energy arbitrage at this Mexican node clearly does not generate a sizeable income stream.

8.1.2.2 Primary frequency regulation and inertia replacement

Primary frequency regulation, also known as *frequency response*, is understood as the capability of generators, loads and storage units to respond to changes in system frequency in a matter of a few seconds and stabilize the frequency, generally at a new value different from the nominal one [1]. With traditional generators, frequency response is performed by adjusting fuel injection (an action generally referred to as *governor response*). This response is normally facilitated by a prior action known as *inertia response*. Inertia response stems from the extraction of kinetic energy from the rotational motion of grid-synchronous generators, which buys some time until the response from control loops of the generators (the "governors") kick in. Modern wind turbines can emulate the inertial response of traditional rotating machinery, thereby mitigating some of the loss of the system inertia, but other technologies such as solar PV systems featuring no moving parts are naturally without inertia. Modern electric grids with a high penetration of renewable energy technologies are therefore said to be "low inertia." Though this is often decried by grid engineers rooted in traditional design thinking, inverter-based systems actually provide a far superior response compared to the traditional combined action of inertial and governor response; this point is explained in some more detail in Chapter 6 on smart grids.

While the dynamic characteristics of inverters operating in primary generation technologies such as wind or solar PV and BESSs are similar, the distinctive advantage of a BESS is of course its capability to provide a regulation margin without having to permanently withhold generation. Evidently, every generator providing regulation services withholds generation, but the difference between a conventional (fuel-based) generator and a renewable one lies with the fact that the conventional generator also saves on fuel and emissions, whereas a wind or solar generator simply wastes free clean electricity by providing regulation services. This is where a BESS may make a good business case, even if the primary frequency regulation is not a remunerated service. Remuneration of primary frequency regulation may be handled quite differently, depending on the jurisdiction [1]. Whereas in some markets frequency response is tendered through spot markets, and in others (such as Germany) it is procured through auctions, yet other markets may simply require certain frequency response characteristics as part of the interconnection procedure. This is the case for Great Britain [1]; similarly, the Mexican grid code specifies frequency response capabilities for all generators alike, both conventional and renewable.

Even if no remuneration is provided for primary frequency regulation, a good business case may be made for storage, based on avoided energy losses alone. Typically, primary frequency regulation requirements will translate into a regulation margin of, say, 3% of active power to cover, say, 95% of all frequency regulation events. In the absence of knowledge of future frequency variations, the generator will have to withhold this full amount of power, leading to a 3% loss in income. A wind farm with a 40% net capacity factor and a power purchase agreement of US$70/MWh will lose an annual income of US$7.4 k for each MW of installed capacity, i.e. some US$370 k/a for a typical 50 MW wind farm. Building a BESS for providing this service may well be possible for US$1 million or less, all in, so even if operating expenses are figured in, a sound business case may be built based on avoided losses alone.

Though primary frequency regulation is essentially a *power* services, it still requires a finite amount of energy storage to design and operate a practical system. Minimizing the size of the battery bank goes a long way toward this goal. One of the ways of doing this is to coordinate the response of the wind or solar plant with the BESS. The combined response of partially loaded wind turbines and a BESS has been shown to produce a superior response to both the wind farm and the BESS alone [2]. It can, however, also help to use the available storage capacity for upregulating alone and use the wind farm control for downregulating, leading to a more effective use of the installed capacity.

8.1.2.3 Secondary frequency regulation

Secondary frequency control, also known as *frequency regulation*, is generally referred to as the regulation capacity provided by generators and loads which may be mobilized on a time frame of around 5–10 min [1]. As in primary control (or frequency response), this regulation is provided by injecting into the grid or withdrawing from it additional quantities of *active* power. Frequency regulation is designed to correct an area control error (ACE) by computing an *automatic generation control* (AGC) signal from frequency deviations and interchange power imbalances [3]. It is the main goal for frequency regulation to minimize this error and restore the nominal grid frequency as soon as possible. Frequency regulation can be provided by units operating in AGC mode, i.e. following as closely as possible the control signal emitted by the grid operator. Alternatively, the generation may be adjusted manually. Traditionally, generators operating in AGC mode are specially equipped, but as new technologies arise that are flexible by design, such as battery or flywheel storage technologies, new regulations are warranted that harvest the full range of benefits provided by such technologies.

An obvious question is how frequency regulation resources can be procured efficiently. In general, three main types of procurement for regulation services can be distinguished [1]:

- compulsory provision
- bilateral agreements
- tendering or market processes.

In a compulsory procurement of regulation reserves, the participating units have a contractual obligation to withhold the required regulation volumes, in case the system operator chooses to request them. These "stand-by" obligations are easy to handle but are generally not efficient, given that a greater volume than is actually needed at any given point in time must be set apart. In bilateral agreements, a more specific set of conditions may be worked out, responding to the actual reserve volumes required at different moments in time, specifying remunerations, as well as delivery conditions. Though being more flexible, bilateral contracts are generally not transparent and have high transaction costs [1].

In view of the considerations above, (competitive) tenders and market processes are generally preferred, wherever a liberalization (or *de-regulation*) of the electric system has occurred. Examples of such mechanisms abound, since many electric grids (including most of the US and Europe) have installed market mechanisms. Following the usual hierarchy of primary, secondary and other frequency control services already discussed in this chapter and explained in further detail in Chapters 6 (in the context of smart grids) and 9 (in the context of forecasting), most electricity systems have installed markets associated with the different time scales and associated services. In advanced markets, such as the ones operated by the Pennsylvania-New Jersey-Maryland Interconnection (PJM), a regional transmission operator operating in 13 states in the Northeastern US and Washington DC, procurement of regulation reserves is performed through a complex iterative procedure involving the co-optimization of energy and regulation services. An opportunity cost is estimated for each regulation resource and added to that resource's regulation price offer to create an order-of-merit curve, which is then used for the dispatch of regulation services. Given the complexity of this process, the results may not always be intuitive; more importantly, the prediction of future market prices will generally be difficult, particularly in a setting where new entrants, such as storage technologies, have to be considered.

In many grids and jurisdictions, particularly the ones with liberalized markets, it may be possible for storage devices to participate in frequency regulation, but without proper incentives recognizing the specific capabilities of fast-acting storage units the overall economic motivation may be too low for such projects to come to fruition.

8.1.2.4 Flexible demand control

Demand control, also known as *demand response* or *demand-side management,* is referred to as a voluntary load reduction incurred by an electricity user, in exchange for an economic compensation. Two large categories are generally distinguished:

- Tariff schemes recognize the time-of-use value of a given amount of electricity, hoping to trigger consumer action in response. A simple distinction made in many commercial and industrial tariffs around the world is the one between on-peak and off-peak; residential consumers, on the other hand, are generally charged on a flat basis. With the advent of Internet-of-Things (IoT)-based technologies and smart meters time-of-day pricing, and even real-time pricing may be offered to the consumer. In either case, the load response is left to the goodwill of the consumer.
- Incentive-based demand-response programs allow program or grid operators a direct control of certain loads in order to mitigate system stress during peak hours or emergencies. A simple traditional scheme is tariffs allowing for load shedding upon prior (often telephonic) notice; in exchange, the user is granted a substantially reduced tariff. For practical purposes such tariffs were limited to large users. Evidently, providing a flat discount on all energy consumption in exchange for the provision of a stand-by service is likely to be inefficient. Moreover, the limitation to large users fails to capture a significant portion of the market. In the era of IoT not only can such services be extended to the residential and commercial load centers, but also is it possible to establish a market where demand-control services can be offered and traded efficiently.

An in-depth description of demand control and response technologies is provided in Chapter 7. Evidently, if demand control can be conducted by passive or low-cost measures only, such as tapping into the thermal inertia of (properly designed) buildings or adjusting fan speeds to compensate for a higher temperature set point, the additional investments required will be minimal and the economic benefits will be maximal. However, as weather excursions become more extreme, ranging from heat waves to polar cold fronts, mild set point adjustments of the HVAC (heating, ventilation, and air conditioning) equipment may not be enough to mitigate, and the required demand control may drive the system out of the consumer's comfort zone. This is where storage can make a substantial difference.

8.1.3 User services

8.1.3.1 Load shifting for users with hourly tariffs

Load shifting is conceptually similar to energy arbitrage, but operates in a quite different setting. While arbitrage works in a (wholesale) market setting (both BTM and FTM, respectively), load shifting always occurs BTM, i.e. as an internal process of the electricity user. An obvious advantage of this arrangement is the absence of interference by the utility or the grid authority, as long as the general terms of the grid-tie contract are not altered. Though in theory time-of-day variations in the cost of electricity and associated services (such as regulation reserves) on the wholesale market level should be reflected on the retail level, in practice the differences may be very considerable. First of all, most retail tariffs, even for commercial and industrial users, distinguish between peak-hour and off-peak usage, possibly with an additional distinction between baseline and intermediate hours within the off-peak period. The obvious field of opportunity then lies with *peak shaving*, i.e. a reduction of electricity consumption during peak hours. In cases where the user can choose to reduce consumption without affecting critical services, the peak shaving may lead to actual reductions in energy consumption. This is evidently the case which maximizes social and planetary welfare. More often, however, the user will wish to only shift a given activity (say, manufacturing process) from peak to off-peak hours, tapping into the (often considerable) difference in electricity tariffs. Though no actual energy is saved, a net benefit to the system does occur, since a part of the generation during peak hours is usually conducted with generators that are either low-efficiency (e.g. single-cycle gas turbines) and/or highly polluting (e.g. fuel oil-fired steam plants). Shifting load cuts back on pollution and saves fuel, for the same amount of total electricity consumed. In case of storage, however, fuel savings are partially offset by the losses in the charge–discharge cycle of the storage device. The following examples illustrate both the cost and the fuel savings of a load-shifting arrangement for a very simple case. Both the economic benefits for a retail user with an electricity tariff differentiating between on- and off-peak periods, and the environmental benefits will be estimated in a hypothetical, yet realistic case. The numbers are inspired by the conditions in the Mexican electric system.

Box 8.4: Load-shifting—battery storage and diesel generation compared

We start out by stating the baseline information:

Tariff information:

- Off-peak: P_{off} = US\$0.1/kWh
- On-peak: P_{on} = US\$0.25/kWh

Marginal generation units:

- Off-peak: Gas-turbine combined-cycle (GTCC) plant, with a net efficiency of $\eta_{NG} = 55\%$ and operated with natural gas (net heating value = $HV_{NG} = 50$ MJ/kg)
- On-peak:
  - Electric system: Fuel-oil steam plant, with a net efficiency of 25%. Heating value of fuel oil: $HV_{FO} = 45$ MJ/kg
  - Diesel-fueled ICE using for peak shaving: Heating value of diesel fuel: $HV_{DF} = 42$ MJ/kg. Net efficiency of the ICM for electricity generation: 40%

Greenhouse gas (GHG) emission factors (CO_2 equivalents):

- Electric system: 0.5 kg/kWh
- Natural gas: 56 kg/GJ
- Fuel oil: 74 kg/GJ
- Diesel fuel: 73 kg/GJ.

We first calculate the cost savings of a BESS, assumed to have a 1-h equivalent storage capacity, i.e. $T_s = 1$ h, and a round-trip efficiency of $\eta_{RT} = 80\%$. The savings for each kWh relocated from the on-peak hour to an off-peak hour are then readily calculated as

$$r = -P_{on} + P_{off}/\eta_{RT} = -0.125\text{US\$/kWh}$$

Note that the amount $\Delta E = (1 - \eta) \times 1$ kWh $= 0.2$ kWh of electrical energy is lost. This corresponds to an additional emission of GHGs of 0.2 kWh × 0.5 kg/kWh = 0.1 kg $CO_{2,eq}$. This additional emission is, however, offset by the avoided emission from a peak-hour plant, assumed to be a fuel oil steam plant. The difference in emissions is calculated as follows:

- Peak-hour emissions = Electric energy avoided/plant efficiency × emission factor = 1 kWh/0.25 × 3.6 MJ/kWh × 74 g/MJ ≈ 1 kg
- Off-peak hour emissions = 0.5 kg/kWh (the system average)
- Net emissions savings by shifting the load from on-peak to off-peak = 0.5 kg/kWh
- Additional emissions incurred because of charging/discharging losses = 0.1 kg/kWh
- Total net savings in emissions = 0.4 kg/kWh.

From this simple example, it can be seen that the emission savings achieved by load shifting using a storage device with a round-trip efficiency of 80% are roughly equal to the *average* amount of emissions caused by producing the shifted amount of energy. This is of course due to the high emissions caused by peak-hour generation. Consequently, even if some electricity is lost in the process, the net effect on the atmosphere is equivalent to generating this amount of electricity from a nearly clean source of energy!

These calculations can be compared with the results for a diesel-based peak-shaving unit. As seen from the emission factors above, the specific GHG emission factors for diesel fuel and fuel oil are essential, so the only gain in terms of emissions to be achieved is the one associated with the higher efficiency of the peak-shaving unit. The GHG emission associated with the generation of 1 kWh is readily calculated to be 0.625 kg of CO_2 equivalents, leading to a reduction of 0.375 kg CO_2/kWh, similar to the reduction that can be achieved with storage.

On a final note, the cost of operating a storage unit vs. operating a diesel-fueled ICM can be compared. Neglecting the (low) operational cost of the storage unit, the main cost incurred is the cost of 0.2 kWh lost per kWh shifted. Assuming again a baseline cost of US\$0.1/kWh, the cost caused by load shifting amounts to 0.2 kWh $\times$ US\$0.1/kWh $=$ US\$0.02. The corresponding calculation for the ICM unit depends on the cost of the diesel fuel, which may vary wildly depending on the delivery required. Assuming, somewhat conservatively, a cost of US\$1/L or US\$0.0271/MJ, and recalling that the generation of 1 kWh of electricity requires 3.6 MJ/kWh/efficiency of fuel, we end up with a fuel requirement of US\$0.24, which is dramatically more expensive than the cost of load shifting with a storage device.

Let's examine a simplified case for a customer in Coyocan, Mexico City, consuming electricity on the medium tension tariff Gran Demanda en Media Tensión Horaria (GDMTH) in Mexico. The consumption will be assumed to be 1 MW flat. This customer is subject to a fairly common rate structure whereby they will pay a fixed service fee, capacity and demand charges which in this case are based on levels of energy consumed during peak hours, as well as three prices for energy depending on whether it is consumed in the base, intermediate or peak period.

For the purposes of explanation, we will simplify the actual tariff structure. Let's imagine that base energy on this regulated tariff is defined as any consumption between midnight and 6.00 AM. This more or less makes sense, as we saw earlier, wholesale prices do tend to be lower during these low consumption periods and transmission systems generally have fewer restrictions in wheeling excess generation in one node to a neighboring deficit node. Let's also assume that the base tariff is 0.9 MX\$/kWh. The higher price peak hours are defined as falling between 8.00 and 10.00 PM. The tariff during these peak hours is assumed to be 1.8 MX\$/kWh. Again, the logic makes sense as the pricing is structured to provide incentives to consumers to change their habits.

Installing a 1 MW/2 MWh battery would allow our customer to shift all energy he or she currently consumes during peak hours to base hours. Ignoring the impact this will have on other products such as capacity or distribution calculations, the customer would see the following savings during a single 24-h period.

Assuming this tariff structure is applied 365 days a year, our customer will save about 660,000 pesos per year or around US$35,000@20MX$/US$. Our customer likely knows more or less how quickly they need to recover capital investments in their business. Until recently, a 1 MW system would cost around US$1 million, so it would have been unlikely for our customer's board to approve a project that has a recovery of capital in 28 years (1,000,000/35,000) based on these savings alone. However, with the recent drops in system prices, it may be possible to build such a system for half that value. Moreover, in many markets, a battery system designed for this load-shifting will reap benefits in the form of reduced capacity and distribution charges, reduction of reactive power consumed from the grid, ancillary services and other avenues for commercialization that can significantly reduce the recovery of capital and improve the attractiveness of a project to a potential owner. The combination of two or more of these commercialization strategies is colloquially known as value-stacking.

8.1.3.2 Reduction of demand charges

The reduction of demand charges follows a similar logic as the shifting of the energy consumption. For users with a tariff including demand charges, the *measured demand* will generally be the highest consumption peak observed during the peak period (which may be identical to or slightly different from the peak-period used for the energy charges). Therefore, the task of storage is then to provide the peak power required by the user (typically caused by the start-up of heavy electrical machinery), if the process causing the peak cannot be rescheduled. Battery storage units are usually well suited for this task, as their strong point lies with the provision of large quantities of power during typically relatively short periods of time.

An important point to be noted here is the fact that in order to mitigate demand charges the storage unit often may not be required to be able to store electricity for a full hour, particularly of the measured demand is dominated by transient consumption peaks, e.g. those caused by the start-up of electrical machinery. A time-resolved measurement of the load profile may be advisable to determine the best configuration for a given storage unit. BESS systems may often be reconfigured for trade-offs between power and energy capacities, respectively, which may be useful if the load profile changes over time.

In Mexico, all regulated medium tension customers are invoiced for a quantity of capacity based on the following reasoning. First, the following two quantities are calculated:

- The maximum 15-min demand $D_{\mathrm{max,p}}$ during the peak period (from 8:00 to 10:00 PM) of any day in the invoiced month.
- The total quantity of energy E consumed divided by the number of hours H in the month and a regulator-determined constant f_L, i.e. $Q = E/(H \times f_L)$. In case of GDMTH, $f_L = 57\%$.

A customer's billed capacity will then be the minimum of both quantities:

$$C_{\text{billed}} = \min(D_{\text{max,p}}, Q)$$

Going back to the customer in our previous example who consumes 1,000 kW every hour without fail for the month of February. Of course, $D_{\text{max,p}}$ for this customer will be 1,000 kW as their consumption never goes above or below that value. Q is readily calculated to be $Q = 1{,}754$ kW. Consequently, $C_{\text{billed}} = 1{,}000$ kW, i.e. the customer in this case will be invoiced for 1,000 kW of capacity. Now, the price of capacity for February 2020 turns out to be MX$350.97/kW-month. Thus, our customer owes 1,000 kW × MX$350.97/kW-month = MX$350,970 (pesos), or roughly US$18,910 considering an indicative exchange rate for 2019.

But wait! Our customer installed a battery system equal in capacity to his flat 1,000 kW with 2 h of duration. With that battery replacing all energy consumed from the grid during the daily 2-h peak period, the revised min equation is as follows: $C_{\text{billed}} = \min(0, 1754\,\text{kW}) = 0$. Our customer has saved the full US $18,910 of the capacity charges he would have otherwise been billed. This is in comparison to the US$2,831 of energy cost our customer has avoided via this load shift. So our customer's energy arbitrage strategy could more properly be considered a capacity charge reduction scheme or "peak shaving." In practice, peak shaving and energy arbitrage are two of the most commonly "stacked" revenue streams for battery systems.

8.1.3.3 Increasing of self-supply levels

The self-supply of electricity refers, quite generally, to the self-provision of electric energy to a load center, residential, commercial or industrial, by either on-site generation, self-generation mediated by the (public) transmission grid or the configuration of virtual power plants. Only in case of on-site generation does a physically traceable flow of energy from the generator owned or leased by the self-supplier exist; this includes the possibility of generator and load center being geographically separated but connected by a dedicated transmission line. In case of a point-to-point self-supply mediated by the transmission grid, power is injected into the grid at or near the point of generation, under the conditions dictated by the interconnection (such as voltage level) and the geographic zone (ruling, e.g., the price the injection is credited with), to be withdrawn at the point(s) of consumption, again under the local conditions. Access to the grid and the terms of utilization and the corresponding charges may be a problem, both because of actual physical constraints (e.g. the existence of congestion) and political adversity. Grid-mediated self-supply projects require bilateral contracts and lengthy negotiations, often limited by grid access. Virtual power plants, on the other hand, tie into existing (or soon-to-be-built) power plants, through the (potentially dynamic) configuration of a generation portfolio, powered by distributed generation and the IoT.

Self-supply is generally motivated by an intended cost saving and often also by a commitment with clean renewable energy technologies. The Environmental

Protection Agency (EPA) of the US even defines self-supply as a form of "green power" procurement. Even if clean generation is not a prerequisite for self-supply, it does converge with the latter, similarly to the convergence seen between markets and clean generation; see Chapter 11 on markets and renewables for an in-depth discussion of this subject. Taking the case of Mexico as a further example, the self-supply sector started out (by the mid-1990s) based mostly on GTCC projects, but then extended itself quickly to include renewable energy projects, mostly with wind energy as the main driver. Most of the wind power capacity in Mexico up to the energy reform in 2013/14 was built through self-supply arrangement and was allowed to continue to operate under the new law as *grand-fathered* or *legacy projects*.

One of the specific windows of opportunities for storage in self-supply projects comes from the remuneration of excess energy. Excess energy may be a real physical issue (if the self-generation exceeds the total consumption at a given moment of time) or may be created by administrative maneuvers by the incumbent utility providing the transmission service. A common practice in Mexico is the treatment of situations where a (smart) meter installed at a given load center for the purposes of the self-generation project momentarily ceases to be in communication, or even more specifically, fails to respond to a periodic poll. The utility will then take this failure to respond as an excuse to disable the (administrative) assignment of self-supplied energy to that load center and replace it by utility-generated energy. Even though the energy is not lost (under the Mexican self-supply agreement, the energy is "banked" according to its time-of-day value), it may lead to a considerable accumulation of credits in the bank which can no longer be consumed fully in the remainder of the operational year. The excess balance is either lost altogether or is purchased by the utility at a price considerably lower than the self-supply tariff.

Storage can help mitigate these unfair practices, though the solution is not quite straightforward. Given that only the *administrative* flow of energy to the load center is restricted, not its physical delivery, avoiding charges by the utility and accumulation of excess energy in the bank requires enabling on-site storage to reduce on-site consumption. Evidently, such a measure (which would be completely unnecessary under fair practices) is an energy application in terms of Table 8.1 and is likely to be feasible only for relatively small load centers and for short durations of the "administrative outages."

8.1.3.4 Black start of internal grids

Beyond producing electricity, power plants are also consumers of electricity. Internal consumption, also known as parasitic load, can be up to 10% of a power plant's total generation. This electricity will be generated by the plant itself while it is online and can be pulled from the interconnected system if the plant is not dispatched. But what happens if a plant needs to come online and cannot pull the required electricity off the grid due to a local, regional or system-wide shutdown? While these events are rare, they are often the most critical times for an electrical system. Getting plants online as quickly as possible after a system-wide outage,

Table 8.1 Power vs. energy applications

Energy applications	Power applications
Arbitrage	Frequency regulation
Renewable energy time shift	Voltage support
Demand charge reduction[1]	Small signal stability
Time-of-use charge reduction	Frequency droop
Transmission and distribution upgrade deferral	Synthetic inertia
Grid resiliency	Renewable capacity firming

[1]Demand charge reduction can be considered a power application in some instances.

along with avoiding the outage in the first place, is one of the milestones of reliability for consumers.

To get a large power plant online and synchronized, an external source of power is needed. In most wholesale energy markets, this service is known as a black start. New England Independent System Operator (NE-ISO) defines a black start as a "service necessary to facilitate a stable and orderly restoration of the power system in the event of a partial or complete shutdown of the system. The ISO selects and compensates specific participating generators interconnected to the transmission or distribution system at strategic locations that can be called upon to re-energize the transmission system. These units must meet certain requirements, including having the ability to quickly restart without an outside electrical supply." Generally an ISO seeks certain technical capabilities to provide black start similar to what is mandated by the NE-ISO Open Access Transmission Tariff (OATT):

- The resource is located within the ISO Reliability Coordinator Area.
- The resource is metered and dispatchable by the ISO or otherwise subject to operational control by the ISO during the restoration of the system following a partial or complete shutdown.
- The resource is capable of starting and remaining energized without support from either off-site power or another Designated Blackstart Resource until such time as additional load is directed to be picked up pursuant to instruction from the ISO or Local Control Center.
- The resource is capable of closing its output circuit breaker to a de-energized bus.
- The resource is maintained and staffed in a manner that allows it to be brought online and available for loading by the ISO or Local Control Center within a specified time period, with such period being measured from the initiation of the startup instruction from the ISO or Local Control Center.
- The resource has the ability to maintain frequency within a prescribed range and is able to operate in a mode with zero governor droop or the equivalent.
- The resource has an automatic voltage regulator capable of being placed in automatic voltage control mode.

- The resource has the ability to provide lead and lag power factor capability.
- The resource has access to a fuel supply during a shutdown of the system that will allow it to, in accordance with ISO or Local Control Center dispatch instruction, run at full capacity for a specified minimum amount of time.
- The Blackstart Owner maintains a communication capability from the resource to either the Local Control Center or the ISO, as directed by the ISO, that is independent of a public telephone or cellular phone communication network and is confirmed to be capable of operating during a shutdown of the system.
- And the Blackstart Service Commitment is established between the Blackstart Owner and the ISO.

A generator capable of restarting independently of the grid and providing power to other power plants to do the same is compensated not only when called upon but also for being available to do so. Historically, diesel generation has been used as a primary source for black starts. In the event of a system-wide shutdown, a small black start diesel generator (BSDG) can be used to start secondary and main generators in a sequenced way. Once online, a larger plant can begin to bring other plants on the same network online.

Depending on the generation technology, varying amounts of electricity are needed for a black start. Thermal plants generally require significantly more power for a black start then renewable plants, for example.

Black start as a service in a wholesale energy market is generally compensated on a fixed yearly capacity payment. For example in PJM, black start is paid a regulated return of 10% on the following cost structure [4]: ((Fixed Black Start Fuel Storage Compensation (BSSC)) + (Variable BSSC) + (Training Costs) + (Fuel Storage Costs)) $\times$ (1 + Z), where Z is an incentive factor usually taken to be 10%.

BESS systems for black start are a good example of a power application vs. an energy application. A black start is a relatively short duration and uncommon event. Therefore, a BESS system does not require the amount of storage that may be required for an energy application such as energy arbitrage. Thus, the primary cost driver of the system, the batteries themselves, is reduced making it more competitive vs. alternatives such as diesel generators. In addition, events requiring black start are highly uncommon, worst case scenarios. As such, battery systems would be available to provide other power applications (peak power, ancillary services, backup, etc.) to generate additional revenue, whereas it is unlikely a diesel backup generator would be economically competitive to do so.

8.2 Regulations for storage in different jurisdictions

Without a doubt, the first countries out of the gate with respect to battery storage are Korea, the United States, Germany, China and Japan (Figure 8.3). What do these countries have in common? Leaving aside China, which benefits from enormous benefits in manufacturing costs and vast deposits of domestic lithium, the

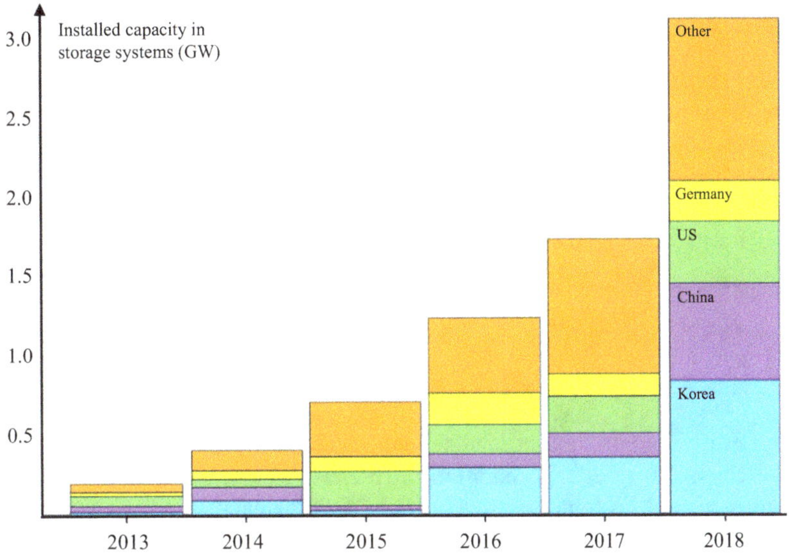

Figure 8.3 Evolution of storage installations worldwide. Source: Modified from [5]

other four countries have portfolio standards, fiscal and technical policies targeted at promoting battery storage and renewable integration into their electric systems.

8.2.1 Portfolio standards

There are many strategies by which regulators draft policy to promote desirable changes to an electricity system. In many markets (NE-ISO, PJM, etc.), renewable portfolio standards (RPSs) have been critical in driving the growth of renewable generation technologies. A portfolio standard is a regulator-issued mandate forcing all load-serving entities to acquire a certain amount of energy from specific zero- or low-carbon sources. In some of these markets, tradable credits or renewable energy credits (RECs) are generated for each MWh of "clean" generation by wind farms, solar farms and other renewable sources. The owners of these plants have their contribution certified and then are able to sell these credits on to contaminating generation sources; they have requirements to offset their carbon contributions. The result is a virtual marketplace whereby these credits are exchanged. The California state RPS is one of the most aggressive to be found, with 50% renewable generation mandated by 2030. But increasingly, other regulators are also using portfolio standards as a way to accelerate the transition to a zero carbon electricity grid.

Box 8.5: Market example—South Korea

South Korea has emerged as one of the initial leaders in BESS, both in manufacturing and the implementation of systems. The reasons for this can be traced back to (1) Korea's RPS [6] and (2) its REC weighting criteria established by the Korean Energy Agency. As can be seen in the table below, from the Korea Energy Agency, Korea has implemented increasing obligations for its renewable service supply ratio.

Year	'12	'13	'14	'15	'16	'17	'18	'19	'20	'21	≥ '22
Ratio (%)	2.0	2.5	3.0	3.2	4.0	5.0	6.0	7.0	8.0	9.0	10

Let us examine how a load-serving entity in the Korean market could have chosen to meet this requirement in 2017. Let us consider a medium-sized industrial customer with about 5 MW of demand. If we simplify this example and assume the customer was consuming electricity at a 90% load factor, this customer will have required 1,971 MWh of RECs during the year (considering an equivalency of 1 REC = 1 MWh). We can see, however, that in 2017 the Korean regulator chose to incentivize certain technologies by applying a multiplier based on various criteria such as size, generation technology and the addition of storage systems; see Figure 8.4.

REC weight for solar PV

Installation Type	Details	REC weights
Install on general site	< 100 kW	1.2
	100–3,000 kW	1
	≥3,000 kW	0.7
Install in existing facility	<3,000 kW	1.5
	≥3,000 kW	1
Floating system installed on water		1.5
On-site generation connected to grid		1
Solar + ESS	2016/17	5

REC weight for other renewables

Installation type	Details	REC weights
IGCC, by-product gas		0.25
Waste, land fill gas		0.5
Hydro, Onshore wind, RDF, waste gasification, marine tidal (with embankment), on-site generation connected to grid		1
Wood Biomass, offshore wind (~5 km of connection distance), hydrothermal		1.5
Fuel cell, ocean energy		2
Offshore wind (over 5 km of connection distance), geothermal, marine tidal (without embankment)	Fixed	2
	Variable	1.0–2.5
Wind + ESS	2015	5.5
	2016	5
	2017	4.5

Figure 8.4 Renewable energy credit weighting in Korea

If we size a system to meet the 1,971 RECs our customer requires, we can see that a stand-alone system of 750 kW will be sufficient assuming a 20% solar capacity factor. That solar production level may be a stretch for Korea, but the principal is clear. This system will generate about 1,314 MWh and benefit from a solar multiplier under 300 kW of 1.5 which will give us the exact 1,971 RECs mandated by law.

But there is another alternative to be explored. By coupling an ESS to our PV project, our system will generate not 1.5× the amount of RECs but

5×. Clearly, the Korea Energy Agency has determined that it is strategically in its interest to promote certain solutions over others. With the addition of the ESS and the subsequent 5× multiplier, we can see that the solar system required is only 230 kW vs. the 750 kW previously needed to generate the same 1,971 RECs. Does this reduced solar capacity size offset the additional cost of the battery system? It will depend on the cost of both technologies. Using indicative prices for 2017, we can compare the two options. We will first consider the cost of a stand-alone PV system considering two capex options, tagged *utility-scale* and *distributed generation scale*; on both cases we will assume the required installed capacity of 750 kW:

1.0	US$/kW utility-scale capex
2.0	US$/kW distributed generation–scale capex
750 k	US$ Total cost—utility-scale
1.5 M	US$ Total cost—distributed generation

Alternatively, we can consider a PV system enhanced by an ESS; as mentioned above, the required system size is now down to 230 kW:

1.0	US$/kW utility-scale capex	Solar component
2.0	US$/kW distributed generation scale capex	Solar component
230 k	US$ Total cost—utility-scale	Solar component
460 k	US$ Total cost—distributed generation	Solar component
650	US$/kW utility-scale capex	BESS component
1.1 million	US$/kW distributed generation scale capex	BESS component
149.5 k	US$ Total cost—utility-scale	BESS component
253 k	US$ Total cost—distributed generation	BESS component
379.5 k	US$ Total cost—utility-scale	System
713 k	US$ Total cost—distributed generation	System

This rough estimation shows that a solar + ESS system will cost about 50% of a stand-alone system generating the same amount of RECs under this system. This holds true regardless of whether the project generating those RECs is a small distributed system, or a larger utility-scale project dividing its RECs among a group of load-serving entities.

As Figure 8.3 shows, in 2018 Korea led all nations in new installed battery capacity with nearly 1 GW of capacity. This may be due to the RPS it implemented in the clear decision the government made to promote storage by granting it additional credits under the renewable energy credit program established. While an effective tool for spurring renewable generation and battery storage technology, portfolio standards and the mechanisms to promote them require constant adjustments. The Korean Energy Agency of South Korea, for example, has reduced the multiplier available to solar +

ESS projects from 5 to 4.5 in 2017. In certain instances, an REC system can overpromote or unfairly favor certain solutions. The explosion of solar + ESS in Korea, and the decision to reduce the 5× multiplier, seems to be a recognition on the part of the Korean Energy Agency that this particular incentive had become overly generous. This may be due to falling solar or battery costs or perhaps a multiplier that was excessive from the start with the intent of "kick-starting" a promising global industry. Regardless, a successful portfolio standard requires active care. Regulators must formulate a long-term plan and constantly evaluate and adjust policy to achieve their objectives.

8.2.2 Procurement standards

An alternative to an REC-driven RPS is simple procurement standards. In such schemes, a regulator will determine minimum levels of deployment needed by distribution companies and other load-serving entities. For example, the state of New York has obligated the New York's six investor-owned utilities to hold competitive procurements for a minimum of 350 MW of bulk-sited energy storage in addition to the 50% renewable generation mandate which utilities are free to achieve through any mix of qualifying technologies. Procurement standards tend to be less common than technology agnostic REC programs as they can be interpreted as being too direct in promoting certain technologies over others.

8.2.3 US market regulations

Why is storage growing in the United States versus a market such as Mexico where there is relatively little development? To understand how regulation affects the emergence of battery storage, we can look at examples from the utility and residential/commercial scale.

8.2.3.1 Utility-scale regulation change: fast-response frequency regulation

PJM provides the most salient example of how technical policy can also be a driver for the integration of renewable generation and batteries. Like many wholesale markets, PJM compensates generators for the service of frequency regulation. It describes this service as "provid(ing) market-based compensation to resources that have the ability to adjust output or consumption in response to an automated signal" [7]. It adds,

> Regulation helps match generation and demand to keep the grid functioning normally by:

- Maintaining a system frequency of 60 Hz
- Tracking moment-to-moment fluctuations in customer electricity use

- Correcting for unintended fluctuations in generation (such as a large generating unit disconnecting from the system)
- Managing differences between forecasted or scheduled power flow and actual power flow on the system.

Historically, frequency regulation services have been provided by spinning gas turbines which can adjust their output in a matter of seconds in accordance with the needs of the independent system operator. Battery storage systems, from a technical and emissions standpoint, are almost certainly better suited than large combined-cycle gas plants for providing this service. This is primarily because batteries can respond to dispatch signals within milliseconds vs. the lag in providing the adjustment with a large-frame gas turbine. PJM recognized the value of this fast-acting service and made a pioneering adjustment. Frequency regulation in the PJM region is now split into the traditional *regulation A* market and a *regulation D* market which mandates certain response times that effectively eliminate certain fossil technologies from meeting the response time criteria [8]. The entire technical justification for regulation D can be found on PJM website and can be debated among electrical engineer and electrical grid experts. What cannot be denied is this single policy change has made an important impact in the investment decision of nearly 2,000 MW of stand-alone battery storage projects that now sit in the PJM interconnection queue.

8.2.3.2 Residential-scale regulation change: HECO smart export program

Given its geographical location, the islands of Hawaii are a natural place for storage technologies to be considered. Historically in such isolated environments, base load capacity was provided using liquid fossil fuels such as heavy fuel oil or diesel. As a state, Hawaii has taken one of the most proactive regulatory positions when it comes to storage. One such example is the Smart Export Program implemented by the Hawaiian Electric Company (HECO) which has been created to incentivize up to 35 MW of battery storage in the system [9].

The Smart Export Program is essentially a net metering program that allows PV and ESS systems in the network to sell electricity back to HECO at defined prices. This not only offsets energy charges accrued by the customer during non-solar/battery hours, but can result in a positive cash flow allowing system owners to recover their investment.

Credit Rates and Export Windows for Interim Smart Export Program for the HECO Companies:

Island	12 AM–9 AM	9 AM–4 PM	4 PM–12 AM
O'ahu	\$14.97/kWh		\$14.97/kWh
Hawai'i Island	\$11.00/kWh		\$11.00/kWh
Maui	\$14.41/kWh	No credit	\$14.41/kWh
Moloka'i	\$16.64/kWh		\$16.64/kWh
Lana'i	\$20.79/kWh		\$20.79/kWh

This net metering program has allowed HECO to be one of the top ten US utilities for storage deployment in 2018, despite a far smaller customer base than competitors such as Pacific Gas & Electric or San Diego Gas & Electric. Across the United States, utilities commissions are extending net metering opportunities to ESS technologies. The result will be a proliferation of projects coming to market over the next 5 years.

Top ten US utilities by annual energy storage capacity (MWh) [10]:

Rank	Utility	State	Capacity (MWh)
1	Southern California Edison	California	154.3
2	Kauai Island Utility Cooperative	Hawaii	102
3	Pacific Gas and Electric	California	73.2
4	Florida Power & Light Company	Florida	56
5	Salt River Project	Arizona	44.5
6	Long Island Power Authority	New York	40
7	San Diego Gas & Electric	California	33.7
8	Connexus Energy	Minnesota	30
9	Hawaiian Electric Company	Hawaii	23.3
10	United Power, Inc.	Colorado	18.2

8.2.3.3 Other incentive examples

As its development has lagged generation, regulation and incentives for storage are relatively less developed. Nonetheless, it is the opinion of the authors that several effective incentives responsible for the growth of PV generation in particular could play a key role in deploying storage solutions on a national and global scale.

Perhaps the single most effective policy in the promotion of wind and solar energy has been the investment tax credit (ITC) in the United States established in the Energy Policy Act of 2005. According to the Solar Energy Industries Association, "[s]ince the ITC was enacted in 2006, the US solar industry has grown by 10,000%...including annual growth of 50% over the last decade" [11]. While this may be largely attributable to falling module prices, increases in cell efficiencies and technological reliability, both perceived and real, the ITC has also been instrumental in improving investor appeal.

As of 2020, the ITC is a 26% federal tax credit that can be claimed by a solar investor against their tax liability. This credit has been instrumental in the United States in supporting the solar industry as it went from concept to scale. Renewable energy investors can reduce their taxable income dollar for dollar on the amount of the eligible investment multiplied by the credit percentage. For example, a solar project with a total investment of US$1 million commencing construction in 2020 would qualify for a $260,000 tax credit. This credit would allow an investor to reduce their taxable income base by the same $260,000 or roll-over this value into the subsequent years. There are several fiscal structures (partnership flip, sale leaseback, inverted lease, etc.) known as "tax-equity" investments that allow project sponsors to partner with outside parties who can make more efficient use of this credit. These structures depend on the tax authority in a given country, such as the

Internal Revenue Service (IRS) in the United States, working closely with energy policymakers to establish clear rules that allow for efficient collaboration between project sponsors and these tax-equity investors.

In fact, the ITC has been so successful that the costs of solar generation have now allowed for the credit to be ramped down from its original 30%–10% in 2022. Like the Korean REC multiplier system, the US federal tax credit shows the power of focused policy to promote growth of emerging technologies and the need for policymakers to make adjustments once these technologies achieve widespread deployment.

8.2.3.4 Accelerated depreciation regulation

Fiscal incentives are one of the most critical sources of value for renewable and battery storage projects. MACRS (modified accelerated Cost-Recovery System) is a program allowing for accelerated corporate fiscal depreciation in the United States. Most renewable energy generation in the United States has been granted a 5-year recovery under this plan. In Mexico, the depreciation of renewable generation technology is even more aggressive. According to article 34 of the Mexican federal law regarding income tax, renewable plants can claim 100% depreciation in the first operational year. It should be noted, however, that having favorable laws alone will not necessarily stimulate investment. Mexico is notorious for a lack of continuity in policy between political administrations, and fiscal provisions may change on a yearly basis, even under the same federal administration. Equally importantly, compliance with tax obligations is low, making fiscal incentives much less attractive than in countries with higher tax ethics. Given the lead time associated with battery storage and renewable projects, regulatory uncertainty can make investors reluctant to base investment decisions on policies that can be changed quickly and without warning. Furthermore, Mexico's accelerated depreciation regulations are not specific as to what technology is included within the definition of renewable energy. BESS, for example, is not explicitly defined as qualifying for these incentives and as of writing there is not a general consensus on whether or not this is the case. Therefore, it is not only for positive regulation to be in place, but for that regulation to be well-drafted so that private enterprise can interpret policies correctly and make long-term investment decisions accordingly.

8.2.3.5 Local incentives

There are a host of local incentives that policymakers have used globally to promote a cleaner generation mix. However, there are still few incentives focused on storage technologies. Nonetheless, it is encouraging that they are increasing. One example is the personal tax credit the state of Maryland [12] launched in 2020. The $750,000 program allows individuals or businesses to receive a tax credit equal to 30% of their installed costs. Residential systems are limited to a $5,000 credit while businesses can receive up to $75,000. Energy rebate programs, common among PV projects, are starting to make their way to storage technologies as well. The New York State Research and Development Authority [12], for example, provides a rebate of \$250–300/kWh of installed energy for systems covered by a 10-year manufacturer guarantee.

8.3 Technologies

There is a wide array of technologies utilizing various mechanical, thermal, chemical, electrochemical and electric phenomena to achieve the bidirectional storage of electricity, as shown in Table 8.2. This allows for great optionality in the selection of the appropriate solution according to the circumstances and application, since there is not yet a "silver bullet" for storage. Some technologies are more appropriate for short- and fast-acting services while others can handle much longer duration, some are mature while others are still in the R&D phase, and some are gaining much mainstream momentum while others are only being used for niche purposes.

In this section, we will cover the main technologies being used around the world, specifying their strengths, weaknesses, relative maturities and approximate penetration. We start by observing that different technologies are differently suited to provide certain services relevant for grid support (Table 8.3) and BTM applications (Table 8.4), respectively. Even though a number of specific differences between technologies are conspicuous in the tables, the dividing line between power and energy applications and the corresponding technologies is particularly notorious in case of grid services (Table 8.3). Battery-based technologies are generally very good at providing power services, such as frequency response and regulation, but are not very effective at applications involving the bulk storage of energy. It is, however, important to note that particularly in battery-based technologies progress is very fast. NaS technology, to mention an example, was classified by reference [13] as a technology contributing little to frequency restoration and energy shifting; however, energy applications with NaS have been

Table 8.2 Energy storage technologies

Type	Technology
Mechanical	Pumped storage
	Compressed air
	Flywheels
	Gravity-based
Chemical	Hydrogen
	Synthetic natural gas
	e-fuels (e-methane, e-methanol, e-Fischer–Tropsch, etc.)
Electrochemical	Conventional batteries (lead acid, lithium ion, NiCad, etc.)
	High-temperature batteries (NaS, NaNiCl, etc.)
	Flow batteries (vanadium redox, zinc bromide, etc.)
Electric	Supercapacitors
Thermal	Molten salts
	Chillers

Table 8.3 Suitability of different energy storage technologies for grid services. Adapted and modified from [13] (CAE, compressed air storage; LA, lead acid; VRLA, valve-regulated lead acid; NMC, nickel manganese cobalt oxide; NCA, nickel cobalt aluminum oxide; LFP, lithium iron phosphate; LTO, lithium titanet; NaNiCl, sodium nickel chloride; NaS, sodium sulfur; VRFB, vanadium redox flow battery; ZBFB, zinc bromine flow battery)

Technology	Enhanced frequency response	Primary frequency regulation	Frequency restoration reserve	Energy shifting
Pumped hydro	None	None	Very high	Very high
CAES	None	None	Very high	Very high
Flywheel	Very high	Very high	Small	Small
Flooded LA	Very high	Very high	Small	Small
VRLA	Very high	Very high	Small	Small
Li-ion (NMC)	Very high	Very high	Small	Small
Li-ion (NCA)	Very high	Very high	Small	Small
Li-ion (LFP)	Very high	Very high	Small	Small
Li-ion (LTO)	Very high	Very high	Small	Small
NaNiCl	Very high	Very high	Small	Small
NaS	Very high	Very high	Small	Small
VRFB	Small	Small	Moderate	Moderate
ZBFB	Small	Small	Small	Small

Table 8.4 Contributions from different energy storage technologies to BTM services. Adapted from [13]

Technology	Self-consumption (small residential)	Community storage	Power quality	Peak shaving	Time of use
Pumped hydro	None	None	None	None	None
CAES	None	None	None	None	None
Flywheel	None	None	Very high	Very high	None
Flooded LA	Very high	Very high	Very high	Very high	Very high
VRLA	Very high	Very high	Very high	Very high	Very high
Li-ion (NMC)	Very high	Very high	Very high	Very high	Very high
Li-ion (NCA)	Very high	Very high	Very high	Very high	Very high
Li-ion (LFP)	Very high	Very high	Very high	Very high	Very high
Li-ion (LTO)	Very high	Very high	Very high	Very high	Very high
NaNiCl	Small	Very high	Very high	Very high	Very high
NaS	Small	Very high	Very high	Very high	Very high
VRFB	Very high	Very high	None	High	Very high
ZBFB	Very high	Very high	None	None	Very high

implemented in several parts of the world, with storage capacities of several hundred MWh, making NaS an option at least for intra-day energy shifting applications.

8.3.1 Mechanical

Technologies based on the mechanical storage of energy transform electrical energy into kinetic or potential energy and vice versa. Mechanical systems account for the vast majority of the deployed storage capacity worldwide, mostly thanks to pumped hydro. However, other technologies have made their appearance and have been subject to continuing development in recent years. With the exception of flywheels, most mechanical storage concepts have a focus on the bulk storage of energy, allowing them to provide services where energy is displaced along the time axis. While typical storage time constants of $T_c = E_c/P_c$ of 4 h are now common in battery-based systems, mechanical systems such as pumped hydro storage (PHS) or compressed-air energy storage (CAES) can be designed for characteristic times of 16 h [14], potentially rendering variable renewable generation plants into dispatchable ones. Whether this makes economic sense depends vastly on the regulatory framework applicable in a given country or region, as discussed at some length above.

8.3.1.1 Pumped hydro storage

The "grandfather" of the energy storage industry, PHS [15] has been around for over 100 years and accounts for over 90% of global installed capacity with over 160 GW deployed [16]. Not surprisingly, it is also one of the most straightforward technologies to understand: basically, they are hydroelectric power plants that can work in reverse. When there is excess capacity (or low prices) in the grid, the hydropower turbine works as a motor (instead of a generator) consuming electrical energy to pump the water back to the upper reservoir where it is stored as potential energy due to gravity. When the energy is needed, the water is allowed to flow back down and the turbines now act as traditional generators transforming the kinetic energy in the water into electricity. PHS is considered quite efficient with round-trip efficiencies that can exceed 80% [16].

As is the case with conventional hydropower, PHS facilities can be massive in scale, with many over 1 GW of power and dozens of GWh of storage capacity. The largest PHS plant in the world so far is the Bath County Pumped Storage Station in Virginia, USA, with 3 GW of power capacity and 24 GWh of storage potential; it was finished in 1985. Evidently, PHS is a great option for longer duration storage as the energy capacity is defined by the height difference between the reservoirs and the size of the upper reservoir, the latter of which can be expanded within geographical constraints. Additionally, PHS is a decent option for seasonal storage (across days, weeks or months) because of these massive basins, and the energy losses (mainly due to evaporation) are relatively small and can be mitigated.

The main drawbacks of PHS are geographic implications and capital requirements. To be cost-effective, PHS plants must be relatively large to take advantage of economies of scale. However, large plants can only be located in certain geographically appropriate sites, with the right topological and hydrological conditions. These large plants typically cause significant environmental impact by affecting the landscape due to flooding of previously dry areas, and disturb the existing aquatic ecosystem. Even so, the costs can be prohibitive if not backed by government investments and incentives, which is one of the main reasons why not so many PHS plants are being built around the world.

PHS will continue to play a key role for energy storage moving forward, as the existing fleet is long-lasting and efficient. Some existing dams that do not have generators installed have been identified as "low-hanging fruit" to apply this technology while mitigating the weaknesses described. Nevertheless, most of the growth of the global storage capacity will come from newer and more flexible technologies.

8.3.1.2 Compressed-air energy storage

Even though compressed air has been used in several forms since the nineteenth century, the first grid-scale storage project was not put on line until 1978, when a 290 MW/580 MWh plant was inaugurated in Huntorf, Germany [13]. CAES uses an air compressor driven by electric motors to store compressed air in caverns or pressure vessels. To utilize this energy, the compressed air is expanded to drive a turbine to generate electricity. When existing salt deposits or depleted gas field caverns can be utilized, the costs of the project can be significantly mitigated, but this also limits the geographic locations available for the technology. High-pressure tanks can be implemented to allow greater siting and capacity flexibility, but the capital requirements also escalate significantly.

One of the main virtues of this technology is that the duration of storage can be massive if located at large existing caverns, and this makes the costs relatively low. However, the technology suffers a great disadvantage in terms of efficiency. When air is compressed to high pressures, it heats up significantly, and cools upon expansion. This residual heat is handled in two ways: it can either be diffused into the environment (diabatic) or it can be reserved in a thermal ESS for use in the expansion phase (adiabatic). Diabatic systems require external sources of energy (mostly natural gas) to reheat the expanding air, and thus typically emit GHGs and are quite inefficient, hovering around 40%–60% of cycle efficiency. Adiabatic systems are more environmentally friendly and a bit more efficient, reaching round-trip efficiencies of up to 70% [13].

CAES remains a relatively immature technology, and the outlook is not as great when compared to other technologies, mostly given the relatively stringent site restrictions. However, research and development is ongoing and it might be the best solution for some niche applications.

8.3.1.3 Flywheels

Flywheel energy storage (FES) is based on a technology with millennia of history: a rotating mass that tends to remain in motion thanks to its angular momentum and can be used for diverse applications such as textiles (spindles), mechanical tools (grinders, mills) and even to stabilize steam engines [13]. Flywheels for electric energy storage convert electricity into kinetic energy by use of a motor to accelerate a rotating mass to high speeds. When the energy is needed, the motor's operation is reversed to act as a generator that converts the kinetic energy into electricity back again, thus decelerating the mass. To mitigate losses due to friction, modern FES typically employ vacuum chambers and advanced mechanical and/or magnetic bearings. Flywheels are typically composed of the following five main components:

1. flywheel,
2. bidirectional motor/generator,
3. power conversion system (PCS),
4. bearings and
5. vacuum chamber.

FES is adequate for short-duration and high-power applications thanks to their quick response times (often in the milliseconds) and their ability for fast charging and discharging. Consequently, they are mainly used for services such as frequency regulation and power quality. The round-trip efficiency of these systems can be relatively high when acting across short time periods, up to 80%–90% [13], but they suffer from relatively high self-discharging rates (idle losses) when used across longer duration. Other advantages of this technology include long lifetimes (as their capacity is less degraded by usage), easy state of charge estimation (based on flywheel speed) and relative maturity due to its history and varied utilization across industrial segments.

However, there are considerable downsides as FES systems have low energy density (thus lower duration), the mechanical bearings need regular maintenance and the magnetic bearings need power to operate (thus reducing efficiency), and lastly they can suffer catastrophic failure due to unexpected physical movements. This technology will continue to be utilized for industrial efficiency and power quality, and will likely continue to find applications in grid (or microgrid) stability.

8.3.1.4 Gravity-based storage technologies

Although PHS also utilizes gravitational potential energy, here we will refer to systems that use massive objects instead of water. Similar to all other mechanical energy storage methods, gravity energy storage (GES) involves the use of energy to drive a motor and then using that motor as a generator to recover the energy. In this case, the motor usually lifts a mass through the use of a winch (or water pump in hydraulic systems) to charge the system, and reverses the operation to discharge. There are many different implementations of this concept; here are a few of them:

1. Cranes and concrete blocks. Energy Vault [17] has created a stir by getting the backing of big industrial and financial players. They propose using several large autonomous cranes to stack concrete blocks around their base to create a tower structure that can be constructed and deconstructed to store and retrieve the potential energy.

2. Hydraulic systems. Companies such as Heindl-Energy [18] and Gravity Power [19] are proposing using water and pumps to raise a very large mass (for the former) or a piston (for the latter). One of the interesting things about Heindl's concept is that the relative system cost decreases with the squared inverse of the radius of the mass, thus promising relatively cheap storage at massive scales.
3. Shafts, ramps and winches. Gravitricity [20] and EarthPumpStore [21] and other concepts such as Mountain Gravity Energy Storage [22] propose using weights in mine shafts, mine pits and mountains, respectively, and bidirectional winches to raise and lower them.

These concepts all focus on the relevant issue of long-term storage, where few alternatives to the incumbent yet limited PHS technology exist. Several of these technologies propose to tap into existing infrastructure (such as abandoned mine shafts [19,20] or even open mining pits [21]), which seems appealing, given the cost savings involved in reclaiming existing wells, and the potentially small footprint, at least in case of vertical shafts. Building dedicated infrastructure such as huge concrete slaps [18] also seems an endeavor readily realizable with existing civil engineering knowledge. More futuristic designs such as the one proposed by Energy Vault [17], which involve the construction of high-precision structures look somewhat more challenging to materialize, but given the vast experience obtained by civil engineers and builders worldwide with the construction of complex high-rise buildings, this proposal does not seem too far-fetched either.

In either case, while being somewhat more incipient compared to PHS or BESS technologies, gravity-based concept seem to be an interesting option for bridging the gap between those two, while offering considerably more flexibility than PHS solutions. Most of the GES systems also boast high efficiencies (up to 90%), long lifetimes and great power and energy capacities. Although relatively immature, these concepts hold much promise if they are successfully developed and commercialized. Evidently, a greater boost to GES technologies, as well as all other storage solutions, will greatly depend on the proper regulations and incentives for storage in the different energy markets of the planet, as discussed in the first part of this chapter.

8.3.2 Chemical storage

8.3.2.1 Hydrogen

Chemical energy storage is mainly done through electrolysis: using electrical current to drive a chemical reaction, a concept dating over 200 years. Lately the idea of using electrolysis to extract hydrogen from water has been gaining much attention as a key piece to decarbonizing our energy infrastructure. However, hydrogen production from electrolysis as a means of storing energy suffers several drawbacks given the low volumetric density of gaseous hydrogen under standard conditions and the poor efficiencies for both electrolysis and the back-conversion to electricity by means of fuel cells. As a replacement for traditional fossil fuels, hydrogen has to be stored at very high pressures, presenting multiple mechanical and safety challenges. However, existing large underground caverns can be used as storage mediums for large-scale applications. The making and utilization of hydrogen for electricity purposes remains relatively inefficient with 70%–80% for electrolysis and about 50% for fuel cell back-conversion into electricity [23], meaning round-trip efficiencies of only around 40%‡. The use of gas turbine combined-cycle plants for H_2 combustion has been demonstrated on a utility scale (16 MW) [24], though, potentially paving the way for the conversion of hydrogen to electricity with efficiencies characteristic of large-scale combined-cycle plants, i.e. up to 60%§.

8.3.2.2 Power-to-X

As pointed out in several of the chapters of this book, the coupling of different sectors is key to a deep decarbonization of the energy sector as a whole. The key role of electricity for driving the renewable energy transition has been emphasized in Chapters 1–3, and the case of a strong sector coupling between urban development and mobility has been made in Chapter 12. Though electricity, both as provided and after cycling through storage devices, as described in this chapter, can be expected to become the dominant provider of energy services in the near future, fuels and sources of heat represent other important facets of the energy sector. Electric power-to-heat, preferably through heat pumps, and Combined Heat & Power (CHP) are two examples of energy sector coupling which have been used successfully over decades. Power-to-fuels processes are another important branch of technology which can be expected to receive strong attention in the next years, as the need for climate action becomes even more urgent.

A power-to-fuel process typically starts with the electrolysis of water, described above, yielding gaseous hydrogen, termed "e-hydrogen," where the prefix "e-" indicates the "electric" origin. Hydrogen can then be upgraded, at a loss, to fuels more susceptible to handling and transporting with the existing fuel infrastructure. Such fuels include e-methane, e-methanol, e-Diesel, e-gasoline or e-jet-fuel. In addition to fuels, chemical feedstocks such as (e-)ammonia (for fertilizer production and other products) and products obtained from syntheses such as Fischer–Tropsch processes can be obtained as well, contributing to the deep decarbonization of industry and agriculture.

When discussing possible business cases for e-hydrogen, often the current cost of *gray hydrogen* (obtained from steam reforming of methane, SMR) is considered a useful reference. It has been shown [25] that "green" hydrogen is already

‡Low round-trip efficiencies may evidently be less of a concern if the alternative is to discard free renewable energy, e.g. through curtailment. Though well-planned transmission grid upgrades are usually considered the most cost-effective way of mitigating curtailment, as demonstrated in Chapter 4, achieving an adequate transmission grid faces a number of obstacles, some of them political.

§Evidently, not everything is about thermal efficiency; combined-cycle plants still need considerable amounts of water for the steam cycle and for dissipating the remaining 40%–45% of fuel energy which was not converted into electricity, which may preclude its wider dissemination in a significant part of the world where water shortage is an increasingly urgent issue. This will still hold true if the water generated in the combustion process is accounted for.

competitive with typical SMR costs of hydrogen which are located in the US \$1.5–2.5/kg H_2 range, if (1) electrolyzers have a high load factor (6,000 load hours/year) and (2) the price of the (renewable) electricity is low, i.e. around US\$20/MWh. Such low costs may have been unrealistic only a few years ago, but thanks to competitive auctions in placing with good renewable resources such as Mexico [26], Chile [27] and Saudi Arabia [28], among others, these numbers have already become the state of the art.

8.3.3 Electrochemical

Also known as battery energy storage (BES), our modern concept of a battery dates back to the time of great historical figures such as Benjamin Franklin, Alessandro Volta and Michael Faraday. All batteries consist of electrochemical cells, each one composed of two electrodes, one positive (cathode) and one negative (anode), and an electrolyte that separates them. A chemical reaction in the battery produces electrons (or positive ions) at one electrode that flow to the other, creating an electric current; this process is reversed when the BES is recharged. The technology has advanced leaps and bounds since its beginnings, and one particular subset, lithium ion, is taking the energy storage industry by storm.

8.3.3.1 Lithium-ion systems

First discovered and proposed in the 1970s, and commercialized by Sony in the 1990s, lithium-ion batteries (LIBs) have benefited from economies of scale and technological advancements due mainly to the demand from modern electronics and electric vehicles [13]. The mass production of the cells owes their rapid rise in the market to a process for a technology that was at that time rapidly declining: cassette tapes. It turns out the coating of film for cassettes proved to be an efficient way to produce the electrodes for the new batteries, a process that remains the gold standard for these batteries. LIBs now permeate all consumer electronic devices and are quickly advancing the evolution and penetration of electric vehicles worldwide. Although they can be considered a relatively mature technology, there are still advancements being developed and commercialized, so costs are bound to keep declining and their performance improving.

One sticking point remains the safety of the technology, as many LIB systems are susceptible to fire and typically require integrated fire-suppression systems and advanced Battery Management Systems (BMS) to monitor the cell voltages and temperatures continuously.

LIB makes up the vast majority of energy storage deployments in the past few years [5], and their ubiquity in electronics and electric vehicles added to the incremental gains in technology and cost declines will mean they will hold the title for the short-term foreseeable future. The technology boasts relatively high round-trip efficiency (80%–90%) and long cycle life (3,000–6,000 cycles). They also have the ability to be sized for many different applications, from residential to grid scale, and can typically be located indoors or outdoors, giving it tremendous versatility.

8.3.3.2 Lead-acid systems

The familiar lead-acid system, common in many volume applications such as in car starters, golf cart and forklifts, as well as in off-grid renewable energy applications, is still considered a viable alternative for certain application in the modern storage world, mainly because of its low cost, the ample manufacturing and recycling experience in most parts of the world, as well as its fair (70%–90%) round-trip efficiency [13]. Its low energy density (50–100 Wh/L) precludes it from more challenging applications such as electromobility, but the numbers of niches where lead-acid batteries can be used is still considerable. Furthermore, a number of manufacturing and systems engineering innovation are being undertaken, including process automation, the use in hybrid applications, e.g. teaming up with flywheels and LIBs, upgrading the negative electrode by integrating a copper stretch mesh, and the deposition of carbon layers onto one or both electrodes with the intent to reduce sulfation [13]. In addition, flow versions of lead-acid batteries have been explored for power systems applications [29].

8.3.3.3 NaS systems

Sodium sulfur (NaS) batteries pertain to a class of high-temperature batteries with well-balanced performance set, including a relatively high (>80%) round-trip efficiency, good discharge capabilities on different time scales and a low self-discharge rate. The fact that NaS batteries only require abundant low-cost materials adds to their attractiveness. NaS batteries have been reported to be suitable for power applications such as the reduction of wind power curtailment in island systems [30].

Pioneering applications of NaS batteries in power systems have been demonstrated in Japan since the 1990s, with a total of more than 300 MW of storage capacity having been installed throughout the country [13]. According to the International Renewable Energy Agency, the cost for NaS battery energy storage (BES) systems ranges between US$263/kWh and US$735/kWh, with typical values being below US$400/kWh [13]. Among the advantages of NaS systems are their long cycle life times; one of their major drawbacks, on the other hand, is their relatively high maintenance cost, estimated to be in the range of US$40–80/kW/year, mainly due to the heat needed to maintain the battery at its operating temperature.

8.3.3.4 Flow systems

As opposed to conventional rechargeable batteries where the electroactive materials are located on the electrodes, flow batteries store them in separate tanks, from which they are pumped into the reaction unit when required [13], i.e. for charging or discharging. The main advantage of this concept is that the capacity can be easily scaled up by appropriately sizing storage tanks and cell stacks (reaction units). The main electrochemical systems used for flow batteries are vanadium and zinc bromine/bromide redox pairs. Costs for such systems still vary widely (values between US$315/kWh and US$1,680/kWh have been reported [13]), although cost reduction potentials of about minus two-thirds have been identified for 2030 [13]. Round-trip efficiencies are still modest (between 60% and 85%) but are expected to improve as well, to 67%–95% in 2030 [13].

8.3.4 Thermal

Thermal energy storage has been used widely in many parts of engineering, with a great number of applications for storage of both sensible and latent heat. A well-established application is the use of phase-change materials (such as molten salt) in solar thermal power plants [31]. However, the use of phase-change materials is not limited to solar–thermal applications. For example, Rao *et al.* [32] propose the use of phase-change materials to store excess power from wind power plants, potentially affected by curtailment, and subsequently generate fuel savings in a combined-cycle power plant.

A very ambitious approach to thermal storage, termed ETES (Electric Thermal Energy Storage), has been proposed by Siemens Gamesa and is being demonstrated with a 30 MW/130 MWh pilot facility at Hamburg harbor. At this facility, hot air at 800°C is blown into a storage unit filled with volcanic rock; the accumulated heat can then subsequently be used to generate steam for power or industrial applications. Though the pilot facility was purpose-built, Siemens Gamesa views the largest potential for this technology in the utilization of soon-to-be-decommissioned coal power plants, following the planned phase-out of coal-fired generation in Germany [33] and possibly elsewhere [34].

8.4 Conclusions

Energy storage is a key element in the renewable energy transition, particularly at high levels of penetration. However, for storage solutions to gain traction it is vital for regulations, incentives and remuneration schemes to be appropriately designed. Though generally agreed upon to be a vital ingredient for the energy transition, storage developers still face a number of hurdles to make a clear business case. Even though commercially viable opportunities have been shown to exist, a clear long-term vision has to be laid out in individual countries and through international collaboration. Suitable storage technologies abound, and solutions once thought to be marginal are now moving into the mainstream. The cost reduction potential in most technologies is considerable, and it is likely that these reductions can be realized in a relatively short term if an appropriate market pull exists. To create such conditions is of course mainly the task of policymakers, but the participation of businesses as well as the civil society is also very valuable. Appropriate stimuli for storage may allow to increase the momentum of the renewable energy transition, direly needed to meet the $\leq$1.5 degree goal, and allow storage to become a main driver of sustainable economic development.

References

[1] Pandurangan V, Zareipour H, and Malik O. Frequency regulation services: A comparative study of select North American and European reserve markets. In: 2012 North American Power Symposium (NAPS); 2012. pp. 1–8.

[2] Tan J and Zhang Y. Coordinated control strategy of a battery energy storage system to support a wind power plant providing multi-timescale frequency ancillary services. IEEE Transactions on Sustainable Energy. 2017;8(3): 1140–1153.
[3] Bolun Xu, Dvorkin Y, Kirschen DS, *et al.* A comparison of policies on the participation of storage in U.S. frequency regulation markets. In: 2016 IEEE Power and Energy Society General Meeting (PESGM); 2016. pp. 1–5.
[4] Hauske T. PJM Black Start Fuel Storage Compensation Education. PJM; 2017. Available from: https://www.pjm.com/~/media/committees-groups/committees/mic/20170607/20170607-item-12a-pjm-black-start-fuel-storage-compensation-education.ashx
[5] Munuera L, Fukui H, Toril A, Lucchese P, Bennett S and Goodson T. Tracking Energy Integration/Energy Storage; 2019. Available from: www.iea.org/reports/tracking-energy-integration/energy-storage
[6] Renewable Portfolio Standards; 2020. Available from: www.energy.or.kr/renew_eng/new/standards.aspx
[7] Regulation Market. PJM Learning Center; 2020. Available from: www.learn.pjm.com/three-priorities/buying-and-selling-energy/ancillary-services-market/regulation-market.aspx
[8] Hsia E. Energy Storage in PJM: A Perspective. PJM Inside Lines; 2020. Available from: www.insidelines.pjm.com/energy-storage-in-pjm-a-perspective/
[9] Smart Export Program. Public Utilities Commission of Hawaii Website; 2020. Available from: ww.puc.hawaii.gov/wp-content/uploads/2017/10/Hawaii_PUC_Smart-Export_CGS_Fact_Sheets_FINAL.pdf
[10] Utility Energy Storage Market Snapshot. Smart Electric Power Alliance (SEPA); 2019. Available from: www.roedel.faculty.asu.edu/sec501f19/pdf/SEPA-2019-Utility-Storage.pdf
[11] Solar Investment Tax Credit. Solar Energy Industries Association (SEIA); 2020. Available from: www.seia.org/initiatives/solar-investment-tax-credit-itc
[12] Program Overview. Database of State Incentives for Renewable and Efficiency (DSIRE); 2019. Available from: www.programs.dsireusa.org/system/program/detail/22076
[13] IRENA. (2017). Electricity Storage and Renewables: Costs and Markets to 2030, International Renewable Energy Agency, Abu Dhabi. Available from: https://www.irena.org/-/media/Files/IRENA/Agency/Publication/2017/Oct/IRENA_Electricity_Storage_Costs_2017.pdf
[14] Mongird K, Fotedar V, Viswanathan V, *et al.* Energy Storage Technology and Cost Characterization Report. Pacific Northwest National Laboratory. 2019;(July):1–120. Available from: https://www.energy.gov/sites/prod/files/2019/07/f65/Storage Cost and Performance Characterization Report_Final.pdf.
[15] Menéndez J, Fernández-Oro JM, Galdo M, *et al.* Efficiency analysis of underground pumped storage hydropower plants. Journal of Energy Storage. 2020;28:101234. Available from: http://www.sciencedirect.com/science/article/pii/S2352152X20300323

[16] Hydropower Vision. Department of Energy; 2016. Available from: https://www.energy.gov/eere/water/articles/hydropower-vision-new-chapter-america-s-1st-renewable-electricity-source
[17] Innovative Solution for Long Duration Utility Scale Energy Storage. Energy Vault; 2020. Available from: https://energyvault.com/
[18] A New Solution for Large Scale Energy Storage. Heindl Energy; 2020. Available from: https://heindl-energy.com/
[19] Grid Scale Energy Storage. Gravity Power; 2020. Available from: http://www.gravitypower.net/
[20] Fast, Long-life Energy Storage. Gravitricity; 2020. Available from: https://www.gravitricity.com/
[21] Gravity Storage System Using Earth Materials. University of Nottingham, World Society of Sustainable Energy Technologies; 2019. Available from: https://www.wsset.org/docs/articles/deep-storage.pdf
[22] Hunt JD, Zakeri B, Falchetta G, *et al.* Mountain Gravity Energy Storage: A new solution for closing the gap between existing short- and long-term storage technologies. Energy. 2020;190:116419. Available from: http://www.sciencedirect.com/science/article/pii/S0360544219321140l0
[23] Fuel Cell Efficiency. Elsevier; 2020. Available from: https://www.sciencedirect.com/topics/engineering/fuel-cell-efficiency
[24] Hydrogen Fuelled Combined-cycle Gas Turbine Power Plant Inaugurated by Enel; 2010. Available from: https://www.power-eng.com/2010/07/14/hydrogen-fuelled-combined-cycle/
[25] Power-to-X: The Crucial Business on the Way to a Carbon-free world. Siemens Gas and Power; 2019. Available from: https://new.siemens.com/global/en/products/energy/technical-papers/download-power-to-x.html
[26] Deign J. Updated: Mexico's Energy Auction Just Logged the Lowest Solar Power Price on the Planet. Greentech Media; 2017. Available from: https://www.greentechmedia.com/articles/read/mexico-auction-bids-lowest-solar-wind-price-on-the-planet
[27] Chile's Auction Concludes with Average Price of $32.5/MWh; 2017. Available from: https://www.pv-magazine.com/2017/11/03/chiles-auction-concludes-with-average-price-of-32-5mwh/
[28] Saudi Auction for 400 MW of Wind Attracts 4 Bids; 2018. Available from: https://renewablesnow.com/news/saudi-auction-for-400-mw-of-wind-attracts-4-bids-621088/
[29] Bates A, Mukerjee S, Lee SC, *et al.* An Analytical Study of a Lead-acid Flow Battery as an Energy Storage System. Journal of Power Sources. 2014;249:207–218. Available from: http://www.sciencedirect.com/science/article/pii/S0378775313017564
[30] Rodrigues EMG, Osório GJ, Godina R, *et al.* Modelling and sizing of NaS (sodium sulfur) battery energy storage system for extending wind power performance in Crete Island. Energy. 2015;90:1606–1617. Available from: http://www.sciencedirect.com/science/article/pii/S0360544215008725

[31] Yu Q, Li X, Wang Z, *et al.* Modeling and dynamic simulation of thermal energy storage system for concentrating solar power plant. Energy. 2020. p. 117183. Available from: http://www.sciencedirect.com/science/article/pii/S0360544220302905

[32] Rao AG, van den Oudenalder FSC, and Klein SA. Natural gas displacement by wind curtailment utilization in combined-cycle power plants. Energy. 2019;168:477–491. Available from: http://www.sciencedirect.com/science/article/pii/S0360544218323302

[33] Oei P-Y, Brauers H, Beirwirth A, *et al.* Phasing out Coal in the German Energy Sector. GTM; 2019. Available from: https://coaltransitions.org/publications/phasing-out-coal-in-the-german-energy-sector/

[34] Wilson IAG and Staffell I. Rapid fuel switching from coal to natural gas through effective carbon pricing. Nature Energy. 2018;3(5):365–372. Available from: https://doi.org/10.1038/s41560-018-0109-0

Chapter 9

Forecasting of renewable energy generation for grid integration

Michel Rivero[1], Alberto Reyes[2], Mauricio Escalante[3] and Oliver Probst[4]

The transformation of the electricity grid, from one which is essentially based on fossil-fuel, nuclear energy, and large-scale hydro, to one driven by renewable energy (RE) sources (many of which are variable in nature), requires a fundamental change in paradigms. Traditional grid dispatch is closely linked to the idea of stacking different types of near-constant generation from the lowest to the highest marginal operating cost, and regulating load variations with a limited amount of flexible (and generally costly) generation. In an electricity grid with abundant amounts of nearly free and nonpolluting energy sources, this is clearly no longer a good idea. Taking advantage of free electricity whenever possible and filling in the gaps by other means, e.g. hydroelectricity, generation from storage technologies (see Chapter 8), and limited amounts of electricity generated from sources of bioenergy, is evidently the approach that minimizes both the total marginal cost of electricity generation and polluting emissions.[*] It is also clear that the amount of regulation needed to cope with variations of the *net load*, loosely defined as the difference between the real load and the variable generation, will increase in a grid with a high penetration of variable generation. Therefore, the cost associated with regulation services will also increase, but this increase can easily be shown to be very small compared to the huge cost and pollution savings brought about by replacing fossil-fuel plants with renewables.

Reducing the cost of regulation (by downsizing the required regulation capacity and minimizing its utilization) is the ultimate goal of forecasting. Even with a perfect knowledge of the future net load, the need for regulation evidently does not drop to zero (because of variations of both the actual load and the generation) but the

[1]Instituto en Investigaciones de Materiales, Unidad Morelia UNAM, Morelia, Michoacán, Mexico
[2]Control, Electronics and Communications Department. Instituto Nacional de Electricidad y Energías Limpias, Cuernavaca, Morelos, Mexico
[3]Facultad de Ingeniería. Universidad Autónoma de Yucatán, Yucatan, Mexico
[4]Tecnológico de Monterrey, School of Engineering and Sciences, Monterrey, Nuevo Leon, Mexico
[*]A comprehensive account of the considerations surrounding this grid transformation is provided by Mark Diesendorf in Chapter 2. A quantitative demonstration of this design thinking for the case of the European electricity is given by Crespo del Granado *et al.* in Chapter 4.

uncertainties in forecasting are still larger than the actual variations of the net load, at least at current levels of penetration in most grids, making improvements in forecasting a critical task. Accurate forecasting is needed on different time scales, corresponding to the ones used to control and regulate grid operations.[†] Meteorological phenomena are also characterized by a hierarchy of temporal and spatial scales, and properly understanding those phenomena goes a long way toward more accurate predictions. In the present chapter, an overview of the topics will be given, providing the reader not only with an account of the scope and capabilities of the different methods available but also an understanding of the importance of forecasting as a grid service.

9.1 Introduction

REs are clean and abundant sources of energy, and their production costs have become increasingly competitive in the recent decade [1]. REs differ from fossil fuels in their diversity, abundance, and potential in any part of the planet, but especially in the fact that they do not produce greenhouse gases or polluting emissions. Given the urgent need to keep the planet from crossing certain tipping points beyond which changes to the climate become irreversible, transitioning to an energy sector based on REs should become the one top priority of decision-makers worldwide. Due to dramatically decreasing costs, several RE technologies, particularly wind and solar photovoltaic (PV) systems, are already cheaper than conventional, polluting technologies.

In 2018, RE technologies had the highest growth in generation, with an increase of about 7% or 436 TWh in worldwide generation (as illustrated in Figure 9.1). REs represented more than 25% of the world electricity generation, just below coal, and around 48% of the global increase in electricity generation. Electricity generation from REs is a key driver for deep decarbonization of the world's energy sector, making it imperative to maintain the momentum in RE deployment observed over the last decades and not to wane in the face of obstacles, both technical and political, arising on the way.

According to the International Energy Agency (IEA) [2], renewable power capacity is expected to expand by 50% (about 1,220 GW in the main case) by 2024, mainly by solar PV, which is predicted to represent around 57% of this growth, as can be observed in Figure 9.2. Wind energy (both onshore and offshore) is expected to increase 352 GW, representing about 29% of the total expected growth.

9.1.1 The case of Mexico

Mexico is not the exception in the current trend.[‡] After the enactment of the Energy Reform in 2013, there have been several changes in terms of energy regulation.

[†]A detailed discussion of grid operations is provided by von Meier *et al.* in Chapter 6 on smart grids.
[‡]Or at least not until recently. Most of the growth in RE installations has been supported by private developments and investments, which are now considered detrimental to the new government's policy to reinstate a privileged role for the incumbent state utility, the Federal Electricity Commission (CFE in Spanish). This policy is oblivious of the fact that CFE is actually making good business by purchasing cheap renewable energy from private generators and selling at about five times its value to the end user.

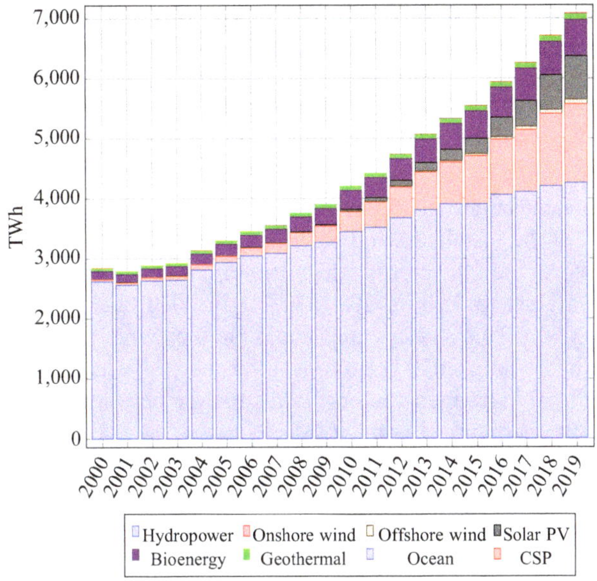

Figure 9.1 Worldwide renewable power generation by technology. Data source: IEA [2]

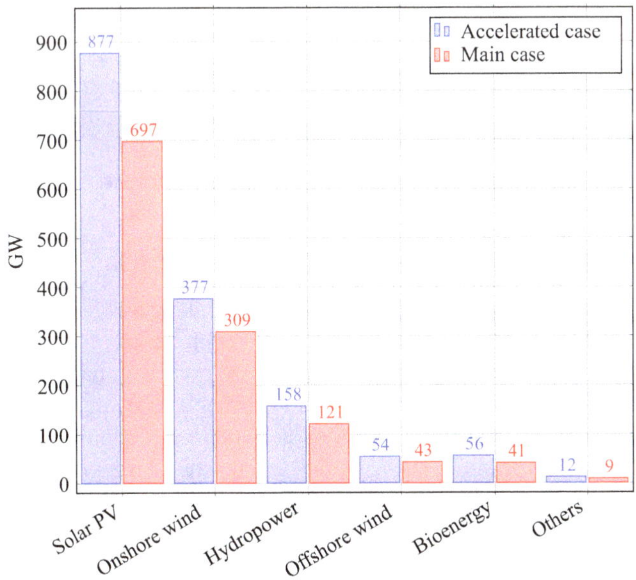

Figure 9.2 Worldwide renewable capacity growth between 2019 and 2024 by technology. Data source: IEA [2]

One of the main objectives of this reform was to enhance development with social responsibility, increase the competitiveness of power generation, and develop mechanisms for fostering clean energy.

According to the Secretary of Energy (SENER), Mexico has become the sixth most attractive economy for RE investments. In 2020 Mexico is expected to add 5 GW of clean energy to its current capacity,[§] the generation of which is roughly equivalent to the consumption of Mexico City. This addition is also equivalent to a 170% increase in the installed wind capacity compared to the last 18 years. The goals established in the transition strategy to promote the use of cleaner technologies and fuels are set to 35% in 2024, 37.7% in 2030, and 50% of total power generation in 2050.[¶] This strategy also states the objective to reduce greenhouse gas emissions by 30% in 2020 relative to a business-as-usual scenarios. For this reason, 70% of total investments in new infrastructure between 2017 and 2030 was expected in clean energy development [3].

According to the Mexican RE prospective 2018–32 [3], wind energy generation was expected to increase about fourfold during the forecast period. This growth prediction was made based on the high competitiveness of wind power in Mexico, considering cost reductions in wind turbine manufacturing, supply chain, and construction, as well as competitive financing. The availability of competitive financing can be readily attributed to (1) the creation of a considerable market pull, achieved mainly through competitive (reverse) auctions for RE, a transparent process, and long-term investment guarantees, at least through the end of 2018. Similar arguments apply to solar PV systems, which have also heavily tapped into the potential created by the auction process.

Despite the very favorable response by the RE development and investment community, the 2018–24 administration stopped the auction process shortly after taking office. The modernization and expansion of the electric grid is also expected to experience a slowdown, since a number of important transmission line projects have been canceled, which adds to existing backlog from recent years. The Mexican National Development Plan of the present administration continues with the goal of moving from 25.6% of electrical energy produced through clean sources in 2018 to 35.8% by 2024. However, there is no specific intention to increase RE capacity after 2024; rather, most initiatives focus on strengthening the utilization of natural gas, with the associated impacts on climate change.

Even though the policies of the current government are not favorable to the large-scale deployment of RE technologies, Mexico will continue to experience a significant expansion in the power generation through renewable sources, mainly solar and wind energy, as projects with signed interconnection contracts are gradually coming into operation. The inherent variability of the generation from RE sources evidently implies a number challenges for their integration in the national

[§]Literally all of these capacity additions come from projects initiated well before the current administration took office. A similar growth is expected to happen in 2021 but in the following years the addition of new capacity is likely to be marginal.

[¶]As of March 2020, these goals have not been altered.

electric system, given both the considerable lag in transmission grid upgrades and a dispatch philosophy at the grid operator largely dominated by traditional thinking.

9.1.2 Forecasting as a cost-effective flexibility resource

A high penetration of RE faces two main challenges: *variability* and *uncertainty*. Though variability is not a new issue in power systems operations, where stochastic considerations have been used for decades—because of the stochastic nature of load variations, the varying availability of fossil-fuel power plants, and the seasonal and annual variations of rain fall impacting on the availability of water for hydroelectric generation—variability does increase with the presence of wind and solar generation, among others. Variability has always been managed in power systems operations with a cascade of regulation measures (see Chapter 6 on smart grids and Chapter 8 on storage). Therefore, integrating high shares of variable renewable generation into the grid is about improving and refining existing approaches, rather than about re-inventing the wheel. It is also worth stressing that the increase in variability in solar and wind resource is fundamentally in the short and medium term, ranging from fractions of minutes to weeks, and to a much lesser degree on interannual time scales. This important distinction will be discussed in greater detail below.

Evidently, the smaller the uncertainty associated with the prediction of the *net load*, commonly defined as the difference between the actual load and the variable generation, the smaller are the regulation reserves required to mitigate a possible variability event. Accurate forecasts of power from wind and solar are therefore a key element for a deeply decarbonized electricity grid. As illustrated in Figure 9.3, forecasting is also one of the cheapest types of intervention available to the grid operator for increasing grid flexibility. Although actual costs are system dependent, it is clear the RE forecasting implies a lower cost compared to other types of interventions such as energy storage, power grid expansions, and flexible generation, among others. Many authors have reported that a decrease in the uncertainty associated to the forecast may lead to an increase in the penetration rate of RE [4–9]. Uncertainty forecasts can provide additional quantitative information for decision-making processes.

From a power systems perspective, prediction tools for wind and solar energy are highly valuable since they enable a better dispatch, scheduling and unit commitment of conventional generators, hydroelectric plants, and energy storage units [11]. They also enable a more competitive energy market under power ramps with solar and wind [12–16]. Although energy storage represents a solution to variability of renewable sources, its use is still limited by current costs [17,18], and more importantly, inadequate remuneration for balancing and regulation services.

Wind presents variations on different *time scales*, ranging from a few seconds (turbulence) to months (meteorological phenomena such as "El Niño" or ENSO). It is also associated with *spatial variations* from a few millimeters (turbulence) to several kilometers (tropical storms), as illustrated in Figure 9.4. Time scales of wind variations can be represented by their spectral distribution, often inspired by

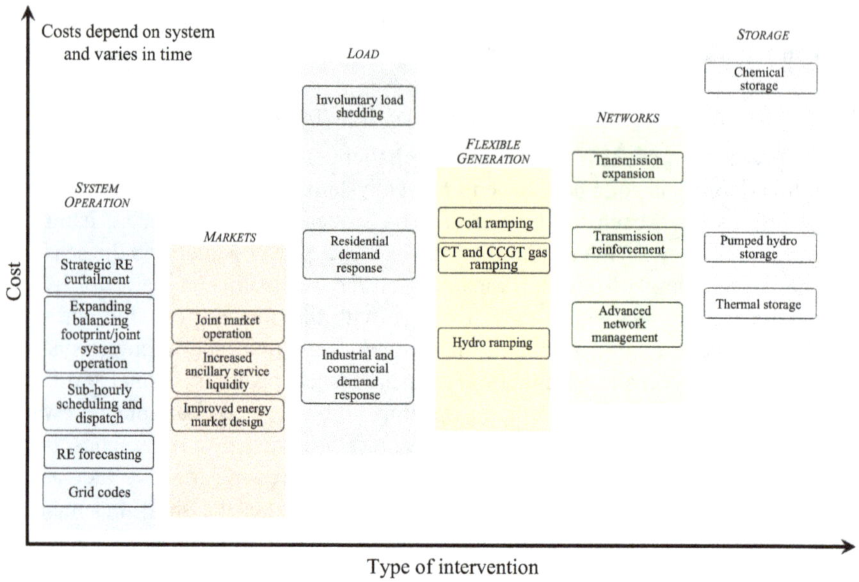

Figure 9.3 Example of integration options to increase grid flexibility. Relative costs are illustrative, as actual costs are system dependent. Modified from [10]

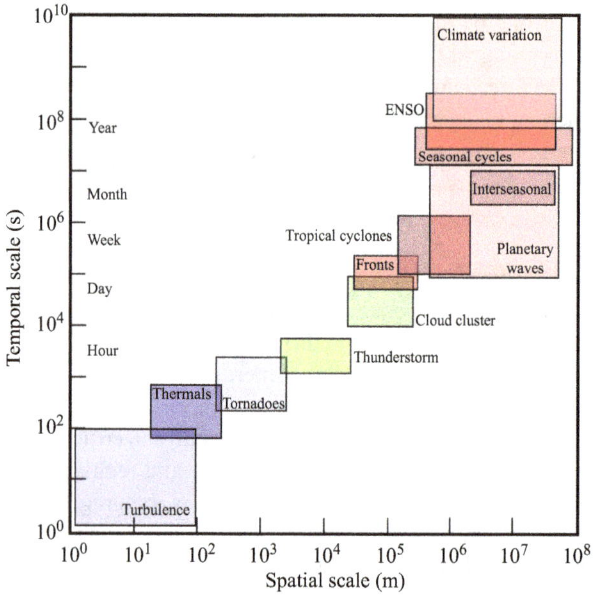

Figure 9.4 Temporal and spatial scales in the atmosphere. Modified from [21]

the results described by Van der Hoven [19], although the specifics of the distribution depends on the land characteristics and other factors [20]. As a result of the interaction between the scales (temporal and spatial), wind patterns are very complex and chaotic, and its prediction is not trivial. Considering that the main variability factor in solar irradiation is cloudiness and that advection and cloud formation are affected by the wind velocity field, there is also a coupling with the temporal and spatial scales for solar energy.

9.2 Power systems

The balance in the power generation systems is maintained by continuous adjustment and control of the generation capacity in response to demand variations. Table 9.1 illustrates the time scales, corresponding applications, and end users in power systems operations and control. Very short-time predictions are useful for power generation systems and frequency control. Short-term predictions (30 min–6 h) are used for scheduling and commitment of power generation units, as well as in the energy market. Medium-term predictions (6 h–10 days) are relevant for the scheduling of maintenance events of the transmission grid and generators. Long-term predictions contribute to scheduled installations of generation units, both renewable and conventional, and to scheduling grid expansions.

Table 9.1 Users of wind and solar forecasts by time scale

Time scale	Application	Stakeholder
<30 min	Electricity market clearing	Traders
	Primary control	Plant operators
30 min–6 h	Trading on intraday energy market. Control of curtailment due to negative market price	Traders
	Influence of renewable production on market price	Speculators
	Balancing. Unit re-dispatch	Grid and independent system operators. Load dispatch centers
	Curtailment of power plants	
6–48 h	Trading on day-ahead energy market	Traders
	Participation in regulation market	
	Influence of renewables on market price	
	Unit dispatch. Load flow calculations. Day-ahead congestion forecast (DACF)	Grid and independent system operators. Load dispatch centers
	Day-ahead planning of maintenance	Plant operators
2–10 days	Trading on long-term markets	Traders
	DACF. Week-ahead planning	Grid and independent system operators. Load dispatch centers
	Medium-term planning of maintenance	Plant operators

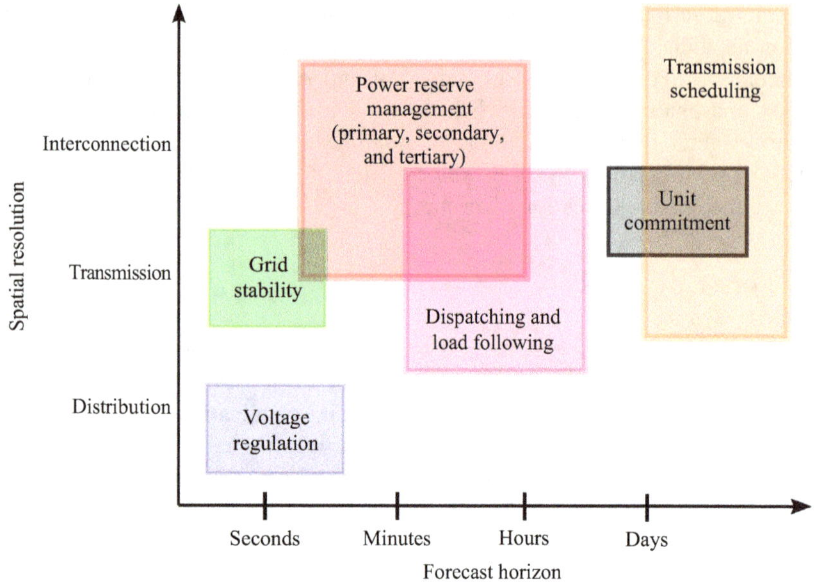

Figure 9.5 Temporal and spatial scales for power systems planning and operation. Modified from [22]

Figure 9.5 illustrates different power system operations and the corresponding temporal (from seconds to days) and spatial (from interconnection to distribution) scales. Only the main applications are listed for brevity, such as voltage regulation, power reserve managing, unit commitment, dispatching and load following, and measures used to guarantee grid stability.

Renewable and conventional power systems must be continuously adjusted to match generation with demand and maintain the frequency within the limits established by the current standard. Frequency response is one important indicator of grid stability. A recurrent concern regarding high-penetration scenarios for RE, at least in countries where renewables are still considered emergent, is the reduction in *system inertia* because of the displacement of conventional generators featuring rotating masses. This inertia contributes to arresting the decline of frequency (a phenomenon often referred to as frequency response) in a loss-of-generation event before the primary frequency regulation (the so-called governor control) kicks in.[||] Frequency response and regulation in the presence of high shares of renewables are nowadays well understood and managed (see Chapters 1, 2, 6, and 8), but may still represent a barrier to grid integration of wind and solar in systems

[||]Though inertia reduction in grids with a decreasing number of directly grid-coupled synchronous generators is a fact, decrying it fails to account for the fact that modern power converters present in wind turbines, solar PV plants, and battery storage systems are capable of providing frequency support in a way which is often much superior to the response provided by the combined action of rotor inertia and governor response.

with little experience with renewables or a general resistance to change.** Similarly, short-term variations of wind or solar power on time scales relevant to frequency regulation can be managed actively on the power plant level, without requiring major interventions from the grid operator. A novel scheme, based on short-term forecast of the (wind) power production, to effectively mitigate wind power fluctuations at almost negligible energy cost has recently been proposed by [23].

High penetration of RE cannot be conceived without appropriate transmission infrastructure. Considering that potential wind sites are often located in areas with weak grids, they might not be able to properly transmit the generated power to the load centers. Therefore, prediction plays a major role in the expansion of the transmission grid planning. Due to their variability, wind and solar farms must be continuously monitored by the control centers. This allows the operators to adjust the power generation or even temporarily disconnect the renewable portion from the grid to ensure stability. Hence, a robust prediction system would bring the possibility of better grid operability, and further benefits already experienced by many economic sectors which rely on prediction models.

9.3 Power forecasting

Wind and solar power forecasting vary in their applications and use may therefore require different types of approaches. For example, a power forecast may be applied to a single system or to a large number of those systems that might be aggregated in a restricted area or might be spread over a wide geographic area. Forecasts are mainly focused on power generated, but there is also an increasing interest in forecasting specific features, such as ramping events. The use of forecasting methods depends also on the available information, such as historic and real-time power generation, data from weather stations, and information from on-site measurements devices (such as anemometers, SOnic Detection And Ranging devices (SODARs), and Light Detection And Ranging devices (LiDARs)), among others. Figure 9.6 presents a classification of power forecast methods based on different aspects [24].

In what follows, emphasis will be laid on two types of approaches: forecasting based on output data and forecasting based on different models, being the latter the wider category.

9.3.1 Deterministic/probabilistic forecasts

With the increasing penetration of REs into the power systems, forecasts are required to provide valuable information to end users, as well as utility managers, at a desired time horizon. In fact, at the beginning of the incorporation of variable REs (solar and wind) into power systems (i.e. in systems with low RE-penetration), forecasting was restricted to providing a single value at the forecast horizon. This is known as point or deterministic forecast. Nowadays, forecasts include, in addition

**It is of course also a convenient pretext for blocking grid access by competitors of the incumbent utility or the grid operator, in case of incomplete unbundling of electricity systems.

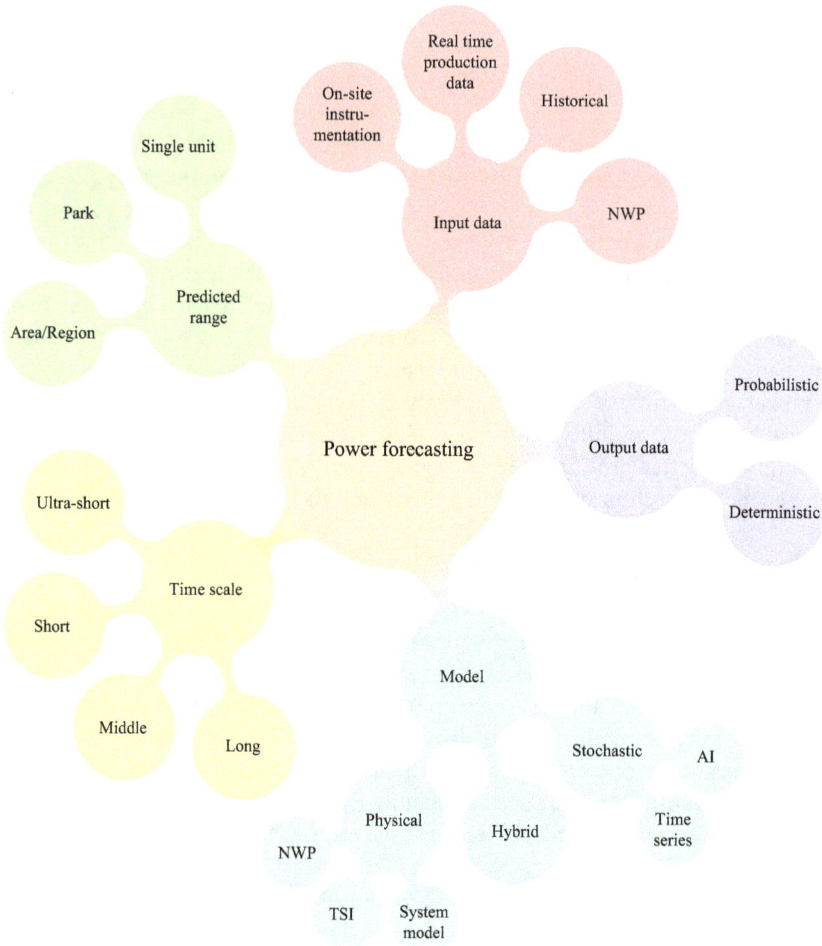

Figure 9.6 Classification of power forecasting methods

to the point forecast, the upper and lower bounds of possible forecasts, or the percentage of confidence of that forecast. This is known as a probabilistic forecast. This kind of forecast can be obtained with robust prediction methodologies such as ensembles. Probabilistic forecasts are found to be very useful for activities where uncertainties are inherent, such as risk managing, and in balancing generation and demand in the market. The use of this kind of predictions results in better allocation of power reserves and greater economical revenue when compared to point forecasts [25].

9.3.2 Stochastic approaches

Stochastic methods are techniques used to forecast nondeterministic processes, i.e. processes where the next state is determined either by a predictable cause or a random event. Stochastic forecasting methods by nature require efficient and robust techniques capable of correlating historical data and inferring the dependency of a desired variable between past events to predict future events. Although in countries like Mexico public data are difficult to obtain, the acquisition and control systems (SCADA) of wind or solar farms produce huge and constantly increasing volumes of historical data which can be used to construct more robust models. Among the most important stochastic methods are the statistical and artificial intelligence (AI)-based techniques described below.

9.3.2.1 Statistical methods

Statistical methods are those where the prediction is based on time-series analysis. Among the most popular methods are autoregressive (AR) time series and some of their variants such as autoregressive moving average (ARMA) or autoregressive integrated moving average (ARIMA) [26].

AR models are commonly used for time-series forecasting since they are capable of capturing persistence in a time series. In simple terms, an $AR(p)$ model relates p past observations to the current value x_t as

$$x_t = \mu + \sum_{i=1}^{p} \varphi_i x_{t-i} + \varepsilon_t \tag{9.1}$$

where μ is the mean value, φ_i is a coefficient reflecting the influence of each past observation x_{t-i} on the current value, and ε_t is a perturbation, drawn from a given distribution of random numbers.

ARMA and ARIMA models describe a univariate time series as the relation between actual observations x_t and their p AR components, and include a moving average (MA) process. Generally speaking, an $ARMA(p, q)$ model is defined as

$$x_t = \mu + \sum_{i=1}^{p} \varphi_i x_{t-i} + \sum_{j=1}^{q} \theta_j \varepsilon_{t-j} + \varepsilon_t \tag{9.2}$$

where φ_i and θ_i are weight coefficients which reflect the influence of past p observations and q stochastic perturbations on the current value.

ARIMA models are an ARMA extension to deal with nonstationarity in data by making data ergodic through d differentiation steps; then, an ARMA model is estimated and used. Such a model is denoted as $ARIMA(p, d, q)$.

Figure 9.7 shows a flow diagram for a typical forecast system. In the diagram, two stages can be clearly observed: model building and forecasting. In the first stage of AR models, parameters can be estimated using past data to iteratively refine the resulting model. The second stage is related to the model performance, in which the prediction is compared to the observations of the phenomena to compute an error. If the model is stable there is nothing else to improve, otherwise the model will need to be re-specified and estimated. Given that, essentially, AR models extend the behavior of time series into the future, an impression remains that the resulting prediction resembles the real-time series over time. A typical example of

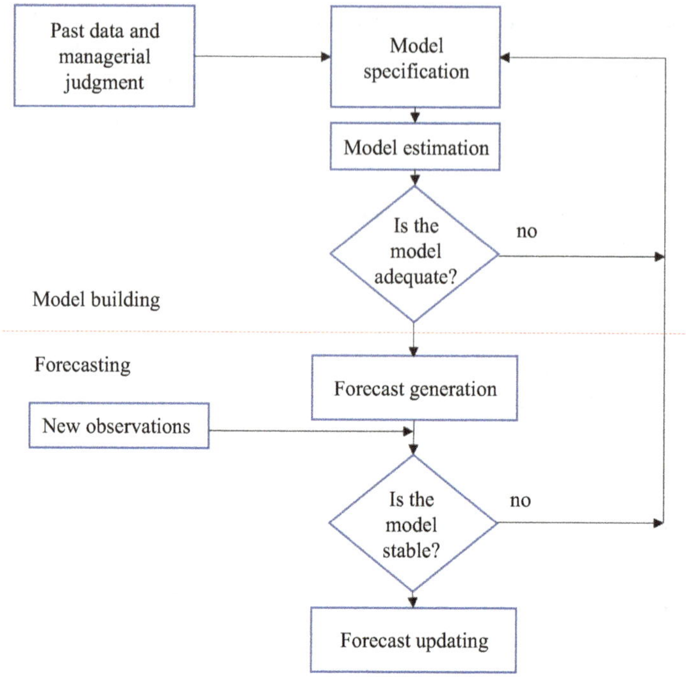

Figure 9.7 Forecasting system: model building and forecasting stages

prediction using AR models, in addition to wind forecasting, is the use of random walk theory for stock market actions.

There are other statistical techniques that are not precisely AR, such as Markov chains, artificial neural networks (ANNs), and support vector machines (SVMs, also known as support machines networks). Although Markov chains might implement a small predictive model by mean of reinforcement learning methods, they are considered more useful for planning problems. Given that SVMs and ANNs are supervised learning models, they are discussed in Section 9.3.2.2.

An application of statistical methods in short-term wind speed forecasting can be found in [27]. A set of advantages and disadvantages of statistical methods are illustrated in Table 9.2.

9.3.2.2 Artificial intelligence

AI methods build models using machine learning algorithms, expert knowledge, or a mixture of both. In the case of wind power prediction, these methods have demonstrated good results in short-term horizons (<2–4 h). AI algorithms are capable of predicting the production of a set of wind power generators or the entire wind farm production. They are also suitable for both modular approaches (weather forecast + a generation model) and direct forecasts from power generation information. For some cases, where prior knowledge is available, AI methods might

Table 9.2 *Advantages and disadvantages of different forecasting approaches*

Method	Advantage	Disadvantage
Statistical methods (AR, ARMA, ARIMA, etc.)	Easily programmable Simple structure Correction capacity of local trend in the data Provides confidence intervals in the predictions	Requires large amount of historical data Nonlinear effects are difficult to model Punctual prediction (one location) Better results on short-term time horizons (1–4 h) Nonlinear patterns due to the chaotic nature of wind cannot be correctly captured
DBN	Ability to manage noise in data Techniques that allow making inferences about processes with uncertainty They are knowledge-based techniques and their models can be approximated with data Delivers probability distributions and not point values	Depends on the user experience or the quality of the data Discrete nature
ANN	Obtain knowledge from the data used for training It is not necessary to specify a mathematical model Little sensitive to errors in the data Adapted to real-time measurements	Requires a training procedure Requires a large amount of data
SVM	Better performance than the ANN approach	They depend on an adequate setting of parameters Complex optimization Longer training times
FLS	Suitable for difficult modeling systems	Prolonged processing, depending on the number of rules
NWP models	Applicable to long-term time horizons Better results on time horizons larger than 4 h Provides information of a wide geographical area Further meteorological variables that might be relevant in the operation of electrical power systems can be obtained	They do not capture short-scale phenomena Not adequate for short-term time horizons Depending on the configuration, it requires high computational capacity Computing time can be very long Depends on initial and boundary conditions, and employed methods

offer better results than purely statistical methods. Some of the most popular techniques are ANNs, SVMs, dynamic Bayesian networks (DBNs), fuzzy logic systems (FLSs), and others.

The ANN model [28,29] is inspired by neurons and their connections in the brain. A neural network is a representation, among others, used in machine learning. Some typical applications of ANNs are pattern recognition, function approximation, and classification. Given that forecasts can be viewed as an underlying pattern (spatial–temporal) recognition problem, ANNs are useful to predict solar and wind resources. The fact that there are certain spatial–temporal patterns that an ANN can deal with better than using other methods, there is no guarantee that one point forecast done by these techniques (for a particular time-horizon) offers better results than using simple persistence methods. This is because the patterns recognized correspond to structures observed as averages of long intervals or large spatial scales.

An ANN consists of a network whose nodes are arranged in layers. The arcs resemble the interconnection between neurons (the nodes) in the brain. The left graph of Figure 9.8 exemplifies this method. The first layer is the input layer, with the number of nodes being equal to the number of input variables. The output layer may produce a number of output variables of the network; in the present case, as illustrated by Figure 9.8 there is only one output signal, typically the predicted power generation value. An ANN may contain none or several hidden layers. Figure 9.8 right represents the jth neuron of the p layer. This neuron receives n inputs from layer $p-1$ and produces one output that will be connected to neurons in layer $p+1$. The output level is given by the sum of all the inputs multiplied by a weight w_{ij}^p, and modified by a sigmoidal function.[††] These weights are defined

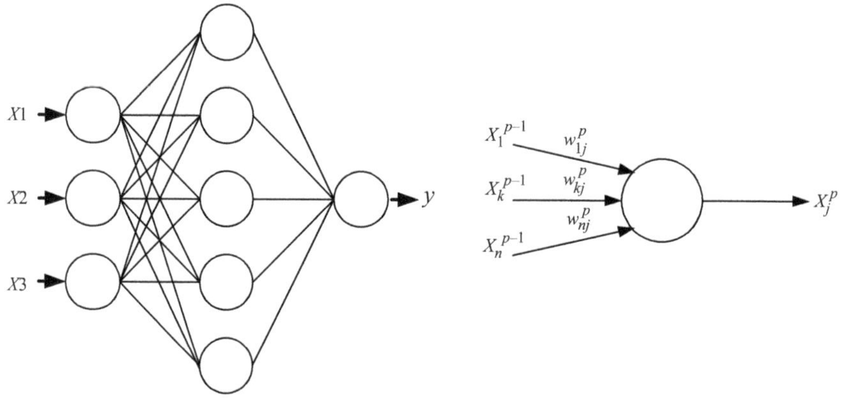

Figure 9.8 A basic neural network. Left: a typical network. Right: a perceptron or basic neuron

[††]A sigmoid is an S-shaped function defined by $1/(1+e^{-x})$.

through a learning process during a training phase. To train the network, several examples of a learned concept are presented and several examples of what is not the concept are also presented. Consequently, the weights are defined so that future candidate input vectors lead to a similar conclusion. This approach has been used in a number of forecasting studies, such as in [30], where a short-term wind speed forecasting method using ANNs has been presented. In [31], a new method for local forecasting of daily global solar irradiation using ANNs was introduced. In that work, the ANN-based model was designed to predict the local Global Horizontal Irradiation (GHI) based on daily weather forecasts provided by the US National Oceanic and Atmospheric Administration (NOAA) for four neighboring locations. In [32], the ANN methodology was applied to predict the power generation of a small-scale (750 W) solar system. The machine learning process (input propagation and error distribution) used training data from the PV system for different time horizons. The ANN demonstrated to be able to predict the generated power with good accuracy for forecast horizons from 30 to 300 min.

In machine learning, SVMs are supervised learning models with associated algorithms that analyze data used for classification and regression analysis. Given a set of training examples, each marked as belonging to one or the other of two categories, an SVM training algorithm builds a model that assigns (predicts) new examples to one category or the other.

SVMs are a relatively new pattern classification scheme introduced by Vladimir N. Vapnik [33]. An SVM classifies an input vector into one of two classes, with a decision boundary developed based on the concept of structural risk minimization (of classification error) using statistical learning theory. The SVM learning algorithm directly seeks an optimal separating hyperplane defined as a maximal margin classifier with respect to training data. For nonlinearly separable data, the SVM uses kernel methods to transform the original input space, where the data are nonlinearly separable, into a higher dimensional feature space where an optimal linear separating hyperplane is constructed. Successful applications of SVM to wind and solar forecasting have been demonstrated by a number of authors, e.g. in [34], where the authors present a systematic study on fine-tuning of the least-squares SVM model parameters for one-step ahead wind speed forecasting.

Most AI-based forecasting techniques appear to be based on ARIMA, ANN, and SVM approaches [35], but a number of challenges remain. A discussion of the results and challenges from the use of such techniques suggest the need for considering conditional dependencies among variables, the uncertainty implicit in the natural resource, the use of expert knowledge as an aid to formal methods, the need to express forecasting in terms of probability distributions, and the use of attribute selection methods and the regularities in the process.

Bayesian networks (BNs) [36] are directed acyclic graphs developed to represent the probabilistic relation between variables in a process. They represent parametric and structural expert knowledge and were created to deal with incomplete or missing information (uncertainty). BNs express their results in the form of probability distributions and are capable of excluding irrelevant attributes. BNs deal with the state of the world in a given instant so that they have to be extended for dynamic applications. DBNs [37] represent temporal relations by replicating each variable for every time instant in the temporal range of interest, including dependency relations within and between temporal intervals. In general, DBNs, as most stochastic approaches, follow two basic assumptions:

- Markovian process: Each variable only depends on variables from the previous and current time steps.
- Stationary process: The structure and parameters of the model remain the same for all time steps.

To ensure that the conditional dependencies are constant (at least in structure), the structure of a dynamic network for the times t and $t+1$ (transition model) is learned first. This net segment defines $P(X_t|X_{t-1})$ for all t in the process. Subsequently, the resulting structure is copied until time n; this is referred to as the network being *unrolled*. The last layer no longer has causal relationships with the next time step. Once the entire structure is formed, the parameter learning process can be carried out. For example, if we define an hourly time slice and we need to model 5 h in advance, the DBN is unrolled to complete six slices. Figure 9.9 shows the resulting six-slice dynamic model. In DBN, model overfitting depends more on the portion of training data so that it is used to splitting the universe of data in equal parts for both training and testing. Although generally the variables of interest are wind speed and direction, the network already trained allows the consultation of the forecast of any variable, every hour and in this case up to 5 h ahead.

In an early example of the DBN approach [38], a 1-h-ahead wind speed prediction model using a Bayesian approach was constructed. To test the method, wind speed data of 2 years from a weather station were modeled as an AR process. The methodology was compared to the persistence method and the results obtained

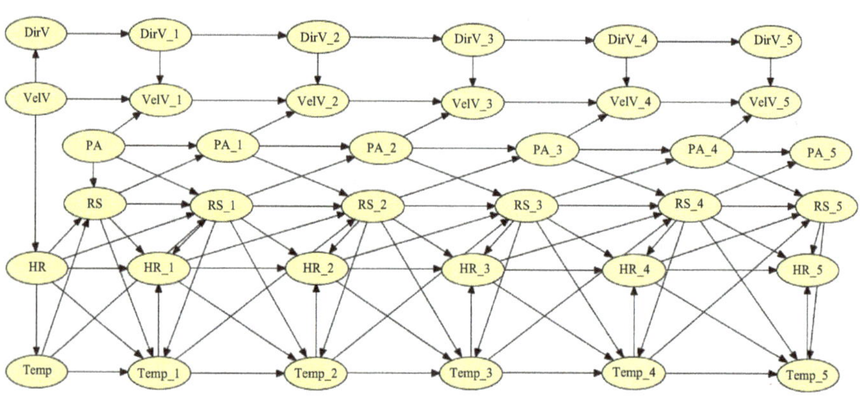

Figure 9.9 A six-slice DBN for a multivariable 5-h-ahead forecast. The variables considered in this model are wind direction and speed (DirV, VelV), relative humidity (HR), ambient temperature (Temp), atmospheric pressure (PA), and solar radiation (RS). The subindex in each variable represents the time slice

indicate that Bayesian inferencing can be a useful tool in wind speed/power prediction, particularly due to the flexibility inherent in the methodology. In [39], the authors use Bayesian inference for solar radiation modeling. For that particular application, it was shown that the use of Bayesian methods instead of classical statistical techniques led to a less biased model. Ibarguengoytia *et al.* [40,41] reported on the application of a DBN to model for wind forecasting. In contrast to other approaches, the proposed model considers all the relevant variables in the domain and produces a probability distribution for the predictions. The method was evaluated experimentally with real data from a wind farm for a time horizon of 5 h, showing superior performance to traditional time-series prediction techniques.

Like DBNs, FLSs are computational intelligence techniques considered as alternatives for uncertainty management. For the same reason, they are suitable for contending with stochastic processes. FLSs are not considered machine learning techniques since they do not build knowledge models from training data but rather through human experts. The natural application of FLS is process control; however, due to their characteristics, they have been used as an auxiliary approach for forecasting applications (e.g. irradiation modeling). Unlike SVMs and ANNs, FLSs are considered white boxes where all their content is exposed to both the designer and its users.

FLSs allow working with information with a high degree of imprecision. The concept was conceived by Lofti A. Zaded in 1965. Fuzzy logic allows partial membership in sets, which in contrast to classic sets are called fuzzy sets [42]. In [43], fuzzy sets theory is applied to relate global solar irradiation to both the daily average and the daily amplitude of air temperature. A set of advantages and disadvantages of AI methods is shown in Table 9.2.

9.3.3 Physical approaches

As stated in Section 9.3.2, stochastic methods are based on statistical relations between available data (measurements, simulated data from physical models, etc.) with power output. In contrast, the physical approach refers to the forecasting of power generation using the fundamental equations and parametrizations describing the relevant physical processes in the atmosphere and the bounding surfaces (land and water bodies).

9.3.3.1 Numerical weather prediction

Numerical weather prediction (NWP) uses numerical modeling techniques to understand the mechanisms that are involved in the complex physical phenomena occurring in the atmosphere. For example, the physics of clouds, including their dynamics and microphysics, needs to be modeled to understand cloud formation and evolution in time. Although such models have been proposed as early as a century ago [44], practical uses have become possible only thanks to the increase in pervasive computing power, numerical modeling, schemes for data assimilation, as well as methods to properly address uncertainties.

NWP models are based on the representation of the atmosphere's behavior through the solution of the set of equations that govern it. For most of the NWP models, those equations include cloud motion (advection) and formation (microphysics), radiation, planetary boundary layer, and the interaction between surface/water with atmosphere. For the numerical treatment of the atmosphere–land–ocean system, it is necessary to discretize the underlying equations. This is achieved by dividing the atmosphere, land, and ocean into cells (as illustrated in Figure 9.10). In order to solve the NWP equations, static input information is required, including the surface elevation (orography), waterbodies (lakes, rivers, gulfs, etc.), and the land-use type; such information can generally be obtained from worldwide or national databases whose resolution varies from 10 m to 1 km, depending on the data source. Given the growing impact of humanity on the land, land-use classifications have to be used with care and are subject to frequent updates. Since orography is often complex, the use of the Cartesian z coordinate is not practical. Instead, terrain-following coordinates are used, such as the so-called σ-coordinate [45], as illustrated in the inset of Figure 9.10. This is possible because the σ-coordinate is defined in terms of the hydrostatic pressure. It is important to mention that when this coordinate is used then the horizontal components of velocity, as well as other variables, are defined along the σ planes, in contrast with spherical or Cartesian coordinate systems. For wind energy applications, these coordinates have to be converted to distances above surface at the desired height.

Phenomena in atmosphere, land, and ocean range from micrometers or even less (nucleation, photosynthesis, radiation interaction, water permeation in land, rate of rain/snow, heat/mass exchange of vegetation, atmospheric emissions, etc.) to tens of kilometers (hurricanes, Rossby waves, thunderstorms, etc.). Solution of all those phenomena (some of them coupled) is technically impossible since their requirement

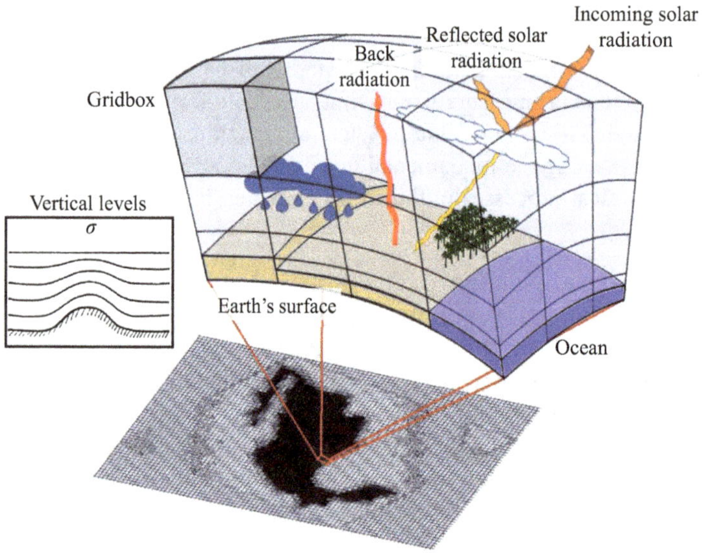

Figure 9.10 Diagram of NWP modeling

in mesh refinement spans several orders of magnitude ($10^{-6}-10^{-4}$m). In order to cope with this challenge, subgrid phenomena must be parametrized. The parametrization process in NWP implies the representation of complex phenomena occurring on scales smaller than the characteristic cell length (such as nucleation or radiation interaction) as a simplified process than can represented by an empirical equation which is tuned to observational results. Given the number of subgrid-scale processes, this implies the existence of several parametrizations for different types of physics. Since these parametrizations are an approximate representation of the physical phenomena, they are a source of uncertainty in the numerical model. Additionally, since different physical processes are interrelated (for example, solar radiation affects the evaporation rate, which in turn is linked to cloud formation, and cloud formation is tied to the incoming solar radiation at the Earth's surface), NWP is highly sensitive to selected parametrizations, as well as to the initial conditions. In order to deal with these later aspects, it is a common practice to produce ensembles (described next in this section).

When working with NWP models, it is important to keep in mind that the output depends on the parametrizations used. This means that, in addition to the common variables (as velocity, pressure, temperature, among others), it is possible to obtain additional variables selecting a specific parametrization (for heat fluxes, hail, rain, etc.). Different parametrizations may be suitable for different climates conditions, namely, a parametrization might be valid for mid-latitude location but not for high-latitude places or vice versa. The misuse of a parametrization may lead to significant error in the prediction.

In accordance to the spatial scales of a given application of NWP models, NWP can be classified as follows:

- *Global models*: These models are used at a planetary scale as well as in climate change studies. The models are also referred to as global circulation models (GCMs). An example of these models is the Global Forecast System (GFS) produced by the National Center for Environmental Prediction (NCEP) and run by the US National Weather Service [46]. The GFS is run four times a day and produces forecasts for up to 2 weeks in advance. The typical spatial resolution of this model is about 13 km with a temporal resolution of 1–3 h, although reanalyses of these data can be obtained at different resolutions.
- *Regional or mesoscale models*: These models are used to investigate mesoscale atmospheric phenomena, as hurricanes, and thus are employed in forecasting in a specified region. Spatial dimensions range from several kilometers up to several hundred kilometers. Although there are different implementations worldwide, the most used models are the Weather Research and Forecasting (WRF) [47] and the European Centre for Medium-Range Weather Forecasts (ECMWF) [48] models. Some of these models have been modified and improved for solar energy applications, such as the WRF-Solar [49]. Mesoscale models use the output of GCMs as boundary conditions and for initializations; since their output can be considered a refined version of the input provided by the GCM, this process is often referred to as *downscaling*.

- *Microscale models*: These models where developed to simulate superficial phenomena and turbulence, relevant to wind energy applications given that wind turbines are located in the first few hundred meters of the atmospheric boundary layer (ABL). This has led to the implementation of computational fluid dynamics (CFD) methods, which are capable of modeling complex flows at complex sites. One example of CFD tools that has been used for resource studies is OpenFOAM [50].
- *Downscaling to the microlevel*: It is important to mention that the temporal integration of NWP models is done with a time step as small as a few seconds but output variables are written every 30 min or more. Additionally, given the resolution of global and regional models, they are not able to correctly capture surface effects or reproduce near-surface phenomena, which are key aspects for wind energy applications. In order to model wind flow or solar radiation on smaller length scales, yet another downscaling process has to be employed. Such approaches are collectively known as *meso-micro coupling* schemes. Methodologies based on this general idea have been demonstrated for both wind [51–53] and solar [54] resource applications. In recent years also, buoyancy effects, leading to different atmospheric stability conditions, have been found to be relevant for the flow conditions in the ABL and therefore for wind energy applications [55,56].

9.3.3.2 Ensembles

Any given NWP model is uncapable of producing continuous, consistent, and precise forecasts. Given the chaotic nature of atmosphere, a single forecast is sensitive to initial conditions, boundary conditions, and models. These three sources of uncertainty are responsible for forecast discrepancies. The longer the time horizon, the larger the uncertainty associated with the forecast. The rate at which forecasting error grows is "flow dependent", leading to variations in modeling accuracy from day to day. An *ensemble* forecast consists of a set of different single forecasts with the same initial time but different configurations (models, data inputs, parametrizations, etc.). These single forecasts, called ensemble members, are then postprocessed to obtain a mean value and an associated uncertainty. Ensemble members are selected in such a way that the set of ensembles be an unbiased and good representation of the forecast distribution. Ensemble forecasts provide an explicit, consistent, and objective representation of the possible future scenarios based on currently available observation and the inherent uncertainties of the model itself. Figure 9.11 illustrates an ensemble forecast. This figure includes the uncertainties associated with the initial conditions as well as their evolution during time. At the target forecasting time horizon, the spread of the ensemble members represents the uncertainty in the forecast value. It has been found that a combination of NWP models results in better forecasting accuracy [57]. For a single wind farm in Europe, Lange *et al.* [57] found the root-mean-square error (RMSE) to vary from 7% to 13.5% through the year, while the mean absolute error (MAE) varied from 4% to 8%, presenting the lowest error during spring. Additionally, the larger the simulated region, the lower the relative forecasting error.

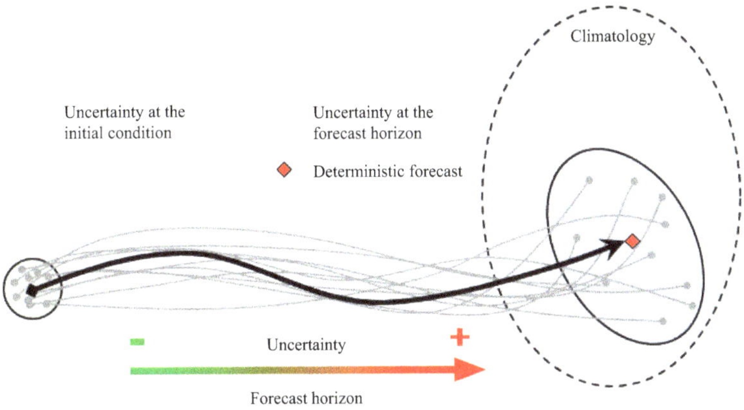

Figure 9.11 Illustration of an ensemble forecast and the corresponding deterministic forecast

Thanks to the advances in computing power, numerical modeling, and schemes for data assimilation, as well as in methods for treating uncertainties, ensembles are nowadays widely used and routinely run in order to provide ensemble forecasts including stochastic [58] and/or physical approaches. It is important to highlight that ensemble members can be produced by one model with a different parametrization or initial conditions, or by a completely different model. This latter approach has been the most widely reported one in literature. Another approach to construct ensembles is by the *comparative method*, in which the data set is divided into several smaller sub data sets (as in an analog ensemble [25,59]) each of them forecasted individually, and a final forecast obtained by combination of all of them. As has been seen, there are a wide variety of forecasting methodologies. In Table 9.3, different forecasting models are compared with respect to five criteria: noise tolerance, ease of implementation, data requirements, forecast horizon and spatial resolution. Values were assigned based on reported literature. Conversely, in the *competitive method* ensemble members are obtained by using the full data set. The main advantage of the comparative method is their lower computational cost, which is why such methods are normally used for very short-term and short-term forecasting [60]. The degree of success of the models

Table 9.3 Comparison of different forecasting models

Method/ criteria	**Noise tolerance**	**Ease of implementation**	**Data requirement**	**Forecast horizon (Table 9.1)**	**Spatial resolution**
Statistical	Low	High	Medium	Flexible	Point
DBN	High	Medium	High	Hours	Flexible
ANN	Medium	High	High	Hours	Flexible
SVM	Medium	Medium	High	Hours	Flexible
FLS	High	Low	Low	Flexible	Flexible
NWP	Medium	Low	High	>30 min	Wide area

depends on the site characteristics, the forecasting time horizon and the implemented forecasting models and methodologies. A set of advantages and disadvantages of physical methods is shown in Table 9.2.

9.3.3.3 On-site instrumentation

On-site instrumentation has been widely used in order to develop new robust predictive tools that help to understand the underlying physics of solar PV and wind parks and reduce energy production costs. Instrumentation located *in situ* allows for the validation and verification of existing and new models. The instruments need to accurately measure the quantities of interest for RE applications, such as wind speed and direction, turbulence intensity, and solar irradiance. For power production forecasting, on-site measurements can be classified into two categories: (1) methods that measure atmospheric variables (wind, radiation, temperature, etc.) within the RE plant and (2) those that measure the power output of wind turbines or solar panels. Measurements help in identifying physical phenomena, but also (and more relevant for our discussion) in testing models for these phenomena.

On-site measurements of the atmosphere variables can be obtained by different methods and instruments, such as cup anemometers, ultrasound anemometers, LiDAR, Spinner LiDAR, SODAR, scanning radar, anemometric towers, pressure probes, sky cameras, particle image velocimetry (PIV), constant temperature anemometer, radiosondes, ceilometer, and microwave radiometers. The measured variables can be used in physical, stochastic, or hybrid methods. For example, for wind energy applications the data obtained by SODAR and/or LiDAR equipment can be used directly incorporated in the power prediction process of stochastic methodologies, or can be assimilated by physical methodologies as NWP or CFD. The use of real measurement in simulations contributes to the improvement of numerical results in the nearby of the measurement site.

Total sky imagery

In order to maintain the stability of the power system in regions with high degrees of RE penetration, power system operators require real-time information. Large-scale and distributed generation of solar PV systems located in relatively small areas is subject to a coherent response to variations in solar radiation on short-time scales (on the order of a few minutes or even less), e.g. due to the passage of a cloud simultaneously affecting large arrays of solar panels. Forecasts based on on-site instrumentation have a great potential for addressing this type of problem by providing reliable predictions of solar ramps.

The variation of solar radiation is both deterministic and stochastic. *Deterministic variation* refers to changes in the radiation due to the rotation (diurnal) and translation (year) of the Earth. This component of radiation can be calculated in a precise way using semi-analytic models with varying complexity [61]. Shadows cast by neighboring obstacles (tree, buildings, etc.) can be easily determined if the geometrical properties are known. *Stochastic variations* are referred to as changes in radiation that are not associated with deterministic variations, namely, dust, moving shadows such as the ones cast by animals, suspended particles, fogs, contrails, and clouds. Among these factors, clouds are the most relevant. Clouds are one of the most important regulators of the Earth's radiation budget, and still not fully understood. Their effect on solar radiation reaching the surface depends on their shape, size, and composition.

Solar radiation has two components: beam and diffuse. Beam radiation is one received from the Sun without altering its direction by interaction with objects. Diffuse radiation refers to solar photons affected by the components of the Earth's atmosphere, experiencing changes in their direction due to optical phenomena such as refraction, reflection, and dispersion. The total solar radiation received by an object on the Earth's surface is given by the sum of beam and diffuse radiation and is called global radiation.

Cloud composition and altitude, as well as its properties, varies from one type to another [62]. In general, clouds can be classified by its altitude into three levels: low (Cumulus, Stratocumulus, Stratus, and Cumulonimbus), mid (Altocumulus, Altostratus, and Nimbostratus), and high (Cirrus, Cirrocumulus, and Cirrostratus). Each level is defined by a range of vertical heights at which each type of cloud typically appears. It is important to notice that the thickness of the atmosphere varies with latitude, being larger at the equator and smaller at the poles. Table 9.4 presents the cloud classification for different polar regions.

As seen, clouds occur in a wide variety of time and space scales, at different levels in the atmosphere and with a varying ice/water content; therefore, cloud prediction is a big challenge for solar power production modeling. To cope with this challenge, three approaches have been reported: NWP (described in Section 9.3.3.1), satellite imagery, and ground-based or sky imagery. NWP models produce a three-dimensional (3D) forecast for each grid point at which temperature, water vapor, cloud water, and cloud ice are reported. Based on these variables, it is possible to infer the existence of clouds. The main advantage of NWP is that they can simulate cloud formation, development, dissipation, and movement. However, (1) the grid size (about 5 km or more) is typically too large to capture small-scale features and (2) cloud patterns are not accurate enough due to the lack of dense measurements networks. As a result, cloud patterns predicted from NWP are currently not reliable enough. In satellite-based imagery, cloud features are extracted from visible or infrared images obtained by sensors located on satellites. Its main

Table 9.4 Cloud classification and main characteristics in accordance to the National Weather Service for different polar regions

Level	Tropical	Temperate	Polar	Characteristics
High	6–18 km	5–13 km	3–8 km	Typically thin and white, but can change its appearance at sunset and sunrise
Middle	2–8 km	2–7 km	2–4 km	Composed primarily of water droplets, although they can be composed of ice crystals when temperature is sufficiently low
Low	<2 km	<2 km	<2 km	Composed of water droplets

advantages are the large dimensions of cloud patterns that can be detected. The main drawbacks are (1) their spatial resolution, ranging from 1–4 km, which is not sufficient to capture small-scale structures or small clouds and (2) the temporal frequency at which images are taken is only about 30 min. Sky imagery (described in more detail later in this section) is in principle capable of detecting the extent, structure, and motion of clouds in the vicinity of the place where the instrumentations have been installed. Sky imagers can take pictures at repetition rates of some seconds (or even less), depending on the specifications and configuration of the camera, which enables their use for the development of very short-term forecasts (of typically a few minutes), which coincides with the temporal frame in which relevant solar radiation fluctuations occur. The main drawbacks of this technique are the following: (1) it does not account for the cloud physics (formation and dissipation), (2) the spatial scale is relatively low (about 10–20 km, depending on the field of view of the camera used), (3) different cloud types (at different altitudes) are not easily distinguished, and (4) there can be saturation of the image due to sunlight. These characteristics limit their applicability to forecast of up to around 1 h, although the specific horizon depends on the cloud type and velocity. For a given velocity, the higher the cloud, the lower the forecast horizon. As can be observed, these three methods complement each other in their temporal horizons.

Total sky imagery (TSI) has been used to obtain cloud cover on solar generation sites. This technology has been found to reduce the forecast error by 50%–60% on a 30-s horizon compared to persistence forecasting [63]. It has been shown that TSI is also useful in the forecasting of solar radiation at a time horizon of 15 min. The setup of a TSI system consists basically of a camera (normally with a high field of view), a battery, a processing unit, and a storage unit located in an isolated container. The storage unit can be omitted if the images are directly transferred to an external node by any communication port. The camera has optical access to the sky by a transparent dome. For robustness, the system is equipped with a refrigeration/heating system. Different configurations have been reported in literature [64–66] with varying components, specifications, and prices. The camera acquires images at user-defined intervals. The images are stored internally or externally (exported and/or transmitted for processing on an external computer), and subsequently analyzed.

The main aim of TSI is to determine the cloud motion in order to forecast solar irradiance, and then power production. The methodology to estimate the cloud motion can be roughly divided into the following three processes:

1. *Cloud detection*: In this step, the taken image is processed to detect which pixel corresponds to a cloud image and which not. For this purpose, the original image (Figure 9.12(a)), which is normally taken in color space (as RGB), must be converted into another space that offers a better contrast between pixels (Figure 9.12(b)). If a proper threshold value is selected, the cloud region can be identified (Figure 9.12(c)). It should be noted that, although in the original image two different types of cloud can be clearly identified, this procedure cannot differentiate them. In some cases, the Sun is occluded by the cloud leading to a saturation in the circumsolar region of the image (Figure 9.12(b)),

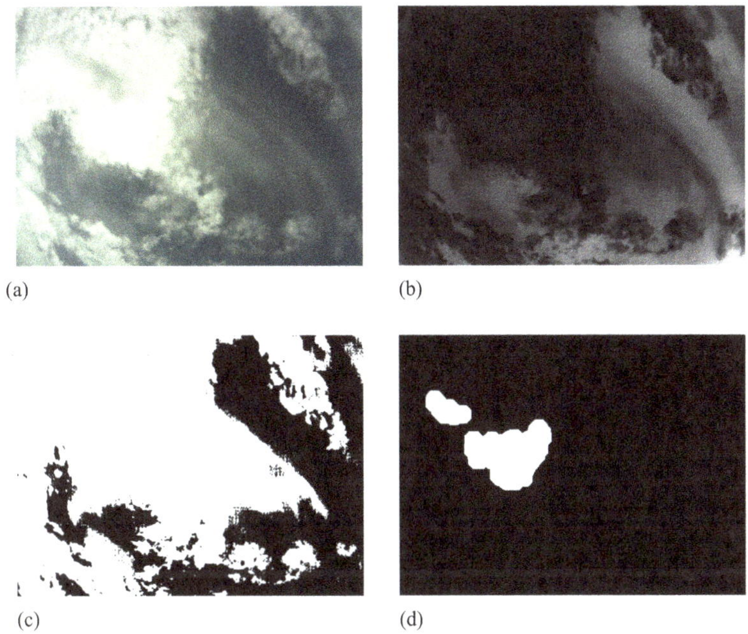

Figure 9.12 *Generic procedure for cloud detection: (a) color image, (b) color space conversion, (c) cloud identification, and (d) Sun occlusion region*

which can be used to divide the image for its use with a more suitable threshold value for cloud identification.

2. *Image features extraction*: Feature extraction is a process in which a set of image characteristics (features) that better represent relevant information is obtained. Suitable image features can be obtained from color-based (in their different color spaces) [67] and/or gray-scale images. This is important because atmosphere color, including clouds, depends on it composition [66]. Thus, image features can be obtained from suitable color spaces expressed in terms of the color layers (RGB), color differences (R-B, etc.), or color ratios (R/B, etc.) which can be used to identify clouds. One of the helpful features in cloud motion estimation are *corners* within the image, which can be obtained by different methods, being the Harris and Shi-Tomasi methods the most widely used. *Corners* are relevant because in order to track the movement of a cloud, there is a need to select reference points in the image.

If we consider that each type of cloud possesses its own characteristics, they can be used to not only identify if a pixel pertains or not to a cloud, but also to classify them [68]. For this purpose, different methodologies have been proposed [62,69]. These classifications may use statistical methodologies or AI, and might use one or several of the features extracted from images. Textural features should be obtained [66,70] as well, in addition to some other features that might be relevant for further analysis. The cloud type provides additional information such as composition, altitude range, and other characteristics that might improve solar irradiance forecasting, and thus power production. In this sense, feature extraction can be used for image segmentation, which is the process of image partition into multiple areas with similar features. Segmentation is a powerful tool that can help image processing. The main goal is to detect and label objects (cloud of the same type) and borders within the image. This process can be implemented in color or grayscale images [67].

3. *Estimation of the cloud trajectory*: Motion estimation has been widely used in recent years [71–74]. In order to estimate the motion of the cloud, it is necessary to assume that pixels within consecutive images do not change, but only move from one position to another. Motion forecasting is based on techniques commonly used in experimental fluid dynamics such as particle image velocimetry (PIV) or particle tracking velocimetry (PTV). PIV is based on the determination of the displacement of particles seeded in the flow from the correlation of two images taken at different times with a known time interval. For cloud tracking, *corners* play the role of seed particles; the PIV technique then provides a velocity vector field. Knowing the speed and direction of the cloud points (corners), it is possible to forecast the position at a given future time. In contrast, PTV "follows" the trajectory of seed particles (corners) in a consecutive series of images. If we know the position in the images and the time at which those images were taken, the velocity of each particle can be computed. Since the position and velocity are defined in term of pixels, there is the need to convert them to desired unit system. This is only possible if a reference dimension is known. This reference dimension can be determined by a stereographic method [75], which requires the use of at least two sky imagers located about 1–2 km apart. Using this method, it is possible to determine the cloud base height (CBH), as well as some information about the 3D shape of the cloud. Additional methods to determine the CBH are available such as radiosondes and ceilometers, but they are relatively expensive and their temporal and spatial resolution is limited. It is also possible to determine the cloud top height by thermal infrared satellite imagery, with the limitations that have been addressed above. In addition to the described methodology based on image processing, it is possible to determine the cloud velocity by means of sensors arrays [76,77], such as the cloud shadow speed sensor [78].

The variables obtained by TSI are cloudiness related, namely, cloud type and cloudiness index, among others, which are used in two different ways for power forecasting:

- The data are processed to obtain the value of the solar radiation which is fed into a mathematical model of the PV system. Mathematical models vary in complexity, depending on the final purpose. For forecasting applications, these models tend to be simple, while complex models are used for power electronics development or power system analysis.
- The data are assimilated by other forecasting models, which may include NWP or AI models. Data normally include, but are not limited to, those image features that better correlate to power productions. The assimilation of on-site measurements into stochastic models may improve their performance.

This methodology can be significantly improved if some phenomena or effects are considered in the forecasting process, such as the following:

- Solar forecasting by sky imagers, as discussed earlier, requires the geolocation of clouds in order to estimate the characteristics, motion, and dynamics. TSI for power forecasting can be either enhanced by the incorporation of mathematical models to determine solar irradiance under clear sky condition (clear sky models) and solar geometry, or used for the determination of atmosphere/cloud properties [79], which are relevant for several application.

  Different clear sky models have been reported with different degrees of complexity but similar structure, but one of the most widely used is the Bird–Riordan model [80] in which the spectral solar irradiance is defined in terms of the extraterrestrial irradiance, a correction factor for the Earth–Sun distance (variation of solar irradiance along the year might be significant depending on the location), and a set of transmittance coefficients for the components of the atmosphere such as ozone, aerosol, and water vapor, among others. TSI can be used to estimate those transmittance coefficients, although their spectral dependence is very limited. On the other hand, including these models into power forecasting enables the possibility to map clouds to a plane that in addition to the solar geometry allows to estimate shadowing effects of clouds on the surface [81]. In the ideal case, the obtained shadow maps will provide the percentage of PV plant that is shaded. If an electrical model for the plant is available, power production can be forecasted. It must be noted that model of PV plants with partial shading are a hot topic in actual research due to the challenges it represents.
- Incorporating stochastic methods, as machine learning, at different stages of the power forecasting process (which constitutes a hybrid model), aiming to construct smart real-time and robust forecast methodologies. Different stochastic methods have been reported in literature [82,83], but there is a huge effort in using them for cloud classification, which represents a key issue for the TSI method.
- For locations where clouds exhibit complex dynamics, the forecast accuracy is low. In order to improve the accuracy, methods for high-resolution cloud assimilation into NWP models have been developed and implemented [84]. Although satellite imagery was used at the work cited to incorporate clouds as initial conditions, there is a possibility of assimilating data from sky imagers into high-resolution NWP models which might have a spatial resolution of the same order magnitude as the resolution in TSI systems.
- Although solar radiation is normally the variable used in solar forecasting tasks, the use of the clearness index has been reported as well [85]. The clearness index is the ratio of the incoming solar radiation measured on a horizontal surface and the extraterrestrial solar radiation; that is, the clearness index represents the fraction of the extraterrestrial solar radiation that reaches the Earth's surface at a given instant. It is important to highlight that during the year, solar radiation varies due to the variation in the angle of incidence of the solar radiation that reaches the surface. In winter, this angle is greater and thus it reduces significantly the energy per unit area on a horizontal surface. This can be observed in Figure 9.13, where

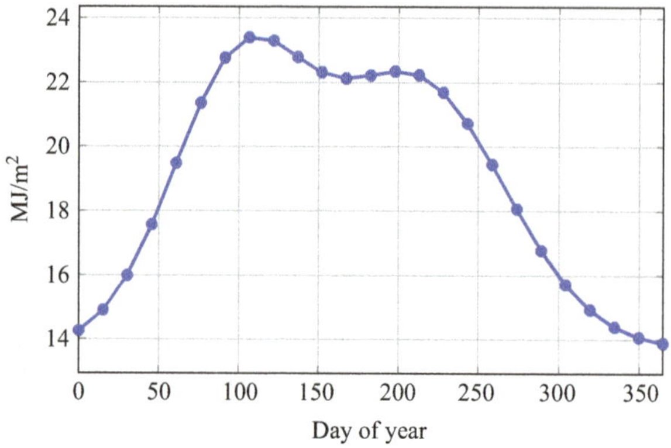

Figure 9.13 Estimated daily global solar radiation through the year at Merida, Yucatan, Mexico

the estimated daily solar radiation at Merida, Yucatan, is plotted along the year [86]. It can be observed that solar radiation reaches its maximum between May and August, while its minimum during December and January. Most important is the fact that there is a reduction of about 40%. This implies that a similar sudden variation in solar radiation during May represents a different percentage than in December. In this sense, clearness index detrends (normalizes) solar radiation behavior. The clearness index can be defined in terms of instantaneous, daily or monthly radiation with analogous definitions. In fact, that the estimation of solar radiation by means of the clearness index has been reported to be more accurate in one study [87].

9.3.4 Hybrid methods

Individual models pose limitations due to the way in which each technique processes data. In order to overcome the limitation of one given model, it is common to combine techniques to take advantage of the individual strengths to achieve better accuracy. Such combinations of models are known as hybrid models. Ensemble methods explained in Section 9.3.3.1 fall into this category. Different ways to combine different models have been reported in [88], but there are still opportunities for improvement.

Seeking to make a good decision, people usually resort to the opinion of several experts.[‡‡] In machine learning (AI), each induced model can be seen as an

[‡‡]Or at least ideally. The importance of irrational drivers for decision-making has been studied for psychologists for a few decades by now, following the pioneering work of Kahneman, Tversky, and others. Often, a subconsciously made decision is later *rationalized* by making it fit into a framework where a technical justification can be constructed. Of course, when it comes to make a machine or a technical process work, intuitive reasoning is generally a bad counselor, unless your common sense has been educated by extended practice on the matter.

"expert", so that a more reliable prediction is produced if the predictions of several models are combined. This idea can be extended to build prediction systems by combining different forecasting methods (physical and stochastic). The implementation of hybrid models represents several benefits:

1. improvements in prediction quality (precision, performance),
2. versatility and robustness,
3. greater number of applications, and
4. greater range of temporal and spatial scales.

Hybrid methods can be classified as follows:

- *Classifier assemblies* (Figure 9.14): Strategies: Majority vote, heavy vote, a new classifier decides the best result. In [89], the authors present a methodology to build wind power forecasting models from data for an Ecuadorian wind farm using a combination of ANNs and DBNs.
- *Parameter selection and optimization* (Figure 9.15): AI is used for selection of relevant variables and the determination of optimization parameters.

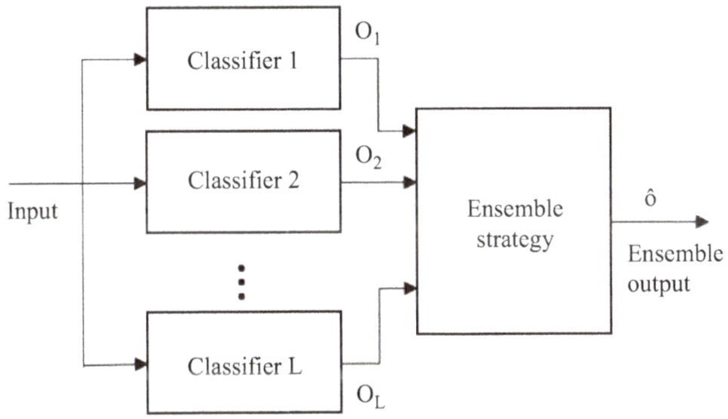

Figure 9.14 Classifier assemblies

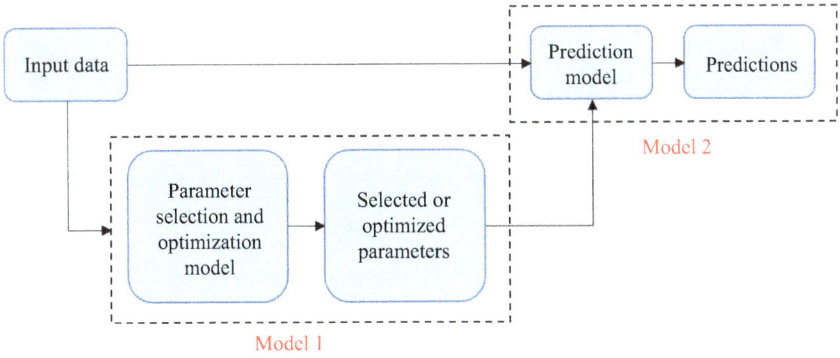

Figure 9.15 Flowchart of a hybrid method using optimization and parameter selection techniques

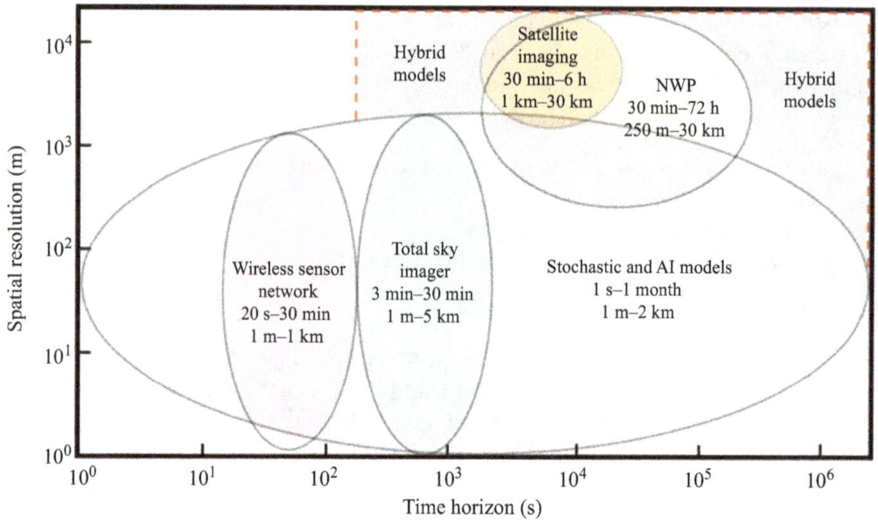

Figure 9.16 Comparison of time horizon and spatial resolution. Solid lines indicate current limits of techniques while the dashed red lines indicate the future progress of hybrid models. Modified from [91]

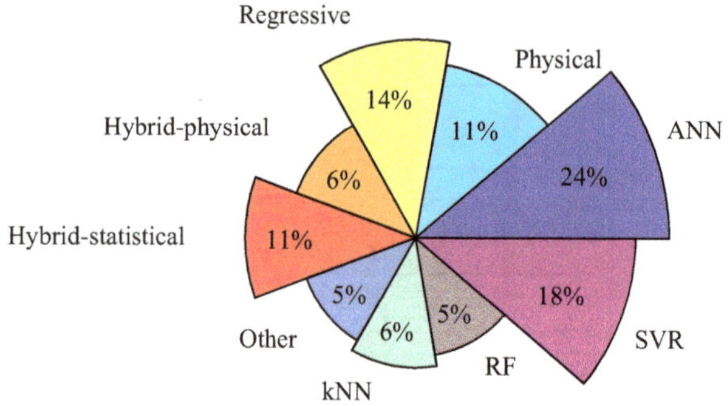

Figure 9.17 Distribution of studies in PV power forecasting, differentiated by their implemented techniques. Physical 11%, stochastic 72%, and hybrid 17%. Data taken from [88]

- *Pre-processing of input data through clustering techniques* (data set decomposition).
- *Complementarity in spatial and temporal resolution*: Figure 9.16 shows a comparison of the time horizon and the spatial resolution for different techniques. AI techniques include local, mesoscale, and global meteorological data which allow for both shorter time horizons and greater areas of interest. In addition, recent trends in NWP suggest that shorter time horizons will be available through the development of WRF models. In [90], a hybrid

forecasting approach, which combines Ensemble Empirical Mode Decomposition (EEMD) and an SVM scheme, is proposed for wind forecasting.

Figure 9.17 presents a distribution of PV power forecasting studies differentiated by the techniques they are using [88]. As can be observed, stochastic approaches represent almost 72% of the implemented methodologies, with physical methods only representing 11% and hybrid models 17%.

9.4 Forecasting evaluation

Forecast verification is the process required to assess the quality of a given forecast. Here, the forecast is compared against an observation (or measurement) or to a good estimation of the expected outcome. In order to obtain information about the forecast error a qualitative (*how does it look like?*) or quantitative (*how accurate is it?*), evaluation may be conducted.

Forecast evaluation is a valuable process since it provides the basis for monitoring the forecast (how accurate is it over time), improving it and comparing it (how a given forecast behaves with respect to another one, under similar conditions).

For decision-makers, it is important for a forecast to have quality, consistency, and value. *Quality* is understood as the degree to which a forecast reproduces the real value, while *consistency* refers to the degree at which the forecast corresponds to the best judgment of the forecaster under different situations. The *value* of a forecast refers to the degree at which the forecast helps in the decision-making process in order to obtain some technical or economical benefit. For wind and solar power forecasting applications, it is desirable that a forecast model possesses these three characteristics. It is worth noting that the scientific literature focuses mostly on the quality of the forecast. Nine attributes are closely related to the quality of the forecast [92]:

- *Bias*: a measure of the correspondence between the mean values of the forecast and observations.
- *Association*: a measure of the linear relation between forecast and observation.
- *Accuracy*: the degree of agreement between the forecast value and the real one. Their difference is defined as the forecast error. Lower errors imply a greater accuracy and vice versa.
- *Skill*: the relative accuracy of a forecast with respect to a reference forecast. In general, the reference forecast is taken to be the persistence model, although a different model can be selected as well.
- *Reliability*: the average agreement between the forecast values and the observed values. When all forecast are considered, then the reliability corresponds to the bias.
- *Resolution*: the ability of the forecast to resolve a set of events into subsets with different frequency distributions.

- *Sharpness*: the capability of the forecast to predict extreme values, such as ramps in wind and solar power.
- *Discrimination*: the ability of the forecast to discriminate among observations.
- *Uncertainty*: the variability of the forecasts with respect to the observations, increasing with the forecast horizon. Uncertainty can also be interpreted as the variability of the observations. In this sense, the larger the uncertainty of a process, the more difficult its forecast.

In order to evaluate the quality of a model, several metrics can be used, including the mean square error (MSE), the normalized mean square error (NMSE), the RMSE, the MAE, the mean absolute percentage error (MAPE), the skill factor, the continuous ranked probability score (CRPS), and the quantile score (QS), among others. Each one of these metrics provides a quantitative measurement of one of the attributes mentioned above. Thus, the performance of a forecasting model depends upon the chosen metric, which must be selected according to the attribute to be analyzed. Metrics used during the evaluation of AI methods (which normally are defined by the MSE or MAE) need to be re-evaluated in terms of their validity in forecasting tasks since they only provide information on the distribution of the error rather than in the instantaneous error of the forecast, which may be more relevant from the operator point of view.

9.5 Economical impact of power forecasting

The economic revenue of wind and solar power plants can be significantly affected by the quality of forecasting methods. Energy markets require a precise day-to-day forecast to match the estimated consumption and the daily energy purchases, and commit the power producer with a certain quota. If the producer fails to meet the arranged energy delivery, the missing fraction will be most likely supplied by a conventional plant at a much higher price than the original one. This economic loss would have to be paid up by the renewable plant operator, resulting in negative revenue for the company.

On the other hand, short-time window forecasts (of the order of 30 s to a few minutes) result in special interest when dealing with abnormal behavior of wind turbines. For example, wind variability is prone to cause two main types of electric phenomena: flickers and frequency distortions. The former is a type of voltage perturbation caused by sudden changes in wind speed, while the latter is the result of both wind variations and voltage and current distortion caused by nonlinear loads in the system. This abnormal behavior can be prevented by precise forecast methods that anticipate both wind gusts and restraints. If a wind turbine is out of a specific voltage and frequency tolerance range due to sudden wind speed variations, the system operator will disconnect it from the grid and there will be a time to restore synchronicity. This event will cause the turbine to stop and generate an economic loss, since it will fail to meet the committed energy delivery. Current methods are quite successful at estimating wind variability in the long and medium term, but there are significant research gaps in terms of wind data analysis in the very short term and in the frequency domain. An optimum forecast system yields better mechanical control of pitch and yaw that will improve turbine performance in response to extreme wind variability.

9.6 Conclusions

Wind and solar PV forecasting is a key tool that can help in the improvement of efficiency, reliability, and operation of electric power systems with a high penetration of variable REs. Although there are a large number of applications of forecasting in power systems, there is still a need to improve the integration of forecasts into different stages or procedures, and even more so, considering the strong increase in installed capacity expected in the coming years.

Wind and solar PV power forecasting is relevant for several tasks within power systems and has both technical and economic impacts. In general, the following conclusions can be drawn:

- Forecasting power production improves the operability of RE power systems.
- It enables scheduled maintenance.
- It helps in the identification of factors affecting power production.
- It helps energy traders in the placement of bids.
- It reduces costs associated with energy storage systems.
- It better forecasts result in a reduction of the required spinning reserve.
- Hybrid methods, involving the combination of different methods, significantly improve the accuracy of power forecasting.
- Although different metrics have been proposed to evaluate the performance of forecasting methods, there is still a need to establish standard metrics.

In literature, a wide variety of forecast methods have been reported, which can be divided into two large categories: physical and stochastic approaches. When two or more of these models are mixed, a hybrid model is obtained whose main aim is to cope with the limitations of the individual methods and to extend the temporal and/or spatial resolution as well as to reduce forecasting errors. The implementation of different methods and methodologies has been performed with different degrees of success.

End users such as system operators, generators, and traders require the development of new tools and procedures to improve the use of information obtained by forecasts in order to make well-informed decisions in different applications. It is important to continue the development and testing of new methodologies to reduce forecasting errors, as well as to efficiently integrate them into power system operations. In this sense, prediction models will help in changing the paradigm in the operation of power systems as we transition into the RE era.

References

[1] Mauleón I. Assessing PV and wind roadmaps: Learning rates, risk, and social discounting. Renewable and Sustainable Energy Reviews. 2019;100:71–89.

[2] International Energy Agency (IEA). Fuels and technologies; 2020. Available from: https://www.iea.org/fuels-and-technologies.

[3] SENER. Prospectiva de energías renovables 2018-2032. SENER; 2018. Available from: https://base.energia.gob.mx/Prospectivas18-32/PER_18_32_F.pdf.

[4] Pinson P, Chevallier C, and Kariniotakis GN. Trading wind generation from short-term probabilistic forecasts of wind power. IEEE Transactions on Power Systems. 2007;22(3):1148–1156.

[5] Zhang Y, Wang J, and Wang X. Review on probabilistic forecasting of wind power generation. Renewable and Sustainable Energy Reviews. 2014;32: 255–270.

[6] Menemenlis N, Huneault M, and Robitaille A. Computation of dynamic operating balancing reserve for wind power integration for the time-horizon 1–48 hours. IEEE Transactions on Sustainable Energy. 2012;3(4):692–702.

[7] Botterud A, Wang J, Miranda V, *et al.* Wind power forecasting in U.S. electricity markets. The Electricity Journal. 2010;23(3):71–82.

[8] Wang J, von Bremen L, Keko H, *et al.* Wind power forecasting uncertainty and unit commitment. Applied Energy. 2011;88(11):4014–4023.

[9] Bludszuweit H and Domínguez-Navarro JA. A Probabilistic method for energy storage sizing based on wind power forecast uncertainty. IEEE Transactions on Power Systems. 2011;26(3):1651–1658.

[10] Cochran J, Miller M, Zinaman O, *et al.* Flexibility in 21st Century Power Systems. Golden, Colorado: NREL 2014. Available from: https://cris.vtt.fi/en/publications/flexibility-in-21st-century-power-systems.

[11] Ahmed A and Khalid M. A review on the selected applications of forecasting models in renewable power systems. Renewable and Sustainable Energy Reviews. 2019;100:9–21.

[12] Heier S. Grid Integration of Wind Energy: Onshore and Offshore Conversion Systems (Third). John Wiley & Sons, Ltd; 2014. Available from: https://www.wiley.com/en-us/Grid+Integration+of+Wind+Energy{\%}3A+Onshore+and+Offshore+Conversion+Systems{\%}2C+3rd+Edition-p-9781119962946.

[13] Infield D and Freris L. Renewable Energy in Power Systems (Second). John Wiley & Sons Ltd; 2020. Available from: https://www.wiley.com/en-mx/Renewable+Energy+in+Power+Systems,+2nd+Edition-p-9781118649930.

[14] Apt J, and Jaramillo P. Variable Renewable Energy and the Electricity Grid. New York, NY: RFF Press; 2014 Available from: https://www.rff.org/publications/books/variable-renewable-energy-and-the-electricity-grid.

[15] Pardalos PM. Handbook of Power Systems II. Heidelberg, Springer-Verlag Berlin Heidelberg; 2010 Available from: https://www.springer.com/gp/book/9783642126857.

[16] Catalão JPS. Electric Power Systems: Advanced Forecasting Techniques and Optimal Generation Scheduling. Boca Raton, FL: Taylor & Francis Group; 2012. Available from: https://www.taylorfrancis.com/books/e/9781315216683

[17] Kiliccote S, Price P, Piette MA, *et al.* Field Testing of Automated Demand Response for Integration of Renewable Resources in California's Ancillary

Services Market for Regulation Products Akuacom/Honeywell. Lawrence Berkeley National Laboratory. 2012;(April 2012).

[18] Hart EK, Stoutenburg ED, and Jacobson MZ. The potential of intermittent renewables to meet electric power demand: Current methods and emerging analytical techniques. Proceedings of the IEEE. 2012;100(2):322–334.

[19] Van der Hoven I. Power Spectrum of Horizontal Wind Speed in the Frequency Range from 0.0007 To 900 Cycles Per Hour; 1957.

[20] Escalante Soberanis MA and Mérida W. Regarding the influence of the Van der Hoven spectrum on wind energy applications in the meteorological mesoscale and microscale. Renewable Energy. 2015;81:286–292.

[21] Introduction to Tropical Meteorology. 2nd ed. The University Corporation for Atmospheric Research; 2010. Available from: http://www.chanthaburi.buu.ac.th/ wirote/met/tropical/textbook_2nd_edition/navmenu.php_tab_2_page_8.1.0.htm.

[22] Huang CJ and Kuo PH. A short-term wind speed forecasting model by using artificial neural networks with stochastic optimization for renewable energy systems. Energies. 2018;11(10):2777.

[23] Probst O. A new strategy for short-term ramp rate control in wind farms. International Journal of Electrical Power & Energy Systems. 2020;120: 105969.

[24] Mao Y and Shaoshuai W. A review of wind power forecasting & prediction. In: 2016 International Conference on Probabilistic Methods Applied to Power Systems, PMAPS 2016 – Proceedings; 2016.

[25] Alessandrini S, Monache LD, Sperati S, *et al.* An analog ensemble for short-term probabilistic solar power forecast. Applied Energy. 2015;157:95–110.

[26] Box G, Jenkins G, Reinsel G, and Ljung G. Time Series Analysis: Forecasting and Control, 5th ed. Hoboken, New Jersey: John Wiley & Sons, Inc.; 2015. Available from: https://www.wiley.com/en-us/Time+Series+Analysis%3A+Forecasting+and+Control%2C+5th+Edition-p-9781118675021.

[27] Wang MD, Qiu QR, and Cui BW. Short-term wind speed forecasting combined time series method and arch model. In: International Conference on Machine Learning and Cybernetics. vol. 3; 2012. pp. 924–927.

[28] McCulloch WS and Pitts W. A logical calculus of the ideas immanent in nervous activity. The Bulletin of Mathematical Biophysics. 1943;5(4):115–133.

[29] Rumelhart DE, Hinton GE, and McClelland JL. 2. In: Rumelhart DE, McClelland JL, the PDP research group, editors. A General Framework for Parallel Distributed Processing. vol. 1. Cambridge MA: MIT Press; 1986. pp. 45–76.

[30] Cadenas E and Rivera W. Short term wind speed forecasting in La Venta, Oaxaca, México, using artificial neural networks. Renewable Energy. 2009;34(1):274–278.

[31] Amrouche B and Pivert XL. Artificial neural network based daily local forecasting for global solar radiation. Applied Energy. 2014;130(1):333–341.

[32] Izgi E, Öztopal A, Yerli B, *et al.* Short-mid-term solar power prediction by using artificial neural networks. Solar Energy. 2012;86(2):725–733.

[33] Vapnik VN. The Nature of Statistical Learning Theory, 2nd Edition. New York, NY: Springer-Verlag New York, Inc.; 2000.
[34] Zhou J, Shi J, and Li G. Fine tuning support vector machines for short-term wind speed forecasting. Energy Conversion and Management. 2011;52(4): 1990–1998.
[35] Ackermann T. Wind Power in Power Systems. West Sussex, England: John Wiley & Sons Ltd; 2005.
[36] Pearl J. Fusion, propagation and structuring in belief networks. Artificial Intelligence. 1986;29:241–288.
[37] Murphy KP. Dynamic Bayesian Networks: Representation, Inference and Learning [Ph.D. thesis]. University of California, Berkeley. Berkeley, CA, USA; 2002.
[38] Miranda MS and Dunn RW. One-hour-ahead wind speed prediction using a Bayesian methodology. In: IEEE Power Engineering Society General Meeting. Montreal, Que., Canada: IEEE; 2006.
[39] Lauret P, Boland J, and Ridley B. Bayesian statistical analysis applied to solar radiation modelling. Renewable Energy. 2013;49:124–127.
[40] Ibargüengoytia PH, Reyes A, Romero-Leon I, *et al.* Wind power forecasting using dynamic Bayesian models. In: Nature-Inspired Computation and Machine Learning. Switzerland: Springer, Cham; 2014. pp. 184–197. Available from: https://link.springer.com/book/10.1007%2F978-3-319-13650-9.
[41] Ibargüengoytia PH, Reyes A, García UA, *et al.* Evaluating probabilistic graphical models for forecasting. In: 18th International Conference on Intelligent System Application to Power Systems (ISAP). Porto, Portugal: IEEE; 2015.
[42] Zaded LA. Fuzzy sets. Information and Control Elsevier. 1965;8(3):338–353.
[43] Tulcan-Paulescu E and Paulescu M. Fuzzy modelling of solar irradiation using air temperature data. Theoretical and Applied Climatology. 2008;91(1): 181–192.
[44] Lynch P. From Richardson to Early Numerical Weather Prediction. Cambridge University Press; 2010.
[45] Stull R. In: Practical Meteorology: An Algebra-Based Survey of Atmospheric Science. University of British Columbia; 2017.
[46] National Centers for Environmental Prediction (NCEP). Global Forecast System; 2020. Available from: www.emc.ncep.noaa.gov/emc/pages/numerical_forecast_systems/gfs.php.
[47] Skamarock WC, Klemp JB, Dudhia J, *et al.* A Description of the Advanced Research WRF Model Version 4 NCAR Technical Note; 2019.
[48] EU. European Centre for Medium-Range Weather Forecasts; 2020. Available from: www.ecmwf.int.
[49] Jimenez PA, Hacker JP, Dudhia J, *et al.* WRF-SOLAR: Description and clear-sky assessment of an augmented NWP model for solar power prediction. Bulletin of the American Meteorological Society. 2016;97(7):1249–1264.
[50] ESI OpenCFD. OpenFOAM. OpenCFD Ltd; 2020. Available from: www.openfoam.com.

[51] Sanz Rodrigo J, Chávez Arroyo RA, Moriarty P, *et al.* Mesoscale to microscale wind farm flow modeling and evaluation. WIREs Energy and Environment. 2017;6(2):e214.
[52] Allen DJ, Tomlin AS, Bale CSE, *et al.* A boundary layer scaling technique for estimating near-surface wind energy using numerical weather prediction and wind map data. Applied Energy. 2017;208:1246–1257.
[53] Temel O, Bricteux L, and van Beeck J. Coupled WRF-OpenFOAM study of wind flow over complex terrain. Journal of Wind Engineering and Industrial Aerodynamics. 2018;174:152–169.
[54] Pereira RM, Santos CS, and Rocha A. Solar irradiance modelling using an offline coupling procedure for the Weather Research and Forecasting (WRF) model. Solar Energy. 2019;188:339–352.
[55] Schmidt J, Chang CY, Dörenkämper M, *et al.* The consideration of atmospheric stability within wind farm AEP calculations. Journal of Physics: Conference Series. 2016;749:012002.
[56] Subramanian B, Chokani N, and Abhari RS. Impact of atmospheric stability on wind turbine wake evolution. Journal of Wind Engineering and Industrial Aerodynamics. 2018;176:174–182.
[57] Lange M. Best-practice of wind and solar power forecasting second session. In: develoPPP Project Integrating variable renewable energies into the Mexican grid via reliable wind and solar power forecasts, Mexico City, 2017.
[58] Sun M, Feng C, and Zhang J. Multi-distribution ensemble probabilistic wind power forecasting. Renewable Energy. 2020;148:135–149.
[59] Delle Monache L, Eckel FA, Rife DL, *et al.* Probabilistic weather prediction with an analog ensemble. Monthly Weather Review. 2013;141:3498–3516.
[60] Ren Y, Suganthan PN, and Srikanth N. Ensemble methods for wind and solar power forecasting—A state-of-the-art review. Renewable and Sustainable Energy Reviews. 2015;50:82–91.
[61] Rivero M, Orozco S, Sellschopp F, *et al.* A new methodology to extend the validity of the Hargreaves-Samani model to different climates: Case study Mexico. Renewable Energy. 2017.
[62] Tapakis R and Charalambides AG. Equipment and methodologies for cloud detection and classification: A review. Solar Energy. 2013;95:392–430.
[63] Pazikadin AR, Rifai D, Ali K, *et al.* Solar irradiance measurement instrumentation and power solar generation forecasting based on artificial neural networks (ANN): A review of five years research trend. Science of The Total Environment. 2020;715:136848.
[64] Chow CW, Urquhart B, Lave M, *et al.* Intra-hour forecasting with a total sky imager at the UC San Diego solar energy testbed. Solar Energy. 2011;85(11): 2881–2893.
[65] Urquhart B, Kurtz B, Dahlin E, *et al.* Development of a sky imaging system for short-term solar power forecasting. Atmospheric Measurement Techniques. 2015; 8(1):875–890.

[66] Herrera-Carrillo NE, Rivero M, Gómez-Ramírez E, *et al.* Solar irradiance estimation based on image analysis. In: 2018 IEEE International Autumn Meeting on Power, Electronics and Computing (ROPEC); 2018. pp. 1–6.

[67] Garcia-Lamont F, Cervantes J, López A, *et al.* Segmentation of images by color features: A survey. Neurocomputing. 2018;292:1–27.

[68] Ye L, Cao Z, Xiao Y, *et al.* Supervised fine-grained cloud detection and recognition in whole-sky images. IEEE Transactions on Geoscience and Remote Sensing. 2019;57(10):7972–7985.

[69] Gan J, Lu W, Li Q, *et al.* Cloud type classification of total-sky images using duplex norm-bounded sparse coding. IEEE Journal of Selected Topics in Applied Earth Observations and Remote Sensing. 2017;10(7):3360–3372.

[70] Barbieri F, Rajakaruna S, and Ghosh A. Very short-term photovoltaic power forecasting with cloud modeling: A review. Renewable and Sustainable Energy Reviews. 2017;75:242–263.

[71] Chow CW, Belongie S, and Kleissl J. Cloud motion and stability estimation for intra-hour solar forecasting. Solar Energy. 2015;115:645–655.

[72] Bone V, Pidgeon J, Kearney M, *et al.* Intra-hour direct normal irradiance forecasting through adaptive clear-sky modelling and cloud tracking. Solar Energy. 2018;159:852–867.

[73] Wang F, Zhen Z, Liu C, *et al.* Image phase shift invariance based cloud motion displacement vector calculation method for ultra-short-term solar PV power forecasting. Energy Conversion and Management. 2018;157: 123–135.

[74] Zhen Z, Xuan Z, Wang F, *et al.* Image phase shift invariance based multi-transform-fusion method for cloud motion displacement calculation using sky images. Energy Conversion and Management. 2019;197.

[75] Nguyen DA and Kleissl J. Stereographic methods for cloud base height determination using two sky imagers. Solar Energy. 2014;107:495–509.

[76] Bosch JL, Zheng Y, and Kleissl J. Deriving cloud velocity from an array of solar radiation measurements. Solar Energy. 2013;87:196–203.

[77] Fung V, Bosch JL, Roberts SW, *et al.* Cloud shadow speed sensor. Atmospheric Measurement Techniques. 2014;7(6):1693–1700.

[78] Wang G, Kurtz B, and Kleissl J. Cloud base height from sky imager and cloud speed sensor. Solar Energy. 2016;131:208–221.

[79] Mejia FA, Kurtz B, Murray K, *et al.* Coupling sky images with radiative transfer models: A new method to estimate cloud optical depth. Atmospheric Measurement Techniques. 2016;9(8):4151–4165.

[80] Bird RE and Riordan C. Simple solar spectral model for direct and diffuse irradiance on horizontal and tilted planes at the Earth's surface for cloudless atmospheres. Journal of Climate and Applied Meteorology. 1986;25(1): 87–97.

[81] Chu Y, Urquhart B, Gohari SMI, *et al.* Short-term reforecasting of power output from a 48 MWe solar PV plant. Solar Energy. 2015;112:68–77.

[82] Chu Y, Pedro HTC, and Coimbra CFM. Hybrid intra-hour DNI forecasts with sky image processing enhanced by stochastic learning. Solar Energy. 2013;98(PC):592–603.

[83] Dev S, Lee YH, and Winkler S. Color-based segmentation of sky/cloud images from ground-based cameras. IEEE Journal of Selected Topics in Applied Earth Observations and Remote Sensing. 2017;10(1):231–242.

[84] Mathiesen P, Collier C, and Kleissl J. A high-resolution, cloud-assimilating numerical weather prediction model for solar irradiance forecasting. Solar Energy. 2013;92:47–61.

[85] Woyte A, Belmans R, and Nijs J. Fluctuations in instantaneous clearness index: Analysis and statistics. Solar Energy. 2007;81(2):195–206.

[86] Quej VH, Almorox J, Ibrakhimov M, *et al.* Estimating daily global solar radiation by day of the year in six cities located in the Yucatán Peninsula, Mexico. Journal of Cleaner Production. 2017;141:75–82.

[87] Fu CL and Cheng HY. Predicting solar irradiance with all-sky image features via regression. Solar Energy. 2013;97:537–550.

[88] Antonanzas J, Osorio N, Escobar R, *et al.* Review of photovoltaic power forecasting. Solar Energy. 2016;136:78–111.

[89] Reyes A, Ibargüengoytia PH, Jijón JD, *et al.* Wind power forecasting for the Villonaco wind farm using AI techniques. In: Pichardo-Lagunas O and Miranda-Jiménez S, editors. Advances in Soft Computing. Cham: Springer International Publishing; 2017. pp. 226–236.

[90] Hu J, Jianzhou, and Zeng WG. A hybrid forecasting approach applied to wind speed time series. Renewable Energy. 2103;60:185–194.

[91] Inman RH, Pedro HTC, and Coimbra CFM. Solar forecasting methods for renewable energy integration. Progress in Energy and Combustion Science. 2013;39(6):535–576.

[92] Murphy AH. What is a good forecast? An essay on the nature of goodness in weather forecasting. Weather and Forecasting. 1993;8(2):281–293.

Part III

Strategies for the clean energy transition

Chapter 10

The role of regulations for providing certainty to the energy reform and transition in Mexico

Guillermo Zúñiga[1]

For the first two decades of the twentieth century, worldwide influences impelled powerful and complex reforms in energy regulatory structures across different countries regardless of its development status. Governments from all over the world began to experiment a reorganization of public utilities, introduction of competition, the strengthening or development of independent regulatory agencies and the re-regulation of rules in the electricity industry.

In the first part of this work, we will analyze the nature of the changes that give way to a change in those regulatory strategies and models. In the second part, we will use the application of the regulatory reform to a single industry, the Mexican Electricity Industry.

10.1 Regulatory reform principles

10.1.1 What is a regulatory reform?

A regulatory reform applied to the electricity sector is a conceptual tool developed to address significant changes in mode of governance and policy strategies, which can be described as a replacement of the centralized and vertical integrated model, a common strategy applied in the electricity sector for most of the twentieth century. This regime was common to the welfare state model and was characterized by an open *dirigisme* role of the government the main features of which were a wave of nationalizations and the creation of state-owned enterprises as public monopolies to centrally manage a sole utility. As explained by Musacchio and Lazzarini [1], governments in the developing world used nationalization to address several goals, such as national security concerns, overcome coordination problems via investments in basic infrastructure, finance R&D in technologies and address market complexities.

[1]Earthjustice, San Francisco, CA, USA

Under a regulatory reform, the principal aim of the state switches from a centralized control to a focus on improving the efficiency of the industry through market liberalization with the development of regulatory capacity to correct particular forms of market failure (e.g. "natural" monopoly, imperfect information and negative externalities, all of them present in the electricity industry). Therefore, the regulatory reform implies that there is a new set of governmental priorities, which would entail a change in governmental strategies.

In particular, the old assumption that an electricity system should be managed by a single entity was overcome when technological advances, operational models and communication systems provided conditions to have different ownerships along the segments of the industry. Moreover, it was proven that more contestable segments of the market (such as generation and supply) could be owned and managed by different actors. Vertical integration with transmission or system operation to achieve security and reliability was no longer required.

Since there are many different interpretations of regulation, for the purposes of this work, Majone [2] argues that Regulatory policies are aimed at providing conditions for private participation and correcting various types of market failures: monopoly power, negative externalities, incomplete information or insufficient provision of public goods [3].

In the centralized and vertically integrated industry, government intervention for the supply of electricity was applied through a process of nationalization involving direct state ownership of industrial sectors, known as the state-owned enterprises model. These enterprises were traditionally characterized by being vested with a monopoly and pursuing simultaneously a diversity of governmental objectives (often contradictory) such as economic development, technological innovation, income redistribution, national security, controlling price, universal service obligations and the setting of financial targets [3].

These state-owned enterprises model had their ideological basis on the central assumption that public ownership would increase government's ability to regulate the economy and protect the public interest. The economic rationale to use the state-owned enterprises model was considered a necessity in the face of the operational complexities present in the attributes of electricity systems, so government ownership was set in place *in lieu* of the market and nonmarket mechanisms that we encounter in today's legal and economic design [4].

Within the surge of a new regulatory strategy, Vogel argues that the government and the markets are not linked by a zero-sum relationship where the increment of one means a decrement of the other; he finds that a regulatory reform, more than a retreat of the state, means a reorganization of the governmental tools to control private activities [5]. Therefore, the regulatory reform represents the development of new regulatory strategies of very specific characteristics that reflect an intent to improve previous practices.

10.2 Characteristics of the regulatory reform

Loughlin and Scott argue that there are three basic characteristics of governmental activity related to the surge of the regulatory state:

1. separation in the government of policymaking from operational tasks,
2. creation of independent regulatory agencies assigned to the development and enforcement of new regulation (re-regulation) and
3. increase in formal rules as guidance of the government activity [6].

According to these positions, the main characteristics of the regulatory state can be categorized into three separate subjects: (1) an increasing reliance on market forces by means of liberalization policies, (2) the surge of independent regulators and (3) the issuance of rules to enforce the new regulatory policies (re-regulation).

10.2.1 Move to markets: liberalization

As such, liberalization would mean the shift away of government activities toward private firms. In turn, this could lead to at least three outcomes; the substitution of a public monopoly for a private one, the introduction of competition in the sector to be run entirely by private parties or the introduction of competition in the sector where private parties and state-owned enterprises (no longer legally vested with monopoly power) will coexist. This last example would entail having an equal footing among private firms and those firms owned and run by the government.

10.2.2 Independent regulatory agencies

Thatcher states that the basic criteria to consider a regulatory agency as independent is that the agency is organizationally separated from the central government; that its members are not directly elected officials, and that it is given its own powers, responsibilities and funding under public law [7]. In this sense, the driving goal for the creation of independent regulators and delegation of regulatory powers is the need of credibility and certainty as opposed to discretion in regulatory performance of the government.

Moreover, the delegation of regulatory powers to independent agencies can also be considered a strategy to show a commitment to certain liberalization policies that would not be credible in the absence of such delegation.

In line with this idea, there has been an important institutional isomorphism process at an international level, by which many jurisdictions are choosing to address certain or many of the public policy complexities surrounding the energy sector by means of an independent regulatory agency.

The increasing international interest in independent energy sector oversight is illustrated by the rise of international organizations that provide platforms suitable for experience exchanges and capacity-building interactions among regulators, such as the Ibero-American Association of Energy Regulators (ARIAE) and the International Confederation of Energy Regulators (ICER). The rise of organizations with such scope can attest how the regulatory reform in the energy sector responds to international factors that happens on broad territories at a relatively small period of time.

Energy Regulatory Agency Model as an International "Bandwagon" Policy

International Confederation of Energy Regulator (ICER) describes itself as a voluntary framework for cooperation between energy regulators. Its mission is to improve public and policymaker awareness and understanding of energy regulation and its role in addressing a wide spectrum of socioeconomic, environmental and market issues.

ICER's membership includes over 200 regulatory authorities located in 6 different continents, including regional regulatory associations:

AFUR African Forum for Utility Regulators
CEER Council of European Energy Regulators
ARIAE Ibero-American Association of Energy Regulators
OOCUR Organization of Caribbean Utility Regulators
AEMC Australian Energy Market Commission
EAPIRF East Asia & Pacific Infrastructure Regulatory Forum
MedReg Mediterranean Energy Regulators
RERA Regional Electricity Regulators Association of Southern Africa
CAMPUT Canada's Energy and Utility Regulators
ERRA Energy Regulators Regional Association
SAFIR South Asia Forum for Infrastructure Regulation
RAERESA Regional Association of Energy Regulators for Eastern and Southern Africa
Source: http://icer-regulators.net/

10.2.3 New regulatory process

The development of independent regulators is clearly linked to the need to create and enforce a new set of general rules to regulate the newly liberalized markets.

At this point, it is important to note that the introduction of competition is supposed to have a deregulatory effect since it has been stated that competition is incompatible with regulation ("competition where possible, regulation where necessary").

Notwithstanding the above, the electricity industry presents attributes that make it difficult for economic competition to enter, such as high entry barriers, a weak position of users in terms of bargaining power, inelastic demand, information asymmetries and lack of efficient substitutes for the services, among many others. Electricity market events if liberalized still require regulation providing a mix of market and nonmarket mechanisms, including wholesale and retail tariff regulation, capacity obligations, frequency regulation, operating reserves and financial transmission rights, among many other ancillary services.

Moreover, the operational requirements for the system operator also rely on robust regulatory and regulatory policies in terms of open and nondiscriminatory access, grid interconnection, procurement obligations, protocols for non-price rationing and curtailments and other system emergencies.

In these kind of cases, the mere use of competition policies may not be enough, so the use, revision and enactment of specific sector rules are necessary to address market failures and provide that competition in regulated markets remains effective.

10.3 Forces for regulatory change

To analyze the regulatory reform process in the electricity industry, it is important to outline some of the main technical characteristics that make it different from other industry. Electricity delivered to the customer is made up of various components or functions, energy generation (traditionally considered non-storable commodity) and transportation (a service), which include transmission, distribution and system operation. Therefore, the electricity industry can be divided in five segments:

1. Generation. This activity consists of the electricity production from primary fuels (solar, wind, fossil fuels, etc.).
2. Transmission. This consists of the transportation of high-voltage electricity from the power plants to the distribution centers.
3. Distribution. This segment covers the transportation from distribution centers to the final consumers, at a lowered voltage, usually less than 69 kV.
4. System operation. The function is to select the different generation plants that could supply energy to meet the overall demand of electricity through a dispatching procedure. This activity is of particular importance because of the special properties of electricity supply, which include balanced needs and the fluctuations of its demand at various time horizons. Therefore, the system operator is the coordinator of transportation and distribution services in charge to ensuring that the system is constantly in a state of balance in supply and demand.
5. End user supply (the delivery of electricity to end users). It includes the procurement of energy and distribution services and the metering and billing of consumption.

Traditionally on the vertically integrated model, the abovementioned activities were controlled by government monopolies; in this case, the advantages of a vertical integration are that all the costs in coordination are reduced when they are internalized in a single entity such as the state-owned enterprise [8].

Even though the Mexican experience demonstrates a clear example of the revision of vertically integrated monopolies, perhaps some of the first and most important experiences took place in Europe, particularly during the 1980s and the 1990s [2].

The electricity regulatory trend still has an important international record; just in October 2019, the South African Government presented a *Roadmap for Eskom in a Reformed Electricity Supply Industry*[*], in which the state-owned company is set to undertake a deep reform that would revise its monopolistic status [9]. Whereas in Australia, government has issued a *Post 2025 Market Design Paper*[†], which is a discussion paper, presented as the first of a series of documents proposing a framework for an in-depth revision and modification of the National Electricity Market.

10.4 Analysis of the forces for change in the Mexican electricity industry

The application of the regulatory reform to foster competitive markets has been possible because a myriad of factors eroded the efficiency of the old command and control of state-owned enterprises. These factors were related to X-inefficiency and the corresponding stress on public finances, lack of innovation and investment. However, the main driver present in all the liberalization process was the potential of new technology to introduce competition in generation and end user supply, and in recent times and to a certain extent, the network segments of the industry; distribution and transmission.

Even before the raise of new technologies, within the remaining "natural" monopoly segments of the industry (transmission and distribution (T&D)) introduction of some kind of discipline had been possible by means of the use of techniques of monopoly regulation, e.g. franchise bidding and the yardstick regulation [10]. Therefore, at least tariff regulation has been put in place to avoid potential monopolistic prices and open access obligation to those networks.

10.4.1 Forces for change: technology

In the field of technological advances, various innovations have dramatically challenged the natural monopoly assumption in the electricity industry; the electricity markets are much larger now, especially since the expansion of the grid has interconnected several regions, even across domestic boundaries. This expansion allows access of supply from a power generation plant to a geographically distant consumer center; therefore, even the biggest plants seem small compared to the size of the market, eroding the notion that a single generator could exert market power to the point of abusing a dominant position.

The advances of new generation technologies have had a considerable impact on the development of renewable generation, such as wind and solar, which have presented an important trend of lower costs, particularly a very low operational and maintenance costs and powerful scale economics.

[*] https://www.gov.za/sites/default/files/gcis_document/201910/roadmap-eskom.pdf
[†] http://www.coagenergycouncil.gov.au/publications/post-2025-market-design-issues-paper-%E2%80%93-september-2019

A good example can be found on the behavior of the levelized cost of energy (LCOE) of renewable energy vis-a-vis traditional fossil fuel technology. LCOE is a methodology that allows the comparison of different generation technologies (e.g. wind, solar and coal) of unequal life spans, project size, different capital costs, risks, etc., by calculating present value of the total costs building and operating a power plant over a assumed lifetime and dividing it by energy production.

Due to falling costs and mature technological stages, during the last few years LCOE has been decreasing for utility-scale wind and photovoltaic (PV) solar plants to the point that they are now cheaper than conventional generation technologies, such as combined cycle gas turbines (CCGTs) or coal.

Recent studies in the US, such as Lazard's annual LCOE analysis, have shown that the trend is very clear and the competitiveness of utility-scale renewables is gaining important low-price momentum.

The same analysis also shows that scale is still very important for this kind of competitiveness; for instance, price of rooftop residential PV solar is still in the range of 160–267 USD/MWh, although in those cases competition is not against other kinds of technology, but against the electricity tariff charged by the utility (see Figure 10.1) [16].

Notwithstanding the fact that the LCOE is a methodology widely accepted to compare generation costs of different technologies, there has been important price deviation from LCOE reports because energy policies that foster clean energy auctions carried out in various regions of the world are resulting in much lower prices than those presented in LCOE studies. The particular conditions that explain these important differences refer, among others, to the following factors:

1. The competitive nature of the auctions for the awarding of contracts has put pressures on the generators to cut their profit margins to a minimum, in order to be awarded with an offtake contract (condition that has slowed the flow from financing mechanisms).

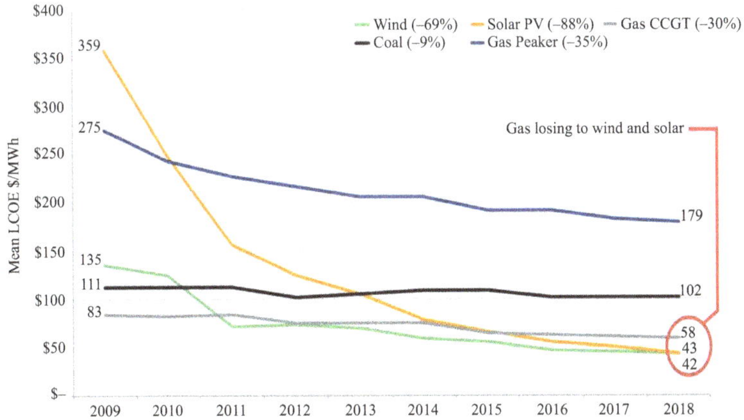

Figure 10.1 LCOE of different generation technologies. Source: Lazard (2018)

2. The contracts resulting from the auction are long term (over 15 years), with that kind of duration the low operational expenditures of wind and solar makes them quite competitive since, once the sunk costs are factored in, maintenance costs are minimal and the input is free. On the other hand, operational and maintenance costs of fossil fuel technologies increase over time while the input (fuels) are subject to market uncertainties.
3. As lenders have become more familiar to the specifics of the renewable technologies, perceived risks have decreased, allowing the flow of financial support, even in the face of record low prices.
4. Technological advances have not only reduced the installation and operational costs of renewable technology, but also new storage technologies have palliated intermittency shortcomings of wind and solar, and interconnections are less of a burden on the network operator.

The reduction of the investment level to build generation units reduces one of the most serious entry barriers to the segment.

Finally, the actual systems of communication and the available technologies of control such as innovative metering instruments dramatically reduce the coordination costs of different market participants in markets, particularly wholesale electricity markets. This is important since the major impediments of coordinating activities were a justification to vertically integrate the industry.

However, making a historical comparison to the telecomm industry, in which it was technology that rendered telephone lines obsolete opened new contestable markets, the most important developments are the ones related to the attack of the natural monopoly assumption in T&D networks. In an electricity context, the surge of "Non-wires alternatives" (NWA) is the electricity equivalent of wireless communications of the telecomm industry, and its outlook promises the same kind of industrial disruption for electricity markets. NWA cover "an electricity grid investment or project that uses non-traditional transmission and distribution (T&D) solutions, such as distributed generation, energy storage, energy efficiency, demand response, and grid software and controls, to defer or replace the need for specific equipment upgrades, such as T&D lines or transformers, by reducing load at a substation or circuit level" [11].

10.4.2 Forces for change: decentralization of the electricity sector

Broadly speaking, the movement toward decentralization challenges the traditional operation of an electrical system which has always been centralized (with the exception of very primal stages of housing illumination, since illumination was born distributed).

Centralization means that all the decision-making capacity, which includes choices related to investment, financing, maintenance, expansion and penetration, among others, is exercised centrally, which results in the various electrical services and products only being provided unidirectionally to the charging centers by the central control.

This model was followed by developments that have progressively diminished the centralized operation of decision-making capacity.

As the degree of decentralization advances, at a theoretical level and for explanatory purposes, it can be said that, together with the centralized scheme, two new schemes have emerged, the decentralized and distributed schemes (see Figure 10.2):

(a) **Centralized systems**. Users in these centralized schemes do not have the ability to consider the electrical system as something in which they can participate or exercise any influence; these are only passive agents who take the electricity supply as a service with fixed costs and qualities, managed exogenously to them.

(b) **Decentralized systems.** The emergence of the process of electrical decentralization implies that the aforementioned decision-making capacity is no longer controlled through a monopolistic exclusivity. Decision-making capacity has reached a certain degree of dissemination and there are actions that are carried out by independent generating units, which coordinate and provide various electrical services to users who belong to one of the groups created around such units, either as a self-user or a client.

(c) **Distributed systems**. In this scheme, the decision-making capacity has been disseminated completely toward the users, who now, with their own generation capacity and independent decision processes, actively determine investment, financing, expansion and penetration actions of the service they provide. Each user is an active agent and their decision-making capacity is fully integrated into the electrical system.

It is important to highlight that the concept of decision-making capacity that we are describing goes beyond the simple initiative to generate. Without a doubt, the emergence of users with generation capacity (including the small users known as "prosumers") is one of the catalysts of electrical decentralization. However, new services and capabilities go well beyond to just adding new generation capacity. For example, a prosumer can provide other services or products such as storage

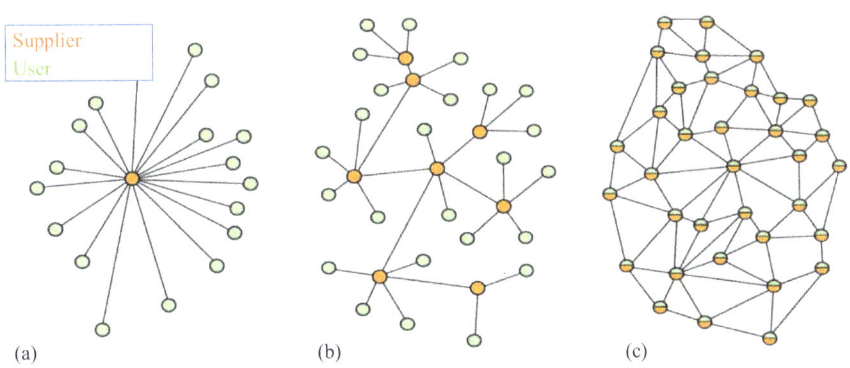

Figure 10.2. Decentralization of electricity systems. (a) Centralized. (b) Decentralized. (c) Distributed. (Model developed by Hyon K. Ra [15])

(particularly by the use of batteries), regulation and demand side management, without forgetting that it is that same user who decided, in a decentralized way, to take the cost and risk of investing and installing this new capacity without any intervention of a central planner.

The opportunity to develop decentralized generation, even to the extreme of having a mix of distributed generation and storage power to carry out off-grid energy schemes (not connected to the electricity grid), is now within the realm of technological advances. Like most sophisticated technological models, early advances on decentralization schemes would require a proper regulatory incentives/structure, taking into account that regulation is one of the most important ways in which governments can diminish the risk of these innovative approaches.

Decentralized generation is particularly appropriate in densely populated areas and where there is limited transmission capacity, since distributed sources are becoming an important substitute of electricity transmission services. Furthermore, the fact the decentralized generation is basically a substitute to new distribution capacity deployment; it is a resource that directly competes with this so-called "natural monopoly."

As Dylan McConnel points out, "The single largest cost of supplying electricity is the network costs (the 'poles and wires') which are regulated, and as such are seemingly insulated from the effects of competition within this paper. Whilst they are described as 'natural monopolies' (and thus must be regulated), this is increasingly no longer the case with distributed generation (such as rooftop solar), and particularly the emergence of cost-effective storage" [12].

"In this context, storage and distributed generation are a direct competitive threat to the network business. Regulating networks on a cost recovery basis makes less sense in the face of such competition."

Thus, decentralization would have an important consumer empowerment effect providing for more consumer alternatives, in addition to enhancing the overall efficiency of the electric system by *literally* challenging one of the most enduring notions of natural monopoly: the electricity distribution grid.

Setting up a new regulatory framework for distribution decentralization in Chile

As of November 2019, the Chilean Senate approved and announced a new legislation to redesign the tariff mechanisms applicable to electricity distribution activities. In this context, this new legislation is also known as the "Ley Corta.[‡]"

Even though Chile was one of the first countries in LATAM to introduce electricity market liberalization, their current framework is no longer considered

[‡]*Official communication of the approval of the "Ley Corta" is available at the Senate's website at* https://www.senado.cl/ley-corta-de-distribucion-electrica-es-despachada-de-la-comision-de/senado/2019-11-06/163240.html

adequate for the new technologies that are pushing for better-distributed resources. In particular, there was a lack of incentives to provide for the entrance of new technologies and business models.

The approved piece of legislation revises the distribution tariff, but it is a part of a broader aim to reform the whole electricity distribution sector, including the unbundling of generation and the commercialization/supply activities, which was carried out jointly by the distribution companies.

The upcoming proposal, also known as "Ley Larga" would also modify the "Ley General de Servicios Eléctricos" and introduce a more generous distributed generation scheme.

Since distributed generation and overall clean energy penetration requires good distribution capacity and the right incentives to invest, this distribution and commercialization separation can foster competition and provide the opportunities of new entrants to offer electricity supply, invest in distributed resources and increase consumer choice by setting a new tariff structure that would effectively identify and differentiate the costs of the transmission and distribution service, provide for an unbundling of these two markets with low contestability and allocate the foundations of structural separation.

10.5 The strength of the forces for change

The applicability of the regulatory reform is directly related to the power of the forces impelling it; in this sense, Bartle argues that the forces of the logic behind regulatory reform are of a lower intensity than in other sectors [13]. Therefore, in contrast to technological advances in generation that began to take place at the end of the twentieth century, at that time there was not a technology that could provide an alternative to the electricity grid infrastructure (at least not until the penetration of distributed resources). Hence, regulatory reform processes are more robust in the generation and supply segment of the market than in the network segments, as has been previously mentioned.

However, this diversity of new technology has led to the creation of new products or services, so there has been a great surge of new actors striving for liberalization. Also, because of the already mentioned attributes of the electricity systems, an important aspect of the regulatory reform process is the design of complex regulatory policies to take care of the most pressing issues hindering such liberalization. Such issues are common generic challenges found in electricity systems previously characterized by vertical integration and some of which are the following:

1. Breaking up the monopoly implies the fragmentation of a market into units, which are numerous enough for effective competition; this creates contractual complexities at the interfaces of those firms with each other and with other economic agents.

2. Concentration (and market power) tends to remain high in liberalized electricity markets, because factors related to entry barriers and important scale economics in the residual grid, among others economic attributes already discussed. Since competition policy is ill-suited to address *ex ante* control of market power, it is specific energy sector regulation policy that can best address those structural measures of market power control, for instance, through asymmetric regulation.
3. Unrestricted and nondiscriminatory third-party access to electricity networks has to remain effective in many markets, mainly because incumbent generators can have incentives to influence grid access in order to eliminate new entry of competitors by denying distribution and transmission services, thus a strong regulatory enforcement to monitor the market and enforce third-party access obligations is needed.
4. Prices have to be sufficient sensitive to supply and demand conditions. The inelasticity in the short-term pricing of electricity is prone to abuse by the generators, which in turn could put pressures to constrain the development of new generation capacity in order to restrict supply and manipulate prices. This is arguably one of the reasons of disasters like the famous California Crisis [17].
5. The boundaries between energy and environmental policies are rapidly converging. Because of the repercussions of energy regulation in environmental concerns, particularly climate change, the electricity markets will be very likely to be prone to government intervention.

These difficulties and complexities in introducing competition and climate concerns into the electricity sector outline the need for a strong regulatory apparatus of the government in order to monitor and control abuses of market power, address the issue of climate change, draw the framework of the contractual scheme and enforce it. There is also the need for the presence of a strong judiciary to resolve controversies and safeguard the new property rights structure. In the case of a regulatory reform, those issues can signify the difference between an effective regulatory regime or an insufficient one.

For the abovementioned reasons, the forces impelling the regulatory reform have had several circumstances that entered into play in its application to the Mexican context. This subject will be developed in the following section.

10.6 The constitutional reform of the electricity industry in Mexico: the fall of a regulatory Chinese wall

The main obstacle faced by any attempt to introduce a regulatory reform in the Mexican electricity industry was the constitutional provision stating that the public service of electricity is an activity to be carried out exclusively by the government, excluding the possibility of any private participation, even by means of concessions or licenses.

Given the fact that the regulatory strands that were drawn at the constitutional level were removed by the 2013 energy constitutional reform, the applicability the liberalization process allowed for the introduction of a good level of liberalization and competition that provided new market entrance and modern regulatory developments.

The tradition of *dirigisme* refers to the emulation of industrial policies that tend to intervene in the operation of the markets with different political aims. In the context of state-owned enterprises, in exchange for its "services," those entities had been delegated a monopoly that included extensive self-regulation and powers in their sector planning. This particular institutional context in Mexico shows that it had characteristics making it hostile to an application of a full regulatory reform.

In this sense, even though, in the 2013, the constitutional reform and its implementing legislation introduced an important degree of liberalization in the generation and supply sectors, the regulatory reform was filtered by the presence of an important institutional factor—the influence of the state-owned enterprise, the Federal Commission of Electricity (CFE). Although CFE was no longer vested with a legal monopoly, it retained the representation of the government, now as a "Productive Enterprise of the State."

10.7 The progress of electric decentralization in Mexico

Ever since the nationalization of the electric industry in Mexico in 1960, the centralized model would be embodied in the Constitution, which is the legal norm of greater hierarchy, reserving to the State any generation for public service purposes.

However, even within that constitutional framework, Mexico began a process of overcoming the centralized system since 1992, with reforms that not only allowed private companies to generate electricity for self-supply, but also other modalities that were not considered for public service purposes: cogeneration, small production and independent production.

What was recorded is that companies with a certain sophistication and scale could "deconcentrate" and begin to exercise decision-making capacity to meet their needs and those of their partners/clients, constituting important private distribution schemes, such as societies for self-supply.

It is important to note that regulatory changes created the opportunity to relocate this decision power that would be exploited by industrial and large commercial companies, in addition to some local governments. For smaller users, the new regulatory spaces went by relatively unnoticed.

Finally, and already within the context of the energy reform of 2013, the Electricity Industry Law of 2014 (LIE) determines that the permits granted under the previous legal framework will be respected on their original terms, so the self-supply societies will continue to operate on a "grandfathered regime," for the remaining years of its title.

10.7.1 Liberalization of electricity generation and supply

The regulatory reform in Mexico began in 2013, when the Congress approved a motion of the government to modify of the Constitution in all energy matters, including the oil and gas, as well the electricity sector. Due to the so-called "2013 energy reform" an important degree of liberalization in the electricity generation and supply was introduced. In broad terms, private investment was allowed in the following schemes:

1. generation (including renewable generation),
2. load-serving activities, also referred to as supply and
3. commercialization activities without actual supply of electricity to load center.

On the other hand, transmission, distribution and the operational control of the electricity public grid were confirmed as exclusively reserved to the Mexican state.

The new legal framework allowed the new private actors to be permitted by the government to follow those activities that were explicitly liberalized. The permits also provide them the right of nondiscriminatory third-party access to the grid, by which the new system operator, a brand-new state-owned enterprise called *Centro Nacional de Control de Energía* (CENACE) had the obligation to allow interconnection with the national power grid. As already been noted, liberalization of the generation sector and the third-party access obligation are the characteristics of a basic liberalization of the electricity industry.

In this regard, it is worth mentioning CENACE's main mandate is to operate the grid independently and offering equal treatment to all the interested parties and market participant, particularly the CFE, entity of which CENACE used to be an internal department before the constitutional reform. The design of the CENACE as a state-owned enterprise, and thus sharing almost identical governmental control with CFE, as well as having a staff comprised primarily of former CFE's employees, have given rise to concern about its neutrality and effective open access rights for third parties.

10.7.2 The emergence of the distributed schemes in Mexico

The practical implementation of the scheme distributed in Mexico emerge in three main areas described in the following.

10.7.2.1 Distributed generation

In general terms, distributed generation refers to a series of technologies, such as PV solar or small-scale wind turbines which have allowed small-scale plants to be installed in the same place where their generation will be consumed. As long as it is already "distributed," said generation does not require transportation. Currently, most of the momentum of distributed generation in Mexico comes from PV systems.

It is useful to note that these types of distributed electricity generation are how the mechanisms for lighting initially emerged. Those of us who grew up in small cities during the last decades of the last century can still remember that the use of oil and kerosene lamps was not unusual. The system was born distributed, and now this generation is having a comeback, in much more powerful ways.

In the legal field, the LIE, in its second article, section XXIII, defines distributed generation as the generation of electrical energy that meets the following characteristics:

1. it is carried out by a generator with a capacity of less than 0.5 MW (therefore exempted of obtaining a generation permit);
2. it is carried out in a power plant that is interconnected to a distribution circuit that contains a high concentration of load centers. In this way, the distributed generation must always be connected to a distribution network, which means that, at least in a legal conception, a generation unit disconnected from the network (off grid) is not legally considered in Mexico as "distributed."

As already pointed out, the LIE requires that distributed generation has a maximum capacity of 0.5 MW; below this value, a generator does not require a government permit to produce electricity, under the terms of section XXV of the same article 2 of the LIE. Undoubtedly, this threshold should be raised to include greater potential for participants.

According to data from the Energy Regulatory Commission (CRE), distributed generation has made relatively rapid progress in Mexico, but still represents less than 0.9% of the country's total generation installed generation capacity. That percentage is much higher in other latitudes, for example, in California that percentage is around 10%, while in Germany the distributed generation has represented at least 22% of the installed capacity for several years.

In Mexico, the legal framework of distributed generation is technologically neutral, an approach that by the way has proven to be virtuous in boosting competition between different technologies; however, PV solar is proving to be the leader in on-site generation issues, because of its manageable scale and low cost.

10.7.2.2 Isolated supply

Article 22 of the LIE defines isolated supply to the generation of electrical energy for the satisfaction of its own needs, without transmitting said energy through the national transmission network or through the general distribution networks. Although the isolated supply corresponds to the distributed scheme, it has the advantage that in Mexico it does allow generation above the 0.5 MW limit, allowing it to be a scheme used by users with very intensive energy consumption.

Other important advantage in isolated supply is that allows the generation unit to be interconnected to the transmission or the distribution grid for balancing and backing up purposes, without any limit of the fraction of energy that can be balanced.

To adhere to the text of the LEI and avoid competition in unequal conditions with suppliers, the requirement that the isolated supply be destined to the satisfaction of their own needs was interpreted by the CRE as the need to have an integration between the unit of generation and load center, demanding that both belong to the same "economic interest group" a concept borrowed from the competition law tradition. If a load center is not part of the same economic interest group, it shall exclusively be served by a supplier.

10.7.2.3 Controllable demand

In accordance with article 2, section XVII, of the LIE, the controllable demand (conceptually a mechanism pertaining to the demand response or demand-side management schemes) is defined as the avoided demand for electrical energy that end users offer through consumption reduction in accordance with the market rules. It is an authentic electricity product that seeks to generate incentives for users to intervene in their own consumption patterns, simply by reacting to a market signal, which are high electricity prices.

Article 49 of the LIE establishes that users of basic supply with controllable demand may offer their demand reduction through a supplier, which provides the necessary legal structure so that the new product can be included among the options to be traded in the wholesale electricity market; however, the final regulation that implements this legal is still being processed.

What are the regulatory options for implementing said policy in the distributed scope? There are at least two, one option being dynamic rates making consumption during peak hours more expensive, thus incentivizing consumption reduction to reduce the electricity bill. Conversely, there is the option to consider controllable demand as an product to be priced by the market, which would provide the offer of the "nega" watt the right to receive a direct payment in exchange for its reduced consumption in certain times of the day where the system reports more stress (think about a Monday around 6 PM, a time in which most regions of the country people return to their homes and daylight begins to diminish).

10.7.3 Distributed generation: the blurring of a natural monopoly assumption

The disruption of renewables technologies has a broad array of manifestations, besides scale economics and diminishing long-term marginal costs, which are factors behind the competitiveness of utility-scale plants. However, the development of cost-effective distributed resources challenges one of the most important assumption of the regulatory design of the electricity industry, the natural monopoly characteristic of the T&D segments.

Because the installation of generation facilities in practically the same place where their energy would be consumed (eliminating the need to transport it) substitutes new transmission capacity in order to bring energy remotely generated. To the extent a service that faces substitution is no longer subject to have a monopolistic status, this assumption could be revised, which would have important consequences to the traditional regulation of T&D networks, for instance, a tariff regulation based on a cost recovery approach.

10.7.4 An independent regulatory agency: the CRE

The CRE was created in 1993 as a consultant office of the Ministry of Energy to advice on the regulation of the liberalized activities introduced in 1992. In 1995, the Congress approved the Energy Regulatory Commission Act, which upgraded it from a consultant office to a higher degree of autonomy, one derived from federal legislation.

With the energy reform, the CRE gained a much higher degree of delegation and independence. However, as the new legal statue provided by the Constitution was that of a Ministerial body ("*Dependencia*")—although highly ranked—it was not given the same full constitutional autonomy as other regulatory agencies in Mexico, such as the Federal Competition Commission or the Federal Telecommunications Commission.

The CRE was situated under the executive power scaffolding, so it was still subject to several budgetary controls from the finance ministry, and political control from the energy ministry by means of a nebulous obligation to "coordinate" with the latter, which greatly affects a true independence and autonomy. In addition, the new corporate design of the state-owned enterprise provided that the chair of the board of CFE was to be occupied by the Energy Ministry and the Finance Ministry will always have a seat at such board.

The conflict of having ministries that were to represent the interest of the federal government as sole owner of CFE at the board could raise concerns taking into account the influence that those ministries could exert into the faculties of the CRE, now in charge of regulating the activities of CFE on an equal footing with the new private entrants.

The demands of private investors for governmental commitment and credibility could be satisfied with the introduction of an independent agency to regulate the liberalized activities free of political intervention and reduce the risk of investment. Even though subjected to all the controls of ministerial bodies, the powers given to the CRE were vast and included the regulation of the activities of generation, permits, a limited intervention in electricity supply tariff setting[§] and dispute resolution mechanisms.

Such a balance normally would create a "shared authority" in the regulation of the sector [7]. In the case of Mexico, where the electricity sector still has a dominant player with a *de facto* monopolistic presence in the provision of the electricity supply, the regulatory authority is distributed between the Ministries (Finance and Energy), the CFE and the CRE.

10.7.5 The re-regulatory process: liberalized activities and its new rules

With the liberalization of private generation activities, a new scheme of permits and regulatory instruments was created. Rules governing the requirements and regulation of the permit holders and development of the specifics of the contractual schemes (e.g. tariffs and specifics of interconnection, power backup and contracts with final users) were developed to regulate the new activities.

Moreover, a new set of market rules to create and guide activities, services and products of the electricity industry was created to regulate the kind of activities that

[§]The tariff-setting power given to the CRE, as set forth by article 139 of the Electricity Industry Act, was subject to be bypassed by the Ministry of Finance in representation of the executive power, which has been the case so far, so the independence of the CRE was limited in this regard.

used to take place only within the former monopoly. The most important content of the market rules is the following:

1. the implementation procedures of the different markets;
2. general provisions to be followed by market participants and CENACE;
3. regulations to be followed for the sustainability of trust in the system, operational procedures in emergencies, minimum communications to be carried out between market participants and CENACE;
4. special schemes for operating small electrical systems;
5. market operations and related services;
6. operative planning;
7. long-term markets;
8. parameters to be registered by electric power plants to provide a unity in market surveillance, and compliance with credit requirements;
9. management of credit for market participants, restriction, suspension and termination of contracts due to financial capabilities;
10. general guidelines for measuring the quality of invoicing.

10.8 Conclusions

The stagnation of some of the regulatory reform forces with others still in motion in Mexico has led to a complex regulatory regime in Mexico. The complexity comes from a coexistence of modern economic regulation by a quasi-autonomous regulatory agency, the CRE, with a *dirigisme* structure represented by a politically motivated body, the Energy Ministry, and the economic influence of a former legal monopoly, the CFE, which is still vested with an active dominance in practically every segment of the electricity industry.

This situation is an accurate example of the prediction about the future of the regulatory state, best to be expressed as some form of hybrid between the regulatory and the welfare state [6]. In the Mexican electricity industry, the forces impelling the change and the application of the regulatory reform clash with the institutional features of the welfare or developmental state. In the Mexican electricity industry, the forces for change to establish the regulatory reform and its institutional mediation has had the outcome of a regulatory regime where the regulatory reform is applied partially.

In this case, the Mexican electricity regulatory structure would fit in Thatcher's definition of "Semi-liberalized public service regime" where there is a continuing direct public role in the supply of services (in this case through the government-owned enterprise), limited liberalization, re-regulation and a "fair and effective competition with a weak or no independent regulatory agency" [14].

A regulatory regime reflects the forces impelling changes, showing which direction the reforms will take. The other part of the equation lies on the institutional context of the specific sector where the regulatory reform is going to be applied. The interaction between those two factors will draw the outcome of a regulatory regime, the questions of which regulatory and welfare state features will adopt depend upon the intensity of the forces and resistances toward changes.

Being the regulatory reform a dynamic process, the forces for change, along with the different circumstances surrounding the institutional context that would mediate them, are inserted in a constant motion, so changes in either force driving the reform, or the institutional endowment, will cause transformations in the regime.

Following our case as an example, if Mexico faces a situation of deep fiscal stress and a high electricity demand, the need for private investment and more technologically complex schemes could derive in a greater dependence on private sector projects. This circumstance would incentivize to lower the risk of investment to private actors, a goal that can be attainable by decreasing the possibility of political intervention (by ministries and the state-owned enterprise), which in turn can be achieved by strengthening CRE's regulatory power, and moving the regulatory regime closer to the regulatory reform principles.

In this sense, the interaction of forces and resistances are inserted in the dynamic on the state's settings of priorities. A greater dependence in efficient regulation would impel the advancement of the regulatory reform, while dependence in the use of the regulatory structure with political and *dirigisme* aims would support the maintenance of welfare state structures.

This balance is always in tension.

References

[1] Musacchio, A. and Lazzarini, S (2014) *Reinventing State Capitalism, Leviathan in Business, Brazil and Beyond* (Harvard, Harvard University Press).

[2] Majone, Giandomenico (1996) "The European Commission as Regulator," in Majone, Giandomenico (ed.), *Regulating Europe* (London, Routledge) pp. 61–79.

[3] Majone, Giandomenico (1997) "From the Positive to the Regulatory State: Causes and Consequences of Changes in the Mode of Governance," *Journal of Public Policy*, 17, 2, pp. 139–167.

[4] Joskow, Paul L. and Tirole, Jean (April 21, 2004) "Reliability and Competitive Electricity Markets," MIT Department of Economics Working Paper No. 04-17. Available at SSRN: https://ssrn.com/abstract=543862 or http://dx.doi.org/10.2139/ssrn.543862. [Accessed 15 January 2020].

[5] Vogel, Steven (1998) "Free Markets, More Rules. Regulatory Reform in Advanced Industrial Countries," First printing, Cornell paperbacks (Cornell, Cornell University Press).

[6] Loughlin, M. and Scott, C. (1997) "The Regulatory State," in Dunleavy, P. *et al.* (eds.), *Development in British Politics* (London, Palgrave).

[7] Thatcher, Mark (2002) "Delegation to Independent Regulatory Agencies: Pressures, Functions and Contextual Mediation," *West European Politics*, 25, 1, pp. 125–147.

[8] Steiner, Faye (2000) "Regulation, Industry Structure and Performance in the Electricity Supply Industry," Economic Department Working Papers, no. 238.
[9] Republic of South Africa (2019) *Roadmap for Eskom in a Reformed Electricity Supply Industry* (South Africa, ABC Printers).
[10] Newberry, David (2001) *Privatization, Restructuring, and Regulation of Network Utilities* Third printing, (Massachusetts, Massachusetts Institute of Technology).
[11] Non-Traditional Transmission and Distribution Solutions: Market Drivers and Barriers, Business Models, and Global Market Forecasts, Navigant, 2017. Available at: https://www.navigantresearch.com/reports/non-wires-alternatives. [Accessed January 18, 2020].
[12] The Conversation, "Energy White Paper Promises Privatisation and Lower Prices: Experts Respond" (April, 2015). Available at https://theconversation.com/energy-white-paper-promises-privatisation-and-lower-prices-experts-respond-39853. [Accessed November 26, 2020].
[13] Bartle, Ian (2001) "Transnational Economic and Technological Forces, Institutions and Policy Change: The Reform of Telecommunications and Electricity in Germany, France and Britain," *Paper Presented in the Workshop: National Regulatory Reform in an International Environment*, Grenoble, France, April 6–11.
[14] Thatcher, Mark (2002) "Analysing Regulatory Reform in Europe," *Journal of European Public Policy*, 9, 6, pp. 859–872.
[15] Sklar, Scott and Rah, Hyon K. (2018) "Reconstrucción de la infraestructura en las islas del caribe tras un huracán: cuando el modelo 'business-as-usual' es riesgoso," Alianza de Energía y Clima de la Américas [Online] http://sp.ecpamericas.org/Blog/Default.aspx?id=140 [Accessed November 25, 2019].
[16] Lazard (2018) *Annual Levelized Cost of Energy Analysis* (LCOE 13.0). Available at https://www.lazard.com/media/451086/lazards-levelized-cost-of-energy-version-130-vf.pdf
[17] Ocana, Carlos (2002) "Regulatory Reform in the Electricity Supply Industry: An Overview," *Working Paper of the International Energy Agency*, September.

Chapter 11

Effective market design for high-renewable penetration

James Bushnell[1]

Since the late 1970s, efforts around the world to transform the electricity industry have featured two simultaneous and overlapping policy goals: (1) the development and promotion of renewable electricity generation and (2) the restructuring and reduction of the role of economic regulation in industry operations. Through much of the last three decades, these two strains of policy goals, often originating from opposite ends of the political spectrum, have had a largely synergistic interaction. The original promotion of small-scale renewable generation helped to spawn what would become a powerful and influential independent (nonutility) power producer sector.

This dynamic continued to build throughout the 1980s and 1990s and culminated in efforts to create transmission markets that were unassailably independent and unbiased and regional efforts to largely deregulate the generation sector. While much of the initial institutional momentum for regulatory reforms can be traced to early efforts to promote renewable energy, as the deregulation wave reached its apex in the mid-2000; in the subsequent period, the policy emphasis has again shifted away from deregulation and toward promoting renewable energy production.

With this last transition, and in part due to the now-sizable share of production claimed by renewable sources, the synergistic relationship between renewable energy policy and deregulation started to break down. Many leading nonutility generation companies are struggling financially to cope with low wholesale energy prices, depressed in part by the large share of renewable energy pouring into wholesale markets.

This paper will survey the recent history of renewable electricity and its impact on power markets and energy policy, with a focus on markets in the US. The discussion will highlight the crossroad at which the industry now finds itself. The industry may be forced to embrace the market implications of a renewable-centric energy policy that, in equilibrium, would result in more volatile and ultimately higher average electricity costs. The rapid influx of renewable capacity that is being

[1]Department of Economics, UC Davis and National Bureau of Economic Research, Davis, CA, USA

layered on top of already adequately resourced systems implies a potentially growing divide between declining wholesale energy prices, and end-use retail rates that must include the costs of the excess capacity.

The increasing share of zero marginal cost production, combined with the inherently intermittent and uncertain nature of renewable energy production, has caused some to question whether markets can support, or are even compatible with, power systems dominated by renewable energy. My general argument is that, while the cost characteristics of renewable energy will almost certainly reshape short run electricity supply, and therefore prices, it is far less obvious that the general design approach of current wholesale energy markets is incapable of transmitting the appropriate price signals and incentives to market participants. Put bluntly, large-scale renewable penetration will lead to extreme prices under our current market designs precisely because extreme prices will be needed for efficient deployment and utilization of those resources. Rather than rejecting the evolution to more efficient marginal cost pricing at the wholesale level, electricity markets need to "double down" on these design philosophies to ensure that appropriate scarcity and opportunity costs are reflected over time and space in future power markets.

11.1 The organization of the electricity sector

In the period from the mid-1990s through the 2000s, a process of industry liberalization, including privatization, and restructuring was adopted in electric sectors in many parts of the world. The experience in the United States is informative because within the US, all the major forms of industry organization are represented in one region or another. As the deregulatory wave crested in the mid-2000s, the organization of the industry across the US stabilized into one of two competing paradigms. The first was that of the traditional, vertically integrated, and regulated utility epitomized by the Southern Company, whose subsidiaries serve much of the Southeastern part of the country. The second paradigm was one of near-complete deregulations of energy provision, epitomized by the market overseen by the Electricity Reliability Council of Texas (ERCOT). In much of Texas, generation is controlled by nonutility producers earning market-based revenues for their production, and retail prices as well are only loosely regulated [1]. However, a deeper look at the different sectors of the power industry paints a more nuanced picture of the state of regulation, or deregulation, across the country. Any categorization of the organization and regulatory status the power industry in the US must consider three distinct dimensions: wholesale transmission operations, generation ownership and remuneration, and retail access and competition. Within the US, one can find individual states occupying almost any combination of organization structure along these three dimensions.

11.1.1 Transmission operations

At the transmission and operations level, the industry has continued to embrace the model of regional transmission organizations (RTOs), independent organizations

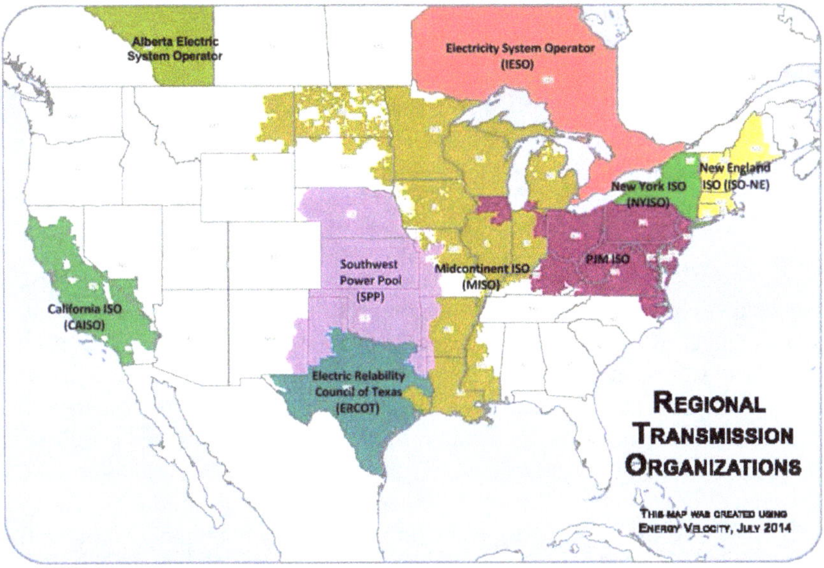

Figure 11.1 US ISOs and RTOs as of 2012

(also referred to as *Independent System Operators*) that operate electricity transmission networks and the markets through which the reliable balance of supply and demand is maintained. Despite a rejection of other aspects of "deregulation" by certain states, the RTO model has come be embraced by an increasing fraction of the country (see Figure 11.1). Indeed, the receding pressures to deregulate the generation and retail sector in much of middle America has made many states hostile to those changes willing to merge their grid operations into regional organizations. In the Western US, a region that, outside of California, has long resisted efforts to merge operations into an RTO style market, efforts to extend the California Independent System Operator (CAISO) spot market into a regional energy imbalance market have secured the participation of seven other western states.*

For unintegrated renewable energy producers, the RTO model is generally seen as more accessible and transparent. The central mission of RTO is to ensure nondiscriminatory access to the grid, and in practice this is accompanied by the operation of real-time balancing markets that are a better economic match for intermittent producers. Outside of RTO regions, intermittent producers are often required to purchase balancing services from the transmission owner and operator at less flexible and competitive prices. The belief that RTO markets such as the one overseen by the California Independent system operator provide a more welcoming environment to grid connected renewables is supported by the fact that in 2010 an

*Utilities operating in Arizona, Idaho, Nevada, Oregon, Utah, and Washington states are in various stages of integration with the Western Energy Imbalance Market. See https://www.caiso.com/informed/Pages/EIMOverview/Default.aspx.

excess of renewable projects throughout the Western US were willing to invest in the capability to "dynamically schedule" there output into the California system, making themselves electrically part of the CAISO system despite being physically located outside of California [2]. The recent broad embrace of the western energy imbalance market after decades of failed efforts at regional integration can also be attributed to the claimed benefits of integration for grid connected renewable energy.

11.1.2 Generation ownership

The most consequential change to result from restructuring efforts of the late 1990s and early 2000s was the shift of the generation sector away from rate-of-return regulation toward a market-based form of remuneration. This was accomplished through both external and internal sales of existing generation capacity from regulated utilities to nonutility generation firms (or affiliates), as well as substantial burst of new capacity investment during the early 2000s that was almost exclusively nonutility generation [3]. While states where this transition took place very rapidly, such as California, Texas, and Pennsylvania, are commonly considered the "deregulated" states, the mix of generation ownership is in fact quite diverse throughout the US. Figure 11.2 illustrates the share of energy produced from nonutility sources as of 2012. Nonutility generation has a nontrivial share even in states such as Alabama and Georgia, not commonly associated with deregulation. In less restructured states, nonutility producers typically operate under either contracts with the incumbent utility or with a small number of large industrial customers.

In the renewable sector, the ownership pattern is less diverse and is dominated by the nonutility sector. This fact illustrates one way in which the origins of

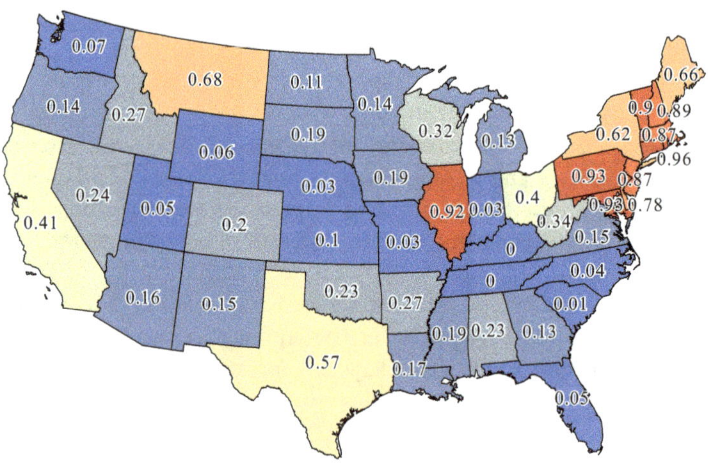

Figure 11.2 Share of output from nonutility generators in 2012 (from [8])

restructuring and renewable generation overlapped. The birth of a substantial nonutility generation sector (and therefore restructuring) can be traced to the Public Utility Regulatory Policies Act of 1978 (PURPA) [4,5]. The PURPA required incumbent monopoly electric utilities to purchase all energy produced by "qualifying facilities," a category that included combined-heat and power, small (less than 50 MW) natural gas production, and renewable generation. Implementation of the Act was left primarily to the states, who responded with widely varying approaches that reflected local enthusiasm for the spirit of the legislation [5,6]. In states such as California and New York, substantial regulatory pressure was placed upon incumbent utilities, and significant incentives were provided to qualifying producers. California's long-standing leading role in renewables development in the US can be traced to this period.

While PURPA provided a market for the output of renewable generation, other incentives were necessary for the financial viability of these plants relative to other qualifying technologies such as Combined Heat and Power (CHP). Within the United States, these incentives largely took the form of production tax credits, and preferred purchase arrangements that required utilities either pay more (e.g. through PURPA contracts) for, or buy a specific quantity of, renewable energy.

Until recently, the renewables industry has been almost exclusively operated by nonutility providers. Changes to the investment tax credit (ITC) in the late 2000s has raised the prospect of ratepayer-backed renewable generation investment and indeed that has spurred some activity in several states. While rising, the share of utility-owned production is still relatively modest as shown in Figure 11.3.

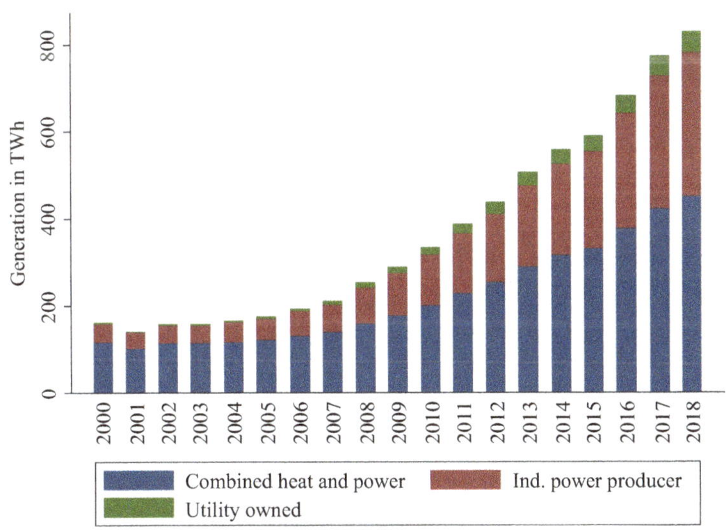

Figure 11.3 Utility-owned renewable generation

11.1.3 Restructuring and reform of retail services

The aspect of restructuring to receive the political attention was the relaxing of the utility monopoly franchise over retailing. The appeal of liberation from one's franchise monopoly was coupled with optimism about innovation in retail services.[†] The place where innovation did seem valuable and likely to occur with retail choice was in financial arrangements: price schedules, payment plans, and options to bundle purchases with complementary products.

More concretely, retail restructuring involved giving customers access to new "energy-only" retail providers who produced or acquired wholesale power for sale to end users. The incumbent utility (and the grid operator) maintained a franchise over distribution and transmission-related functions. In many cases, the incumbent utility was allowed to continue to offer a default "bundled" retail rate for customers who did not switch retailers [9].[‡] Customers who did switch received a bill for "energy-only" service from the third-party retailer they chose, and a separate charge, intended to recover transmission and distribution system investments made by the incumbent utility. The extent to which this transformation has materialized has varied greatly around the US. Texas has far outstripped the rest of the country on the retail competition front, with the only other significant activity clustered in the Northeast [8].

One area where retail competition might be seen as providing more diversity and innovation in services is the provision of so-called "green pricing" plans. Under such tariffs, the retail provider agrees to source a certain percentage of the energy consumed by a green pricing customer from renewable energy sources. Indeed, outside of time-varying pricing, green pricing was one of the most frequently cited niches through which a retail might be able to differentiate itself. There are at least two problems with placing much value on retail choice as a fruitful channel through which to expand the output of renewable energy. First, although such plans for the most part lived up to their commitments to procure renewable energy in sufficient quantities to plausible claim that their customers were "buying" from a renewable heavy mix, this is not the same thing as truly expanding renewable output. To the extent that such renewables were already in existence or in the development pipeline, green customers were simply grabbing a proportionately higher share of this energy at a premium price. This was more a reshuffling of energy consumption than an actual expansion of the total. This situation was more likely to arise in states that had other programs promoting the expansion of renewable energy as such states often had a decent amount of renewable energy already.

The second flaw in the logic of the "choice will promote green" point of view is that green pricing tariffs need not be limited to states or load-serving entities (LSEs) that had deregulated their retail services. As early as 2002, utilities in 32 states serving roughly 20% of the nation's retail customers were offering green

[†]See Joskow's discussion of the potential for new product innovation under electricity restructuring [7].
[‡]The bundled rate combined energy with the incumbent utility's transmission, distribution, and retailing charge. This was sometimes called the "default provider" or "provider of last resort" (or POLR) rate. In some states, the default provider franchisee is selected through auctions overseen by local regulators.

Figure 11.4 Sales under "green pricing" tariffs in 2012

pricing tariffs. While one could make the reasonable argument that competitive pressures played a role in moving regulated utilities to take such steps, most states with utility green pricing programs do not have any retail deregulation measures in place.[§] The US Energy Information Program surveys all LSEs (competitive and regulated) on both their conventional and green electric sales. As shown in Figure 11.4, the greatest green pricing activity is in the upper midwestern and Southeastern US, regions that have never seriously considered retail deregulation.

11.1.4 Restructuring and distributed generation

The relationship between retail competition and distributed generation is not immediately obvious, but in many ways distributed generation is another manifestation of the political dynamic that promoted retail choice. In both cases, there is a strong undercurrent of motivation to gain a degree of independence from an often unpopular incumbent provider. Borenstein and Bushnell [8] also point to another similarity. With both retail choice in the late 1990s and distributed generation in the 2000s, much of the potential apparent benefits to the customer accrue from an opportunity to avoid paying for the sunk or fixed costs of an integrated provider. This trend has led some observers and utility executives to predict a "death spiral" in which a significant number of customers self-generate much of their electricity, forcing the utility to raise rates for the electricity they still sell in order to cover fixed investments, in turn making solar photovoltaic (PV) economic for a larger set of customers who then reduce their purchases, leading to a greater revenue shortfall

[§]Wiser, Olson, Bird and Sweezy, 2004.

and another rate increase, and restarting the cycle. Ultimately, some argue, the monopoly utility disappears. This scenario has triggered widespread debate—both positive and normative—about the future and viability of the utility. The regulator in New York state has even proposed a complete redesign of utility systems that is focused on customers also being generators [9].

The social welfare gain from increasing reliance on distributed PV generation, however, is still far from clear. Even the most optimistic cost scenarios suggest that the full social levelized cost of electricity from residential solar PV is likely at least $0.15/kWh in relatively sunny areas, more than double the full cost of utility-scale solar installations [10]. Distributed PV generation is eligible for the same tax benefits as large-scale solar, a 30% tax credit through the end of 2016 and accelerated depreciation. Borenstein and Notsund [11] estimate that the accelerated depreciation amounts to an additional effective subsidy of about 15%.[‡‡]

Distributed PV generation also benefits from being compensated at retail prices for the power it produces. Under "net metering," which has been adopted in most of the US, customers are credited for all power produced from their PV system by deducting the quantity from the customer's consumption.[§§] In reality, calculations in [12] suggest that less than half of the power produced by a typical household PV system is consumed onsite—actually reducing the customer's retail demand—but net metering treats all power as demand reduction, thereby crediting it at the retail rate the customer would have paid.[**] If the full benefits that residential rooftop solar PV power brings to the market are less than the marginal rate the customer pays, then net metering policies lead to overcompensation of residential rooftop solar production. A simple calculation suggests this is very much the case, but the full system benefits are a matter of some dispute [10] and [13]. What is clear is that retail electricity rates are set in ways that are not closely tied to long-run marginal cost, so incentivizing residential rooftop solar through net metering will conflate solar policy with rate design policy and will have unpredictable effects on the incentive to install residential solar.

11.2 Policies driving renewable electricity

While recent public support for renewable electricity stems largely from the potential environmental benefits, Federal support for renewables was initially motivated by concerns surrounding energy security. During the late 1970s, 17% of US electricity production came from petroleum—up from 7% during the 1960s. In

[‡‡]Actually, the accelerated depreciation benefit is available only if the system is owned by a company, not an individual. This has been a significant factor behind the rapid growth of third-party-owned residential systems in which the third-party owner leases the system to the homeowner or, more commonly, sells the electricity from the system to the homeowner. Third-party owners of these systems point out that this model also greatly lowers, or eliminates, the up-front payment the homeowner would otherwise have to make.

[§§]See http://dsireusa.org/solar/solarpolicyguide/?id=17 for timely information on US state net metering policies.

[**]This is for a system that generates electricity equal to half of the household's demand. The figure could be even lower for a system that is larger relative to household demand.

response to the oil price shocks during the 1970s, policymakers placed an emphasis on reducing the electricity sector's dependence on foreign oil and increasing generation from domestic energy sources. PURPA was the centerpiece of policymakers' efforts. By creating an avenue for independent power producers to supply electricity to the regulated utilities, PURPA not only laid the groundwork for the creation of a competitive energy market, it also opened the door for private investment in renewables.

While PURPA provided a market for the output of renewable generation, other incentives were necessary for the financial viability of these plants relative to other qualifying technologies such as CHP. Beginning in the late 1970s, the US Federal government began implementing a number of tax provisions designed to directly spur investment in grid-scale renewables. The Energy Tax Act of 1978 established the business ITC, which remains in place today. The ITC provides a corporate tax credit equal to 30% of the expenditures made on new renewable investments. The more recent Energy Policy Act of 1992 established the renewable electricity PTC. In 2016, the PTC provides qualified renewable investments—which includes new wind, geothermal, and biomass plants—a tax credit worth $23 for each MWh produced during the first 10 years of operation.

In addition to the investment and production tax credits, firms also benefit from the Federal Modified Accelerated Cost-Recovery System—i.e. accelerated depreciation—which allows firms to recover a large portion of their investment through favorable depreciation reductions.[†††] Federal support extends to distributed renewables as well. Since 2009, individuals have been able to claim the 30% ITC on qualified renewable investments—most notably, solar PV. In addition, individuals leasing solar systems, or purchasing power from the business-owned systems installed on their homes, also indirectly benefit from the favorable accelerated depreciation [10].

Despite the long history of Federal support, renewable generation in the US remained relatively steady throughout the 1980s and 1990s. In 1978, wind and solar accounted for effectively 0% of US electricity production. By 2000, this figure had risen to only 0.16%. During the last decade and a half, however, there has been a surge of investment in renewables. During 2014, wind and solar supplied over 5% of the total generation in the US (see Figure 11.5) and this share is expected to continue to grow.

In large part, the recent growth in production from renewable energy sources has been driven by the widespread adoption of state-level policies supporting renewables [15,16]. Similar to the Federal government, individual states offer a wide range of tax provisions incentivizing investment in renewables [14]. However, the key state-level policy has been the renewable portfolio standard (RPS). An RPS mandates that, over a given period of time, each LSE in a state must produce or procure a minimum share or, in some cases, level of electricity from qualified renewable energy sources. Compliance with the RPS varies from state to

[†††]The Federal Government also provides indirect support for renewables through expenditures on R&D programs. For an overview of the R&D support, see [14].

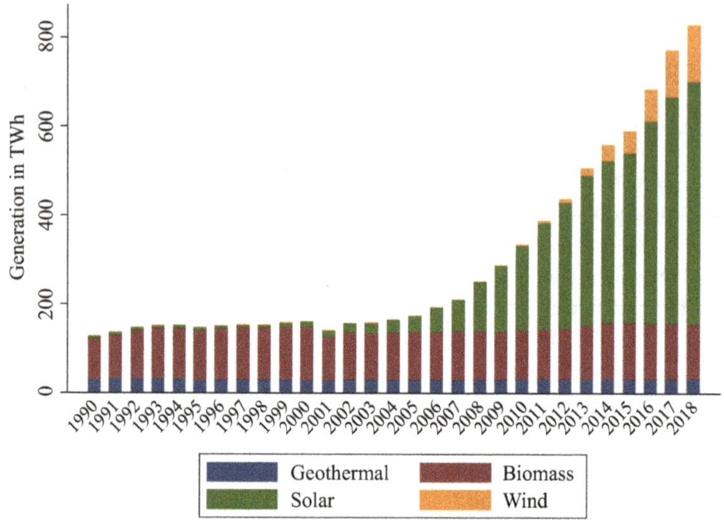

Figure 11.5 US renewable production 1990–2018

state. In states with deregulated wholesale markets, renewable electricity producers typically receive a renewable energy credit (REC) for each MWh of output. LSEs within the state can then meet their renewable obligations by purchasing and surrendering a sufficient number of RECs each period (e.g. annually). In states with vertically integrated utilities, compliance is often achieved through direct contracting with renewable suppliers [17].

Prior to 2000, minimum levels of renewable electricity had been mandated in only two states. By 2016, 29 states had RPSs in place. The required share, and growth, of renewable electricity varies dramatically across states. In many cases, the mandates are quite aggressive. For example, California's RPS currently requires electricity providers to procure at least 33% of their electricity from renewable energy sources by 2020. The recently approved California Senate Bill 350 has set this target to 50% renewable energy by 2030.

Thus, despite the support provided by long-standing, if intermittent, tax credits as well as the demand pull created by PURPA, it appears that state portfolio standards are the strongest driving force in the deployment of grid-scale renewable generation. While state policies are creating the demand, the supply is relatively geographically diverse (see Figure 11.6). However, this diversity is largely limited to wind generation. California, with its natural endowments of sunshine and geothermal potential, combined with its long-standing attractive PURPA terms and newly adopted aggressive portfolio standards, dominates the production from geothermal and solar and is a substantial biomass producer also (Figure 11.7).

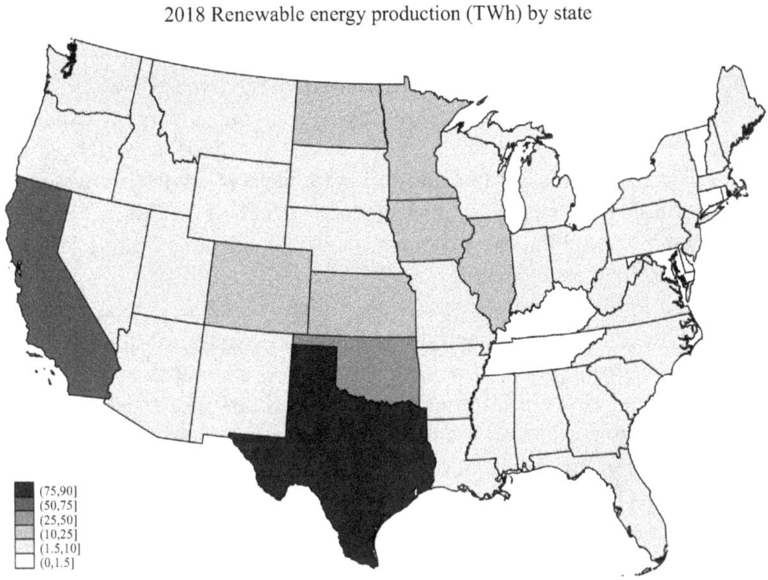

Figure 11.6 US renewable production 2018

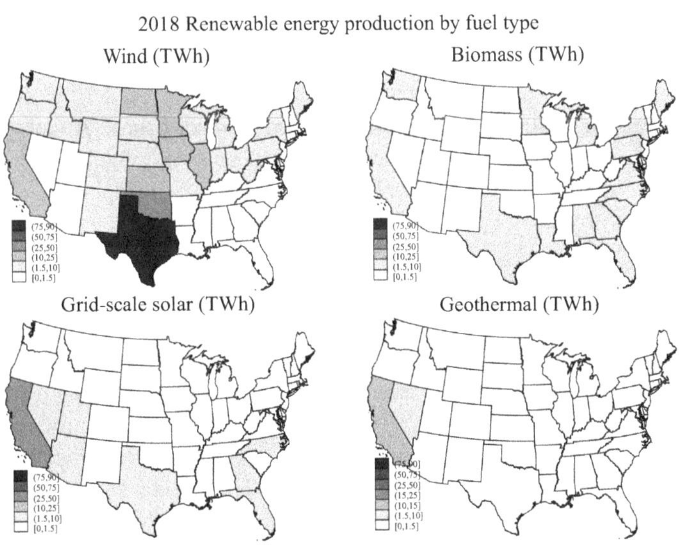

Figure 11.7 2018 renewable production by technology

11.3 The challenges of high-renewable penetration

Much of the transformation of the electricity industry since the early 1980s has coincided with and been largely spurred by the advent and rapid expansion of natural-gas-fired generation. Spurred by the PURPA, the US experienced a rapid expansion of smaller-scale combustion turbine and later combined cycle generation. This technological change in turn transformed the economics of the sector away from highly capital intensive, large-scale base-load generation to less capital-intensive and modular investment.

The advent of increasingly cost-effective wind and solar generation has again tipped the balance toward capital-intensive, low-marginal-cost technology. However, the scale economies of renewable energy installations are nowhere near the level of nuclear, or even traditional coal production. The transition to higher renewable electricity production has led policymakers in electricity markets around the world to question the ability of existing market designs to accommodate these changes [18,19]. These critiques generally coalesce around the theme that short-term wholesale energy prices will be very low if not zero most of the time, and therefore would not on their own be sufficient to support capital investment in a high-renewable world. In the next two sections, I discuss each of these questions in turn, first the short-term price effects and second the long-term implications for investment.

My general argument is that, while the cost characteristics of renewable energy will almost certainly reshape short run electricity supply, and therefore prices, it is far less obvious that the general design approach of current wholesale energy markets is incapable of transmitting the appropriate price signals and incentives to market participants. Put bluntly, large-scale renewable penetration will lead to extreme prices under our current market designs precisely because extreme prices will be needed for efficient deployment and utilization of those resources. Rather than rejecting the evolution to marginal cost pricing at the wholesale level, electricity markets need to "double down" on these design philosophies to ensure that appropriate scarcity, and opportunity costs are reflected over time and space in future power markets.

11.3.1 Short-term market operations

The fundamental paradigm under which pricing in wholesale electricity markets operates is that of a production merit order, in which lower marginal cost production operates with highest priority, and the highest marginal cost production operates only when high demand conditions make it necessary. This principle is illustrated in the left-hand side panel of Figure 11.8. Ideally this supply curve in generation is combined with the principles of efficient congestion pricing, which has its most complete manifestation in locational marginal pricing (LMP).[‡‡‡] When executed accurately, the least-cost security-constrained dispatch of generation in a

[‡‡‡]The framework for modern electricity market pricing is first developed in [20].

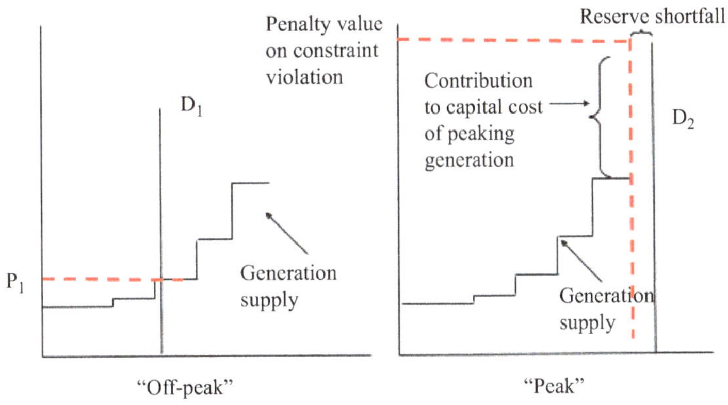

Figure 11.8 Scarcity pricing

manner that honors necessary operating and reliability constraints produces a set of shadow prices that reflect the marginal cost of producing or consuming energy at every time and location in the network. In organized wholesale markets, these shadow prices form the basis for market prices paid by consumers and earned by suppliers.

This basic operational paradigm is remarkably robust to the addition of alternative resources or additional reliability constraints. If an operating constraint can be represented as a constraint to the solution of the dispatch problem, it can produce a shadow price that provides a meaningful incentive to a market participant. However, in practice there have traditionally been barriers erected between the pure shadow values produced by the dispatch optimization and the prices faced by market participants. The most common of these barriers has been a reluctance in many parts of the world to fully reflect transmission constraints in wholesale prices.

Another area where spot pricing of electricity has encountered both practical and political difficulty is in the area of *scarcity pricing*. This is the question of how to set prices when demand exceeds the production capacity of all available (and accessible) generation supply. In principle, the supply demand balance is just another constraint that would generate a shadow price which could be used to set market prices in such periods, as illustrated in the right-hand side panel of Figure 11.8. The recovery of capital costs would come from periods where prices rise above short-run marginal cost. In such conditions, investors either implicitly or explicitly determine the market value of capacity through the concept of scarcity pricing [21]. However, several practical challenges complicate pricing in these conditions.

First and foremost, wholesale market demand remains almost completely inelastic, so finding a price that actually reduces demand to a level that clears the market would be difficult and probably politically unsustainable. This is combined with the inherent vulnerability of electricity markets to supplier market power, particularly under scarce supply conditions [22]. While economists have consistently pointed to the need for more demand elasticity in power markets, the conditions for clearing a market on the demand side have remained elusive. In place of letting demand set the price under scarcity conditions, therefore, market operators deploy "penalty" values that allow for the relaxation of the "power balance" as well as other constraints to allow a solution to the dispatch problem to be reached in reasonable time. These penalty values become in practice de facto price caps on market prices. While there has been some discussion about the proper role and level of penalty values, they have remained largely within the realm of engineers and computer programmers and have not attracted sufficient attention from economists.

A final obvious challenge to scarcity pricing lies in the fact that the mission of power system operators is largely defined by providing reliable supply, and that "scarcity," to the extent it would result in involuntary interruption, is viewed as a failure of the operator and the market organization in general. Therefore, operators take many actions both within the market rules and sometimes outside of them to ensure scarcity almost never arises [23]. One measure some markets have taken to allow for scarcity pricing without impinging reliability is to redefine scarcity to mean a reduction in operating reserves rather than an actual interruption of load. Under this principle, scarcity penalties would affect market prices well before interruptions are triggered. In recognition of the fact that different levels of this type of scarcity likely imply different levels of reliability risk and cost, operational "demand curves" have been proposed [24]. As illustrated in Figure 11.9, these administrative functions result in prices rising steadily as reserve margins are reduced, up to a maximum scarcity price.

This scarcity pricing approach accommodates the fact that a substantial share of consumption would not be price responsive. The horizontal segment of the black (energy) and red (energy + reserves) demand curves captures this fact. One critical parameter in scarcity pricing is therefore the height (or price point) of these segments at which prices would often be set in times of scarcity. This is the penalty value at which prices are set if reserves are insufficient to clear on section where

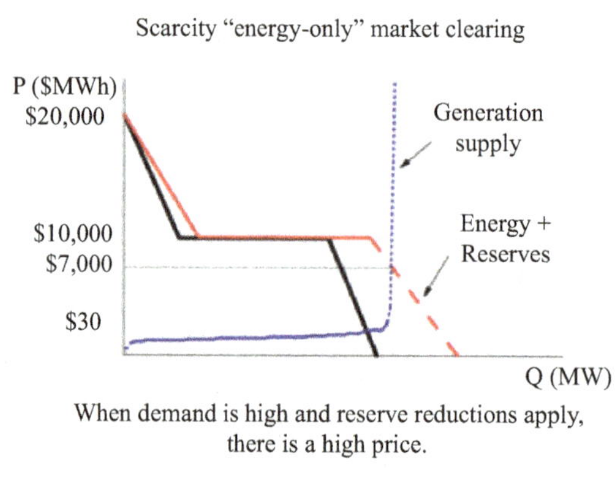

Figure 11.9 Scarcity pricing and reserve deficiency [24]

demand is downward sloping. Hogan proposes this level be set at the value of lost load in order to capture the average willingness to pay for electricity service amongst those customers not actively bidding demand into the market.

11.3.1.1 Implications of renewables for short-term pricing

Despite some concerns to the contrary, an influx of renewable energy does not necessitate a fundamental redesign of the principles of short-term electricity pricing. Renewable production will constitute an expansion of the "first-step" of the merit-order supply curve, one that will fluctuate with weather conditions. At moderate levels of penetration, say 20% of overall production, this rightward shift of the supply curve may depress prices in some hours as higher cost generation is moved outside the margin. At high levels of penetration, there will be some periods where renewable production exceeds current demand and prices will fall to zero or negative levels [25]. In the period from 2014 to 2016, prices in US markets cleared at negative levels in 1% of the hours.[§§§]. The least intuitive scenario is very high-renewable production, 90%–100% renewable energy. It is under this scenario that some have argued short-term prices will be zero or lower at all times [19].

However, this should not be the case under a rational expansion of renewable energy, even to 100%. While prices with 100% renewable production will be, and should be, zero or negative much of the time, prices will and should also be extremely high the rest of the time (Figure 11.10). Under a market with substantial zero marginal cost, but variable, supply, energy scarcity, pricing will play an increasingly critical role. Simply put, when there is enough wind and solar

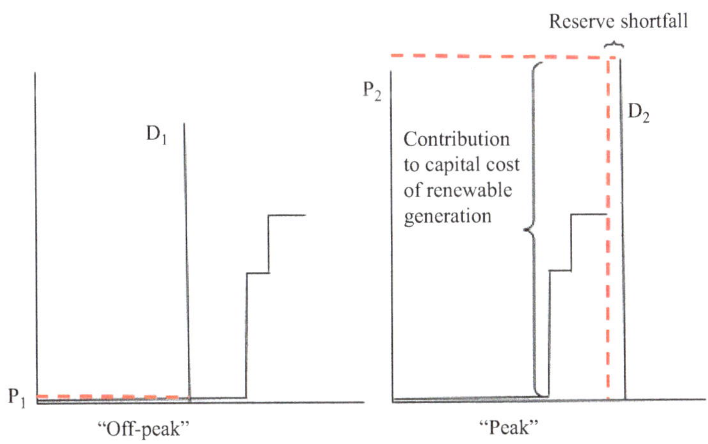

Figure 11.10 Short-run pricing with high-renewable penetration

[§§§]See [26]. There are a number of causes for negative energy prices, including non-convex costs of conventional generation and the lost opportunity cost of tax and renewable credits for renewable generators if they curtailed their output.

production, prices will be zero, but when there is not, prices will be set by demand, storage, or administrative scarcity values.

The only scenario in which short-term prices should be zero all the time is one in which renewable capacity is so overbuilt as to always be in sufficient (or excess) supply. Because of the variable nature of renewables, and the high regional correlation of output from wind and solar facilities, it is extremely unlikely that this scenario would emerge. Even if massive extra solar capacity were built in a system, it still would not provide overcapacity at night. In some areas, balancing wind and solar with geothermal resources, or more plausibly, conventional generation powered by biofuels, may be a realistic pathway to 100% renewable output. However, neither of these dispatchable renewable resources is truly zero marginal cost.

The expected expanded role of storage technologies, particularly lithium-ion batteries, in high-renewable markets signals the importance of a second consideration in short-term pricing: opportunity cost. Under most high-renewable penetration scenarios, storage will often be the marginal source of supply during hours in which variable renewable resources are not available. Storage facilities will base their offers into the market upon the opportunity cost of not being able to use the energy in later periods. Given the trend toward periods of extremely low prices intermixed with periods of extremely high ones, it is natural to expect that the opportunity cost-based offers of storage units will reflect the prospect that the energy being stored could be used in a very high price scarcity hour.

Further complications arise when one considers the intertemporal aspects of conventional power plant production and of storage resources. In most markets, the production of variable renewable resources tends to be concentrated in specific hours, leaving daily cyclical gaps that need to be filled by other resources. Much of the incumbent conventional generation fleet is poorly suited for such cycling however. Inflexible base-load generation, such as conventional steam-based coal or combined cycle gas units, were not designed to either rapidly increase or decrease output or to shut down multiple times per-day. When faced with daily cycles of extremely low prices, such units are forced to either continue operating at a loss, or shut down all together and consequently be unavailable when the renewable energy resources are not operating.

This dynamic is already well underway in markets such as California, where prices have transitioned from peaking at midday to peaking during morning and evening ramp periods. This transformation in pricing patterns is due to the massive influx of grid-scale solar capacity [27]. The result has been a large reduction in output from combined cycle gas and an increased reliance upon more nimble but higher marginal cost gas turbines. As a consequence, while the solar influx has reduced prices overall, it has actually contributed to an increase in market prices during the early even periods after sundown (Figure 11.11).

Rather than ushering in the irrelevance of short-term pricing, renewables create an even more pressing need for electricity prices to reflect marginal conditions. There are several areas where the current pricing structure needs to be further developed in this regard. First, pressure from renewable development is making it increasingly important to properly price congestion. European markets that have resisted LMP

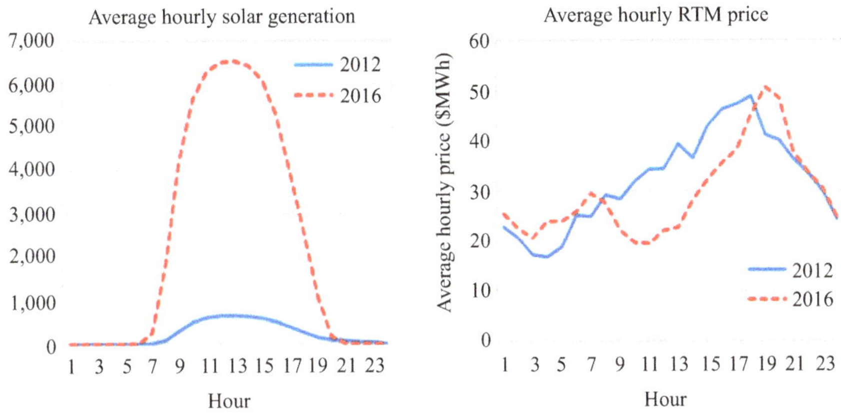

Figure 11.11 Evolution of CAISO solar and prices (from [27])

will need to reconsider their resistance in the face of renewable construction in areas with already limited transmission export capability [28]. Second, daily markets will need to better accommodate the operational needs of storage capacity. Currently in day-ahead markets, individual hours are cleared largely independent of each other except for honoring ramping and start-up limits of conventional units. Market designs need to be adjusted to allow to solve optimally for an inter-hour pattern of charging and discharging from storage units. Third, markets may need to better anticipate ramping needs in the face of growing short-term supply volatility. Markets such as California's are integrating a flexible ramping requirement into their market optimization, essentially setting aside more fast-ramping capacity for stand-by services during periods of higher renewable production volatility [29]. Last, markets may need to raise the parameter values currently applied during scarcity conditions.

11.3.2 Long-term market (investment) challenges

Ultimately, the question of market design and renewable production comes down to how to finance the capital costs of renewable investment. The position that markets need to radically change to accommodate renewables is grounded in the perception that energy prices will be too low to sustain investment in both renewable generation and its complementary production under current designs. These arguments are generally used to make the case that some type of capacity payment will be necessary in high-renewable systems, even if they would not be otherwise [18,19]. This position in turn stems from the perception that energy prices in a high-renewable market will necessarily be lower on average than in today's markets. In particular, prices will be zero or negative during periods of high-renewable production (e.g. midday in solar dependent systems).

In order to explore these hypotheses, we need to also address the fact that many market designs already reflect a belief that energy prices alone are insufficient to support capital investment in generation, and therefore require some form of capacity payment structure. Therefore, the question of whether markets have to change to accommodate renewables also becomes whether one of capacity payments are the change that is needed, and whether those capacity payment structures themselves need to change.

Underlying any discussion of investment and resource adequacy is the fact that the vast majority of new capacity and the bulk of energy is transacted either through long-term contracts such as power purchase agreements or financed internally as part of a vertically integrated utility. In both cases, the revenues from power generation are set by instruments other than the short-term energy prices discussed above, or the capacity payments discussed below. Therefore, any consideration of the interplay of energy and capacity markets with investment needs to also consider the impact of those markets on the underlying markets for contracts through which most generation revenues are set.

11.3.2.1 Renewables and capacity markets

In the United States, capacity requirements take one of two forms: a bilateral resource adequacy requirement that is imposed upon LSEs, or a centralized capacity market [30]. We will use the term "capacity requirements" to encompass both general approaches. Ironically, while markets in Europe that are lacking capacity requirements are speculating on the need for them in order to accommodate renewables, markets in the US that already have capacity requirements are struggling to accommodate renewables within their requirements frameworks.

Capacity requirements may be affected by the integration of alternative resources including renewables, energy storage, and demand response. The shift to these alternative resources impacts power markets in three important ways: their impact on average energy prices; their impact on capacity market prices; and the extent to which their intermittency and energy limitations create new reliability needs that are not satisfactorily addressed by conventional resource adequacy (RA) and ancillary service policies. Each of these concerns have been observed in restructured electricity markets. In California, for example, the penetration of utility-scale solar has helped contribute to low, or even negative, energy prices during the middle of the day. In addition to influencing energy prices, alternative resources (primarily renewable generation) have earned an increasing share of capacity or RA payments.

The magnitude of these changes is illustrated in Figure 11.12. While the overwhelming share of new capacity added in Independent System Operator (ISO) regions during the early 2000s was fired by natural gas, since 2010 roughly half of all new capacity has been powered by renewable sources. If this trend continues as expected, the implication is that not only will conventional resources find it increasingly difficult to recover costs from short-term markets, they will also find more competition and lower prices in the capacity markets. While renewable resources have played a very large role in resource adequacy in California and MISO, to date, demand response has been the most prominent alternative resource in much of the east.

One key policy question, therefore, is whether alternative resources can and should provide a comparable form of capacity as conventional resources. Such questions get to the heart of what has been a central issue with RA policy from the

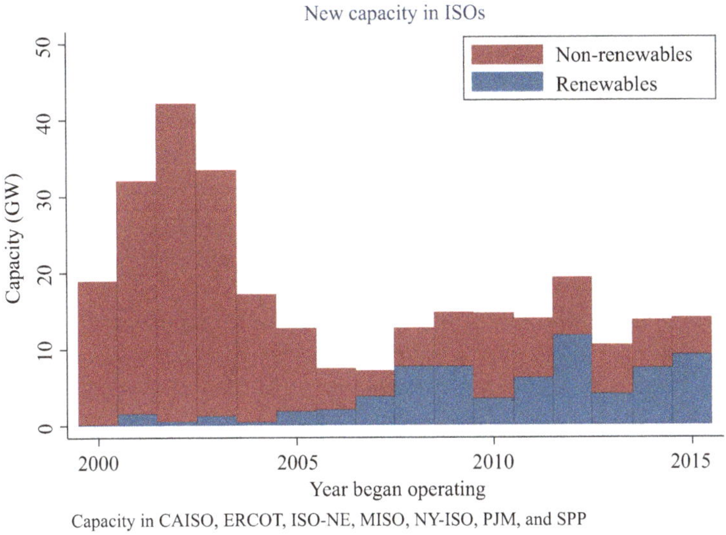

Figure 11.12 Decomposition of new capacity in the US. Reproduced from [30]

start: what exactly constitutes "capacity" under such policies? ISOs have struggled even to define the attributes that constitute the boundaries between conventional and alternative resources. Among the key elements that have been debated and periodically revised are the following:

- Should qualifying capacity be limited to resources that can be made available on demand, or evaluated based upon a probabilistic expectation of performance?
- How location-specific should capacity procurement be?
- What performance characteristics should be required?
- What are the performance obligations of participating capacity?
- Should those obligations be applied uniformly or adapted to specific resources such as energy-limited storage and variable energy resources?

These questions highlight the distinction between an energy-only setting and those with compensation for capacity. Performance in an energy-only setting is simply the sale of energy or ancillary services in a daily market. If a unit is operating and selling into the market, it earns revenue. If it is not, then it earns no revenue. Under a capacity payment paradigm, qualified units earn revenue in advance and can keep those earnings even if the unit is not available under a long set of possible exemptions. When resources were of like type and operated by firms with similar incentives, common assumptions about availability did not distort procurement. However, with a new range of more diverse resources, ISOs are again revisiting their ex ante assumptions about performance and the incentives provided to resources committed through RA markets.

11.3.2.2 Renewables and capacity performance

A capacity market would have no value if resources were not expected to be able to produce energy when the market was tight. Here we examine how capacity markets are being modified to consider incentives and mandates to achieve performance. In their review of the New York Independent System Operator (ISO) capacity market, Harvey, Hogan, and Pope [31] note the following:

> *The larger the total revenues collected through the capacity market rather than the energy or ancillary service market, the greater the concern with the many inherent approximations that appear in the necessary simplifications of the complex problem of constructing forward estimates of resource requirements and defining administrative requirements to provide appropriate performance and investment incentives for capacity suppliers.*

The simplifications and assumptions made in the procurement of capacity were more tolerable when the types of capacity being procured were relatively similar. The assumptions did not bias procurement toward one type or resource or another. These stresses have become more significant with the increase in the use of unconventional resources to meet capacity needs. This has left the designers of RA policies with two choices: (1) further refine and categorize the types of capacity to be required; or (2) increase reliance on short-term energy and ancillary services revenues to provide signals about the characteristics and performance abilities of new capacity.

ISOs are taking a diverse approach to this choice. Harvey *et al.* [31] strongly support an emphasis on short-term market rewards, arguing that "attempting to use capacity market rules to elicit capacity resources with the optimal mix of characteristics to meet load over the operating day has the potential to become more and more difficult as the diversity of the resource mix increases and has the potential to end badly, resulting in both lower reliability and higher consumer cost." In New England, ISO-NE has also shown a preference for strong performance incentives that would be uniformly applied to all resources. The ISO argues that performance incentives are the key to inducing flexible resources necessary to complement intermittent supply: "Changes to the FCM that improve incentives for resource flexibility and availability will provide better incentives for investment in resources that can balance intermittent power supply? [32]."

Conversely, in California the CAISO, in conjunction with California state agencies, has been incrementally working toward a setting with multiple, nested, capacity requirements. In addition to a standard RA requirement that is applied to all participating LSEs, the California Public Utilities Commission adopted a "flexible" Capacity Procurement requirement in 2014. The requirement for the first time explicitly distinguishes types of capacity by operational characteristics. Other RA requirements and capacity markets differentiate resources by location, and reduce their qualifying

The California proceeding highlights many of the difficulties inherent in specifying not just a quantity of capacity, but also a range of operational requirements in an RA context. If fast ramping capability is a key need, must such capability be available for a full hour or smaller intervals? Must resources be available all the time, during peak needs, or during shoulder ramping periods? The difficulties have been magnified by the need to compare dramatically different resources types, including energy-limited storage, conventional generation, and demand response. In addition ensuring capability, California has also experienced problems enticing flexible resources to perform. A nontrivial amount of conventional capacity in the California market bids in such a way that operators must treat it as an inflexible unit. The flexible RA proceeding has therefore been about inducing existing flexible resources to act flexible as well as securing the investment of new flexible resources.

11.3.2.3 Capacity instruments and performance incentives

One reason why even electricity markets that provide significant revenues to suppliers through capacity instruments, such as PJM, also utilize scarcity pricing is that the prices that firms earn for providing electricity (namely, the energy market and ancillary services prices) provide an extremely strong incentive for the performance of generation capacity. Even Cramton and Stoft [33] acknowledge that scarcity pricing, if it were feasible, sets "the economic gold-standard for performance and investment-quality incentives."

Performance incentives pose one of the biggest challenges for RA structures, particularly in the face of an increasingly diverse mix of resource types. Providing missing money alone does not ensure the adequacy or reliable supply, only the adequacy of generation capacity with the potential to provide reliable supply. But reliability is not enhanced if the "adequate" capacity is not operating when it is needed.

This challenge is articulated by Harvey *et al.* [31] in their review of the New York capacity market. They note that

> *The use of a capacity market to make up the "missing money" needed to support the capacity required to meet capacity requirements has the unintended consequence of creating a series of missing incentives relative to an energy-only market as that maintained in ERCOT. The New York ISO attempts to replace these missing incentives with a series of administrative rules and requirements, some of which work better than others.*

In PJM and New England, the policy direction has been toward encouraging flexible participation through the prospect of extreme penalties if units do not perform and additional rewards if they do perform. However, these payments, while significant, are limited to hours in which scarcity, in terms of deficient energy or ancillary service provision, have been identified by the operator. While this approach can focus resource efforts on a relatively low number of scarcity hours, it is less clear if it is suited to, for example, a routine daily need for ramp. To the extent ramping or other constraints lead to volatile prices, but not official "scarcity" conditions, these mechanisms may still fall short of providing adequate incentives and revenues for high-frequency flexible resources.

11.4 Conclusions

The expansion of generation capacity from renewable electricity sources during the last decade has been remarkable. It also foreshadows the changes in market conditions that are likely to envelop electricity systems around the world. Some observers have come to view the coming dominance of renewable supply as a reason to fundamentally reshape electricity markets. I have instead argued here that expansion of renewable supply is not creating *new* challenges for markets but rather sharply exposing issues that have existed for decades. These issues include a continuing shortfall of revenue from energy and ancillary services markets, resistance to pricing scarcity accurately by location or time, and a need to reconcile engineering expectations of capacity performance with the incentives necessary to realize those expectations.

Almost from the first months of organized trading in liberalized electricity markets over two decades ago, there has been concern that short-term energy and ancillary services revenues have not been sufficient to cover the capital costs of new generation. This ongoing "missing money" problem is not unique to the era of renewable generation, but may be exacerbated by it. Ultimately, the vast majority of new generation capacity has been supported by long-term contracts between LSEs and the supplier or by vertical integration of the supplier with the LSE. Capacity payments, where available, are generally not considered to be sufficient or stable enough, or of long-enough commitment, finance new investment. Therefore investment sufficiency has traditionally relied upon a healthy market for long-term contracts. This will continue to be the case in a world with high-renewable penetration. For renewables, this is nothing new. Almost all new renewable investment has been spurred by long-term power purchase arrangements between developers and LSEs.

Regardless of the source of financing for new generation, efficient pricing of scarce energy and transmission will continue to be a critical cornerstone of efficient market operations. While all organized markets in the US have fully migrated to LMP, internationally strong resistance to LMP remains. However, this resistance may become untenable soon in the face of large renewable entry into regions with insufficient transmission capacity. Markets from Australia to the UK to Germany have already begun to examine the need for more effective pricing of scarce transmission capacity. Similarly, markets around the world are being forced to reconsider the proper trade-off between limiting supplier market power and sending efficient scarcity pricing signals. With the emergence of renewable supply and, more importantly, the complementary technologies of storage and demand response, markets are being forced to determine what an appropriate reward for supply during truly scarce conditions should be.

In markets that continue to limit scarcity prices to levels deemed insufficient to support investment, capacity payments will continue to play some role in generation financing. Rather than strengthening the case for capacity payments, the onset of renewables has further clouded an already murky set of design issues. The key question in these markets is whether the capacity markets should change to accommodate renewable resources or whether the renewable resources need to change to accommodate the reliability needs addressed by capacity structures. The crux of the question is the same in either case. What exactly do we expect from "capacity"? Should it be an "on-average" resource in which variable supply gets at least partial capacity credit for being concurrent with system needs? Or is capacity and "on-demand" construct in which generators of all types are expected to be available when needed, and will pay serious performance penalties if they are not available?

In each of these key areas, investment financing, scarcity pricing, and capacity payments, existing markets have made large strides over the last 20 years but remain far short of theoretical, or even practical ideals. The optimistic view is that renewables will force markets to identify and implement solutions to problems they have been able to avoid confronting in the past. The pessimistic view is that these issues will force a reconsolidation to state-ownership and or dominant integrated utilities. Under this latter path, the opportunities provided by liberalization to improve short-run and long-run efficiencies will be abandoned, and the question posed in the introduction—whether renewables and markets are complementary or in conflict—will be answered not by necessity but by choice.

References

[1] Hortaçsu A, Madanizadeh SA, and Puller SL. Power to choose? An analysis of consumer inertia in the residential electricity market. American Economic Journal: Economic Policy. 2017;9(4):192–226.

[2] Bushnell J, Hobbs B, and Wolak F. Opinion on the ISO's Dynamic Transfer Policy for Intermittent Resources. Market Surveillance Committee of the California ISO; 2010.

[3] Joskow PL. Markets for power in the United States: An interim assessment. The Energy Journal. 2006; pp. 1–36.

[4] Joskow PL. Restructuring, competition and regulatory reform in the US electricity sector. Journal of Economic Perspectives. 1997;11(3):119–138.

[5] White MW. Power struggles: Explaining deregulatory reforms in electricity markets. Brookings Papers on Economic Activity. 1996; pp. 201–250.

[6] Kahn E. Electric Utility Planning and Regulation. American Council for an Energy-Efficient Economy, Washington, DC (United States); 1995.

[7] Joskow PL. Why do we need electricity retailers? Or, can you get it cheaper wholesale? MIT Libraries; 2000. Available from: https://dspace.mit.edu/handle/1721.1/44965.

[8] Borenstein S and Bushnell J. The US electricity industry after 20 years of restructuring. Annu Rev Econ. 2015;7(1):437–463.

[9] Makholm JD. The revolution yields to a more familiar path: New York's reforming the energy vision (REV). The Electricity Journal. 2016;29(9): 48–55.

[10] Borenstein S. The private and public economics of renewable electricity generation. Journal of Economic Perspectives. 2012;26(1):67–92.
[11] Borenstein S and Notsund K. The private net benefits of residential solar PV: and who gets them. Energy Institute at Haas Working Paper. 2015;259.
[12] Darghouth NR, Barbose G, and Wiser RH. Customer-economics of residential photovoltaic systems (Part 1): The impact of high renewable energy penetrations on electricity bill savings with net metering. Energy Policy. 2014;67:290–300.
[13] Cohen M, Kauzmann P, and Callaway D. Effects of distributed PV generation on California's distribution system, part 2: Economic analysis. Solar Energy. 2016;128:139–152.
[14] Schmalensee R. Evaluating policies to increase electricity generation from renewable energy. Review of Environmental Economics and Policy. 2011;6(1): 45–64.
[15] Hitaj C. Wind power development in the United States. Journal of Environmental Economics and Management. 2013;65(3):394–410.
[16] Yin H and Powers N. Do state renewable portfolio standards promote in-state renewable generation? Energy Policy. 2010;38(2):1140–1149.
[17] Wiser R and Barbose G. Renewable Portfolio Standards in the United States: A Status Report with Data Through 2007. Lawrence Berkeley National Lab. LBNL Report Number LBNL-154E; April 2008.
[18] Newbery D, Pollitt MG, Ritz RA, *et al.* Market design for a high-renewables European electricity system. Renewable and Sustainable Energy Reviews. 2018;91:695–707.
[19] Helm D and Hepburn C. The age of electricity. Oxford Review of Economic Policy. 2019;35(2):183–196.
[20] Schweppe FC, Caramanis MC, Tabors RD, *et al.* Spot Pricing of Electricity. New York: Kluwer Academic Publishers; 1988.
[21] Borenstein S. Understanding competitive pricing and market power in wholesale electricity markets. The Electricity Journal. 2000;13(6):49–57.
[22] Borenstein S. The trouble with electricity markets: Understanding California's restructuring disaster. Journal of Economic Perspectives. 2002;16(1):191–211.
[23] Joskow PL. Capacity payments in imperfect electricity markets: Need and design. Utilities Policy. 2008;16(3):159–170.
[24] Hogan WW. On an "Energy only" electricity market design for resource adequacy. California ISO. 2005.
[25] Traber T and Kemfert C. Gone with the wind? Electricity market prices and incentives to invest in thermal power plants under increasing wind energy supply. Energy Economics. 2011;33(2):249–256.
[26] Borenstein S and Bushnell J. Do Two Electricity Pricing Wrongs Make a Right? Cost Recovery, Externalities, and Efficiency. National Bureau of Economic Research; 2018. 24756. Available from: http://www.nber.org/papers/w24756.

[27] Bushnell J and Novan K. Setting with the Sun: The impacts of renewable energy on wholesale power markets. National Bureau of Economic Research; 2018.

[28] Burtraw D, Bushnell J, Gambardella C, *et al.* The response of market and policy design to increasing shares of renewables in California and Germany. Resources for the Future; 2019. 19-03.

[29] Abdul-Rahman KH, Alarian H, Rothleder M, *et al.* Enhanced system reliability using flexible ramp constraint in CAISO market. In: 2012 IEEE Power and Energy Society General Meeting; 2012. pp. 1–6.

[30] Bushnell J, Flagg M, and Mansur E. Capacity markets at a crossroads. Working paper WP278. Energy Institute at Haas, April 2017. Available from: http://ei.haas.berkeley.edu/research/papers/WP278Updated.pdf.

[31] Harvey SM, Hogan WW, and Pope S. Evaluation of the New York Capacity Market. FTI Consulting; 2013. Available from: http://lmpmarketdesign.com/papers/Final_New_York_Capacity_Report_03_13_2013.pdf.

[32] ISO-New England. Flexible Capacity Performance. White Paper. October, 2012. Available from: https://www.iso-ne.com/committees/comm_wkgrps/strategic_planning_discussion/materials/fcm_performance_white_paper.pdf.

[33] Peter C and Steven S. The convergence of market designs for adequate generating capacity with special attention to the CAISO's resource adequacy problem. Massachusetts Institute of Technology, Center for Energy and Environmental Policy Research. Working Paper WP-2006-007. April, 2006.

Chapter 12

Regulating the interdependencies of the mobility and electricity sectors

Paul Adrianus van Baal[1] *and Matthias Finger*[1,2,3]

Innovations in the mobility and electricity sectors are increasingly leading to interdependencies, resulting in so-called "sector coupling." This can lead to substantial welfare gains, but only if managed properly as most of the potential lies exactly at the interface of the two sectors. In order to avoid suboptimal coupling and for the potential of sector coupling to be fully exploited, a holistic approach that steps away from the traditional sector-specific approach to regulation is needed. This chapter describes the institutional and technological transformations shaping the mobility and electricity sectors, and shows how these changes can be traced back to the same three global megatrends: decarbonization, digitalization, and deregulation. Using the coherence framework, we then argue that there should be a certain degree of alignment between the technological advancements, located at the interface of the electricity and mobility sectors, on the one hand, and the prevailing institutions on the other hand. Institutions can be an enabling or an inhibiting factor to sector coupling, and unfortunately, without a holistic approach to regulation, the latter will be the more likely outcome.

12.1 Introduction

In this chapter, we highlight the growing interdependencies between mobility and electricity, leading up to the two sectors' "coupling," and discuss their implications for regulation, so that the potential of these interdependencies for both mobility and electricity can be fully exploited. The chapter starts out by describing the current transformation of the mobility sector or industry. It then does the same, but in a much shorter form, for electricity, given that the transformation of the electricity sector is at heart of this volume. In the fourth section, we will crystalize the three underlying forces, namely deregulation, digitalization and decarbonization, that drive the transformations of both sectors. In the fifth section, we will explore

[1]EPFL (Ecole Polytechnique Fédérale Lausanne), Lausanne, Switzerland
[2]EUI (European University Institute), Florence, Italy
[3]ITÜ (Istanbul Teknik Üniversitesi), Istanbul, Turkey

whether, and furthermore argue that, these forces lead to so-called "sector convergence" or "sector coupling," giving examples of potential and already realized innovations. Finally, in the concluding section, we will discuss the institutional dimension of such sector convergence, both in the case of institutions enabling convergence, as well as in the case of institutions preventing it.

12.2 Description of transformations in the mobility sector

The transport or mobility sector is very diverse. It includes a large variety of organizations, technologies, and stakeholders. In this section, we offer a general description of the mobility sector and the elements that are currently under transformation. To structure the text, we use the coherence framework, also called the alignment framework. The coherence framework stems from the theory of co-evolution of technology and institutions [1–3]. It was further developed to evaluate the performance of network industries, such as energy or mobility, as a function of degree of the alignment between the institutions and technology that together constitute the industry [4]. It has been still further developed and improved several times since [5,6].

According to the coherence framework, when institutions and technology are not aligned properly ("mis-aligned"), this creates a new dynamics between stakeholders, often diminishing the overall performance of the network industry, yet at the same time creating a need and space for actors to solve the associated problems in, perhaps, new and innovative ways. Large-scale changes of institutions such as deregulation or privatization will be followed by changes in technology. Inversely, large-scale changes in technology such as digitalization will necessarily have to be accompanied by institutional changes in order to re-optimize the performance of a given network industry. The coherence framework consists of several so-called *building blocks*: technology, institutions, alignment, and performance. In the remainder of this section, we will describe the (needed) changes in the two main building blocks of the transportation sector, i.e., technology and institutions, before highlighting the challenges this raises for performance and alignment.

12.2.1 Technology in transportation

The first building block is technology. The coherence framework considers technology as assets employed in networks for the purpose of providing goods and/or services [43], which implies that any technology used in the network will have certain inherent objectives of use built-in; indeed, technology is not simply "there." Technology can be divided into physical assets (hardware) and any complementary information and communication layer (software). The individual assets make up the system and together exhibit unique network characteristics, such as a specific network topology, capacity constraints, and various types of flows.

The technological assets in the mobility sector are on the one hand the vehicles such as the cars, trams, metros, and bicycles, and on the other hand the supporting physical infrastructure to these regimes: roads, railways, parking spaces, signage, refueling infrastructure and stations, etc. Certain researchers would count the global supply chains supporting the steel-and-petroleum car as part of the regime, as these are essential in the regime's ability to assert its dominance in society [7]. Up to half of the land area of cities can be dedicated to roads, parking spaces, and refueling stations for private cars [8]. These determine the network topology: cars are generally the most flexible mode of transport, as most developed countries have a network of roads and associated infrastructure that can, quite literally, take you anywhere. Other modes of urban mobility generally have much less dedicated land area. Consequently, the networks of these public and unmotorized modes of transport (rail, bus, metro, bicycle, etc.) are not as widespread and these forms of mobility do not provide the same amount of flexibility for users. Nonetheless, private cars have many performance issues related to the network's capacity, which will be discussed later. On top of these physical assets, there is the layer of information and communications technology (ICT) supporting the mobility sector. ICTs have traditionally been essential for managing operational efficiency and safety, for instance, by dynamically ensuring fastest routes and intersection crossings.

There are many technological transformations happening in the mobility sector. First, there are the various new types of private vehicles that are emerging, such as battery electric vehicles, (plug-in) hybrid electric vehicles, and fuel cell electric vehicles. The traditional internal combustion engine is also seeing itself reinvented to accommodate new fuel types, such as biofuels and hydrogen. In public transit, electrification of rail and buses is continuing. This trend was started a long time ago already and is more prevalent in certain countries than it is in others [9,10]. Traditionally "soft" modes of transport are also seeing increased electrification: electric bicycles, (kick) scooters, skateboards, and other similar transport options are all on the rise. Developments and innovation in the ICTs are also stimulating much of the technological transformations happening in the mobility sector. Cars are becoming smarter and more and more like a "smart" home away from home, where smartphone apps can track and control your car settings. Assisted and autonomous driving is furthermore becoming a reality. The electrification of mobility is also making the convergence of the electricity and mobility sector possible: for instance, smart technologies enable the utilization of the flexibility of car energy storage capacity to provide grid services (vehicle-to-grid). In public transport, technological developments allow for better planning and operation of various modes of transport and integrating the various modes at an urban system level in terms of planning, operation, and payment options, as can be seen by the emergence of mobility platforms, public transport planning apps, "smart cards," and "Mobility-as-a-Service" (MaaS) solutions.

Public and private transport modes are also blending together, with many new options of shared mobility becoming possible through various ICT solutions: car sharing, bike sharing, ride sourcing, ride sharing, ride splitting, and many others [11]. Car sharing, bike sharing, and other urban vehicle sharing options can be divided into three categories. The first is so-called "free floating," where users can pick up the vehicle, use it within the service area, and drop it anywhere. This has been a particularly popular options for electric kick scooters ("e-scooters"), but their abundance in public parks and across pavements is considered a nuisance to many, causing a public backlash in some places [12]. The second is "station-based," where users can pick up and drop off vehicles at specific stations spread throughout the service area. This model is very popular for (electric) bikes, but also cars and other types of vehicles. While this offers less versatility for their users, it also generally causes less nuisance to the public as the vehicles are contained to specific sites. The last is "peer-to-peer," where users can offer their personal vehicle for others to use at times when they are not using it. Ride sourcing, also called ride hailing, is usually done through mobility platforms like Uber and Lyft, where people offering rides can connect with people looking for such services on demand. The platforms can offer various ride services such as regular taxi services, drivers using personal vehicles, or more luxury rides. Ride sharing is when normally private rides are opened to others sharing the same destination, for instance a commuter allowing passengers on his way to work or driving between cities. Carpooling is a traditional name associated with this, but the advancements of the ICTs have given ride sharing a new impetus, making it much easier for riders and passengers to connect, even if they are strangers. Ride splitting is a dynamic type of shared mobility which falls somewhere in between ride sourcing and ride sharing. Rides are split between passengers not sharing the exact same route. An algorithm can determine the optimal combination of the two routes in terms of time and money. These are just some of the new ICT-enabled options to shared mobility. Innovations in this field are advancing at an incredible rate. All in all, these technological transformations are rapidly creating new possibilities for passengers to be mobile; they are also challenging the existing institutions.

12.2.2 Institutions in transportation

The second building block of the coherence framework is institutions, commonly defined as the "*humanly devised constraints that shape human interaction*" ([13], p. 3). Institutions can be divided into formal institutions, such as laws, constitutions, and regulations, and informal institutions, such as behavioral rules, customs, traditions, and cultural norms. Both formal and informal institutions, as well as their enforcement mechanisms, are important in shaping the dynamics of socio-technical systems, such as mobility or energy. Markets and how they are shaped and operate are a particular form of institutions and are therefore included in this building block as well.

In mobility, the most important institutions are the various types of markets and regulations. The institutional dynamics are very different at different scales: international, national, regional, and urban mobility, but also for the different types of mobility and vehicles. In terms of markets, rail and urban public transport have traditionally been provided by local or national monopolies. These mobility monopolies were providing a public service that was thought to be a natural monopoly, including the entire value chain. In the late 1980s and early 1990s, the global trend of liberalization (more precisely deregulation) reached international and national rail systems [14], where many systems were unbundled or privatized with the aim of increasing their performance. A critical look at the various mobility systems showed that there was room for introducing more competition as well. This led to the creation of various new institutional regimes, which can generally be categorized into four different configurations of network industries [5]. The different institutional regimes are summarized in Table 12.1. The first is the traditional public monopoly model, where a single mobility operator owns the network and is in charge of operating it. The monopoly can either be a branch of (local) government, an entity owned by (local) government, or a regulated private entity. The public monopoly configuration was more prevalent in the mobility sector before liberalization but still exists in many places for local or urban transport. Metro systems are sometimes owned by cities, but taxi companies are often private oligopolies or even monopolies under concessions. Different regions and cities will have different setups and many variations exist, even between neighboring cities in the same country. The second institutional setup is that of a fully unbundled system, where the infrastructure remains monopolistic, owned and maintained by a single public or private entity, whereas other actors compete for providing services on the network. In mobility, we see this setup for instance in the way railways are in some countries. Again, competition to provide rail services on the network can take many forms. The third institutional setup will be the partially unbundled network with access-based competition, where different mobility services can be competing at various levels using different networks. This is, for instance, evident of local competition between taxi and metro services. This drives a dynamics, whereby the network is no longer a traditional natural monopoly, as various alternatives to the system exist and are competing. The last institutional setup would be full competition in both infrastructure and service. Considering the natural monopoly of most of the mobility infrastructure, this last setup can only really be seen in maritime transport.

Table 12.1 Overview possible institutional regimes in transportation. Adapted from [5]

Institutional setup	Ownership of network	Provision of services	Competition
Public monopoly	State-owned or regulated company	State-owned or regulated company	No competition
Fully unbundled system	State-owned or regulated company	Private companies	Competition at service level, nondiscriminatory access to infrastructure
Partially unbundled system	State-owned or regulated company	State-owned, regulated, or private companies	Access-based competition for network access, intermodal competition between networks
Full competition	Private companies	Private companies	Competition at service and infrastructure level

On top of these different forms of institutional regimes for mobility, there is sector-specific regulation and competition regulation where applicable. Public service obligations are commonplace in the mobility sector and they define terms such as minimum or maximum tariffs, especially in case of monopolies, continuity of services, regularity of services, or minimal capacities. Public service obligations were traditionally a feature of the rail systems first, but many cities have extended these types of obligations to local transport systems as well [15]. In case of regulated monopolies of infrastructure, service, or both, regulation explicitly determines what is and what is not allowed by the transport organizations. Such regulations can be determined at the global or international level, for instance, with regard to air or international rail, at a national level or at a city level. The division of responsibilities between such different levels is not always clear.

There are many ongoing institutional changes in the mobility sector. Liberalization as a macro-trend was already discussed. Liberalization has set in motion many related processes [15]. Publicly owned transport companies are becoming privatized in more and more locations. The use of contracting is spreading, replacing more direct intervention in publicly owned operators, especially in urban and municipal settings. This is related to privatization, where such direct intervention is no longer as easy. The awarding of contracts or monopoly concessions is gradually becoming more competitive across the board, with tendering, for instance, replacing direct award or grandfathered rights.

Shared and ICT-enabled mobility can potentially contribute greatly to solving some of the biggest problems facing mobility today, notably by improving the performance and flexibility of public transport and reducing the congestion and environmental issues caused by single-occupancy private vehicles. Most of the new types of shared and ICT-enabled mobility are operated by new platform actors in the mobility space and rely on nonownership modes of transport, even if they involve private vehicles ([16], p. 24). These new mobilities and platform operators are facing existing institutional restrictions that were not designed for them, such as market access restrictions, operational requirements, and financial disincentives. The rise of these shared and ICT-enabled mobility options, despite these institutional barriers, initially left many regulators and policymakers struggling to find answers. Conflicts between (local) governments and regulators and the new mobility platform providers are often reported and widespread [17]. The question if and how these new shared and other ICT-enabled modes of mobility should be regulated has been answered or is being answered, very differently in different places ([16,18]).

12.3 Description of transformations in the electricity sector

In this section, we are going to describe the transformation of the energy sector, yet focusing only on electricity. We are following the coherence framework already described in the previous section. To recall, this framework considers that infrastructure systems co-evolve between technology and institutions. Let us discuss both technology and institutions separately.

12.3.1 Technology in electricity

Technology in electricity can, like in transport, be divided into physical assets, such as power-generating plants, transmission and distribution grids, and the electricity transported, thanks to the grid. The specificity of electricity, when compared to any other network infrastructure, is that electricity cannot be stored and thus must constantly be balanced: as much electricity must be fed into the grid as is consumed and the other way around, risking substantial costs and blackouts if the balance is not strictly maintained. As with mobility, there is also a software layer on top of the infrastructure in the electricity sector. The ICTs have become increasingly essential for the balancing of the grid, in part due to the grid becoming more and more volatile as a result of the generation of increasing amounts of electricity by way of intermittent renewables.

There are many technological innovations in the electricity sector as of recently. The most obvious and most often mentioned ones pertain to the generation of electricity thanks to renewables. Most of the innovation here is in the form of solar photovoltaics and wind turbines, but technological advances are also taking place in geothermal generation, as well as in electricity generation from biogas and others. The most innovative part here is the fact that electricity is increasingly generated in a decentralized manner in smaller generation units and power plants. Not only does this result in the physical location being changed, it also means the network fundamentally changes. These decentralized assets, unlike large-scale power plants, are not hooked onto the electricity system at the transmission grid level but mainly at the distribution grid and household level, where the grid was originally designed for unidirectional rather than bidirectional flow.

A little less advanced are technological innovations in terms of storage. Energy storage can somewhat be considered as being the other side of generation, or the interface between the demand and supply side. Energy storage can be particularly important in times of excess electricity production, again mainly because of renewables, to better match supply with demand. While many different types of innovation exist in storage technologies [19,20], one of the most advanced is electrical energy storage by way of various types of batteries, which boast fast response times, easy integration in electrical grids, and versatility in deployment [21]. The fast response times of batteries also make them suitable to provide ancillary services such as grid stabilization and frequency regulation.

One of the most talked about innovation in the electricity sector as of recently is certainly smart meters [22]: their innovations is directly related to the ICTs, as they allow just-in-time metering of consumption, and generation in case of self-generation, and consequently a much better balancing of the electricity grid. However, data concerns are prevalent, as these data can hold incredible amounts of detail about the most intimate parts of one's private life [23]. Nonetheless, smart meters also allow for just-in-time demand-side management, opening up the valorization of household load flexibility for the benefit of the consumer and the grid. Various novel business models have been created in the wake of the smart meter rollout and digitalization in electricity [24].

The examples above show substantial transformation to the energy sectors in terms of technological innovation. However, we are not yet as advanced in electricity, and in energy more generally, when it comes to service innovation. There is not yet the equivalent of "energy-as-a-service," as there is "mobility-as-a-service." This is mainly because the integration of the various energy vectors, such as gas, heat, and biogas from waste is less advanced, and no digital platform has yet emerged that could combine all these energy services and serve as an intermediary between them and possible energy customers. Energy is also harder to "platform," as the various energy vectors are not as interchangeable as the various mobility offerings. However, it is very likely that this evolution will take place at some point, as the substantial service innovation potential remains untapped.

12.3.2 Institutions in electricity

Institutions have also significantly evolved as of the past 20 years in the electricity sector, owing, in essence to the sector's liberalization. Indeed, there have been numerous policies that have promoted electricity market liberalization. While there is incredible diversity among states and countries, we can largely distinguish two models of such liberalization: on the one hand the dominant US model liberalizes the sector by allowing qualified producers (power plants) to feed electricity into the grid, forcing grid companies to buy their electricity at regulated prices. The dominant European model is more radical, inasmuch as it mandates unbundling of generation and transport, both high voltage and distribution, thus creating not only a wholesale but also a retail electricity market. Today, electricity markets are liberalized in most industrialized and emerging countries.

Like we see with mobility, a lot of institutional innovation is happening around the new technological developments. The advent of new types of actors in electricity, such as prosumers, storage operators, virtual power plants, demand response aggregators, and energy communities, has hit many institutional barriers. Regulators and policymakers have been struggling to update their institutional framework to accommodate this new reality [25].

The combination of liberalization, decentralized generation, and smart metering does indeed lead to a radically transformed electricity system, which opens the way for coupling it with other sectors, such as transport. This is now where we turn.

12.4 Drivers of the mobility and energy transformations

From the two previous sections, it is clear that both the mobility and energy sectors are undergoing significant transformations. In this section, we show that the transformations in energy and mobility have the same drivers, namely deregulation, digitalization, and decarbonization, also called the "3Ds." Each driver will be discussed separately, explaining their origin and their driving effect on each of the two sectors.

12.4.1 Deregulation

Of the three drivers, deregulation is the one that has been influencing the transformations of the mobility and energy sectors for the longest. Deregulation is a result of the rise of neoliberalism, which emerged as the supporting ideology of globalization during the 1980s. Neoliberalism regards competition as the key to unlocking unlimited economic growth and the overall betterment of society. Conversely, neoliberalism sees most forms of government regulation and intervention as an inhibiting force to competition and economic growth. As a result, many previously state-owned companies and regulated sectors have been deregulated and even privatized since. Liberalization—as the concept that encompasses both deregulation and privatization—has been pushed by governments and policymakers around the world, but it has also been driven by consumer behavior and more recently by technological developments.

As in electricity, deregulation has led to the unbundling of transport systems, to the privatization of (local) transport operators, to the introduction of competitive tendering, concessions, and contracting. Because of such deregulation, independent regulatory authorities had to be introduced to enable markets and to keep them functioning properly. However, liberalization is not only being pushed by governments; rather it has triggered a global dynamic of its own and turned into a kind of global megatrend. Because (long-distance) transport systems are becoming more international, or global even, competition has naturally increased among national transport providers that did not face any or very little competition before. This has in turn led to significant intermodal competition: airlines, railways, and long-distance bus operators all compete for international and long-distance travelers. Feeling the competitive pressure, EasyJet, a UK budget airline operator, has expanded into bus travel under the name EasyBus, initially only for airport transport services but now offering long-distance services across Europe. In an another example, FlixBus, a German long-distance bus operator, has expanded into rail under the name of FlixTrain. Some countries still reserve in-country long-distance travel for their national (rail) champion, but they are increasingly facing criticism and encountering regulatory problems as cross-border travel is possible via a host of other providers. Another example of liberalization driving change is in the urban transport area. Taxi companies are traditionally granted concessions by cities. The advent of ride-hailing apps like Uber, Lyft, or Grab can, at least partially, be attributed to consumer demand because the traditional taxi company offering was not appropriate and characterized by high prices, lack of diversity in offers, and a low service quality. Companies like Uber have not always followed the regulations set to protect the taxi monopolies but in doing so have *de facto* liberalized taxi services. As a result, (local) governments have had to adjust their regulatory frame in order to accommodate this new reality, whether trying to protect the incumbent and status quo or to facilitate the arrival of new players.

In electricity, we see a very similar trend of deregulation as driven by government policies, consumer behavior, and new technologies. The European Union saw a chance in deregulation so as to integrate formerly nationally protected sectors and by doing so create a true single European market. It was hoped that such economic integration would ultimately lead to European political integration [26]. Before the 1990s, electricity trading among EU countries was negligible. But during the 1990s the EU started to deregulate the European electricity sector, forcing all Member States to unbundle their monopolistic electricity companies and to create a European market [27]. Now, electricity can be traded from Finland to Portugal thanks to a single coupled market [28]. The various starting points of the European countries were very different: while countries like France had a single national company generating, transporting, and distributing electricity across the country, federal countries like Germany had hundreds of small regional and municipal utilities in charge of various elements of the supply chain. Therefore, while in some countries deregulation has led to a growing amount of market players, in some other countries the same process led to consolidation. But this process is still far from complete. In any case, deregulation has forced incumbents and competitors to come up with a way to differentiate themselves and thus venture into new business models.

In countries where liberalization has not actively been pushed by governments, we nevertheless see that consumer behavior and technological developments are contributing to the spreading of markets and business innovation. The advent of decentralized energy resources means that almost anyone can now become an electricity producer. Similarly, technological developments in the ICTs, such as blockchain and digital ("smart") meters, have made it possible for consumers and especially communities of consumers to supply themselves without relying on their local utility. Such semi-independent energy communities can exist at various scales, from neighbors trading their locally produced (and perhaps stored) electricity to apartment buildings or even whole neighborhoods, *de facto* liberalizing the consumers' choices of electricity supplier and providing an alternative to the local grid, even when they are in the service area of a regulated monopoly [29]. It will thus become increasingly harder to enforce monopolies in electricity with these and other such technological developments unfolding.

12.4.2 Digitalization

The process of digitalization of network industries such as energy and mobility has several steps. First, the installation of sensors and meters across the infrastructure, so that all network activity can be monitored. Second, the data have to be transmitted from the network back to a central location or locations, where these can be processed. High-speed, high-quality transmission of information is vital, because the data are most valuable if they can be reliably interacted with in real time. Such real-time interaction is necessary for improving operational efficiency, resilience, real-time pricing, and many other applications. Third, the information has to be processed and analyzed. Because the energy and mobility sectors generate such vast amounts of data, artificial intelligence and machine learning are especially important in this aspect. Lastly, the generated insights have to be acted on, either in real time for management of the system, or on longer time scales for planning and strategic decision-making.

Digitalization has been a driver of many transformations in both mobility and energy. In mobility, digitalization has "*enabled a whole array of new transport solutions to take off [...] enabling people to connect more easily to each other, and even in some cases to propose mobility services by themselves*" ([16], p. 24). Almost all of the technological transformations in the mobility sector that were discussed earlier can be linked to digitalization. While most of the new shared mobility solutions have technically been technically possible for a while, the advances in digital technology, smartphones, and mobile Internet have strongly contributed to their development, spread, and wider acceptance [30].

According to Audouin ([16], p. 25), digitalization affects public mobility in three different ways: first, by enabling the creation of advanced travel information systems, which integrate the information about several modes of travel onto a single (digital) platform. This platform can contain routing information, timetables, prices, delays, and many other pieces of vital information. The aggregated information can be used by mobility services providers to operate the system, reduce congestion, and plan for capacity expansion where needed, but can also be of benefit to consumers. Routing and trip planners can be constructed that integrate this information to help travelers use multiple modes of transport more efficiently and increase their alternatives. As the owner of such platforms will have the ability to choose what routing options to show customers, they should be transparent and nondiscriminatory to allow customers to make fair choices. Second is the enabling of the integration of payment schemes. Because of digitalization, mobility providers can integrate their payment schemes, so that multimodal travelers pay for an entire route, rather than separately for each part of the route to the respective mobility provider. This greatly facilitates using multimodal travel. The integration of payment schemes also enables so-called "smart" payment cards, such as London's Oyster card or the Netherlands' OV-chipcard, that allow users to pay for various types of public transport in the service area on-the-go, obviating the need for planning or booking trips and buying tickets beforehand. Lastly, digitalization is enabling the integration of the first two solutions into so-called integrated mobility platforms, which offer mobility "as-a-service," rather than a product, meaning the mode(s) of travel or the route specifics are secondary to the service that is being provided to the consumer: going from A to B most efficiently. These new mobility solutions offer a platform interface for consumers to interact with for all of their mobility services. Most of the added value will go to the owners of the platform, even if they are not the main provider of the services.

In electricity, digitalization has radically changed the way the grid is operated and managed, leading to improvements in cost efficiency and resilience. The various electricity and balancing markets are now cleared and operated at increasingly shorter time periods, at increasingly larger scales, spanning entire continents. However, fundamentally, this has not (yet) changed the electricity system. Nevertheless, digitalization has also changed the way consumers view and interact with the electricity system. Before digitalization, the electricity system functioned in a supply-driven manner. The consumer's only interaction with their electricity supplier was the yearly check of the analog meter installed in their basement in order to correct their monthly billing. Because of digitalization, the electricity system is increasingly operated in a demand-driven manner. Electricity tariffs have become dynamic and time-dependent. Meters have become digital or "smart," with governments stimulating their rollout. Demand response schemes have become possible, not only for large industrial consumers, but also for households. So-called distributed resource "aggregators" are now able to group appliances, electricity supply sources, or demand response assets into so-called "virtual" power plants. Electricity can be traded between peers, with the involvement of their utility or even without intermediaries through blockchain and other ICT-enabled schemes. Numerous other examples abound.

12.4.3 Decarbonization

Decarbonization is the third driver of the transformations in both the mobility and energy sectors. As a result of climate change, societies around the world are increasingly looking toward ways to reduce the emission of greenhouse gases. This is referred to as decarbonization, where "carbon" is short for carbon dioxide, the most prevalent greenhouse gas that stems mostly from burning fossil fuels. The energy and mobility sectors are two of the most carbon-intensive sectors of the current economy. Fossil fuels, in the form of oil, coal, and gas, (still) power the large majority of the mobility and (still) generate the large majority of electricity worldwide.

Decarbonization is more often thought of in relation to the electricity than in relation to mobility, mainly because that is where most public policies are directly aimed at. The subsidization of renewable energy sources, such as solar and wind, has initiated a global technological wave of innovation into these technologies and many of them are now competitive with conventional fossil-fuel power plants in many locations [31]. The variable, intermittent characteristics of these renewables have increased the demand for flexibility [32]. Previously, this was provided for on the supply side, but flexibility can now increasingly be provided on the grid management side through balancing markets and re-dispatching, or on the demand side through storage, demand response, or other types of load flexibility. Governments worldwide are trying to come up with effective and cost-efficient policies to reduce carbon emissions, leading to significant institutional changes, while a power struggle unfolds between businesses that are trying to break through with new, more sustainable business models, and incumbent companies trying to maintain the status quo.

However, decarbonization is also a very prevalent driving force for the transformations of the mobility sector. Electrification of transport across all modes is widely recognized as one of the first and most important steps toward combating climate change [33]. Such electrification also benefits public health, especially in larger cities that struggle with air pollution and smog. Many places around the world are seeing large growth in research and subsidies for and subsequent deployment of electric vehicles (cars, buses, trains, etc.) and associated software and infrastructure [34]. Many of the technological advances in energy storage for batteries, electric vehicles, buses, and trains are the result of the political will to decarbonize. Not only does decarbonization affect the (fossil) automobility regime, it also impacts public transportation thanks to the electrification of buses, trams, and trains. Many institutional changes are also happening in order to push consumers to switch from carbon-intensive private transportation toward greener modes of mobility focused around traditional public transport and new ICT-enabled shared mobility solutions, but also by incorporating a sustainability perspective in urban and regional planning. This has the potential to considerably improve the overall resilience and sustainability of transport and mobility. In this regard, advancing comprehensive mobility or transport integration has an especially large potential for decarbonization [35]. But ultimately only the alignment of the physical system with the institutional regimes, both in transport and electricity, will lead to optimal performance.

12.5 Toward sector convergence and sector coupling?

Most infrastructures have developed independently from one another and constituted self-contained socio-technical systems. This is not only the case of electricity and transport, but also of telecommunications, postal services, water, wastewater, waste management, and others more. Consequently, when regulation of these infrastructures arrived, it was also set up in a self-contained sector-specific manner, and it has largely remained that way.

However, this way of doing things cannot and will not continue into the future, as the three above underlying drivers of convergence will push for sector integration, sector convergence, or sector coupling, of all the network industries. This is, in the first instance, because of the technological and economic dynamics that have been triggered by liberalization, which have led to new technologies, often at the interface of the different sectors (e.g., power-to-gas, power-to-fuel, and vehicle-to-grid), along with corresponding cross-sectoral business strategies. Convergence also results from recent developments of digital networks, in particular of the fifth generation (5G) of wireless technologies, which increasingly act as driver of convergence between sectors. 5G enables vast amounts of information to be reliably transmitted and processed in real time, which is essential for further digitalization of infrastructures, as these simultaneously have high reliability requirements and generate incredible amounts of data. This is leading to cross-sectoral and much more integrated infrastructures services (e.g., MaaS). The take-off of the Internet of things based on 5G networks, which is sometimes addressed as the next Industrial Revolution [36], is expected to accelerate this trend. Finally, climate and other ecological challenges force a direct comparison among different sectors, such as in the case of externalities caused by energy generation (by renewables or by fossil fuels) or by the different transport models. For all three reasons, a more convergent view of the different network industries is needed.

Let us, in this section, focus exclusively at the convergence between the transport and the electricity sectors. Decarbonization is probably the most powerful force driving convergence between transport and electricity in the form of the electrification of all forms of transport [37]. Almost all transport modes are currently relying on engines powered by oil derivatives such as gasoline, diesel, and kerosene. These fuels emit significant amounts of carbon emissions. Decarbonizing mobility is almost impossible without electrification, and even other options such as power-to-fuel require deep integration with the electricity sector [38]. Electrified transport has to either be constantly attached to an electrical power grid, in the case of trains, trams, or metros, or only some part of the time, to charge their batteries, in the case of cars, buses, bicycles, scooters, and various forms of micro-mobility. Batteries are being developed and supported as a fundamental technology for electrifying and decarbonizing mobility, but also as an essential technology for integrating more renewables into electricity grids [39]. Batteries can even cost-effectively serve both purposes at the same time, in schemes referred to as vehicle-to-grid [40,41]. Decarbonization and electrification thus provide the physical underpinning of the two sectors' coupling.

Liberalization, subsequently, will allow for the type of service innovation that will lead to the further integration by way of new business models pushed for and exploited by innovative firms. Electric mobility offers indeed much demand-side flexibility, which can be exploited at a commercial level to the benefit of the electricity producers as well as to the benefit of the electricity consumers, the latter becoming able of participating in the open power and balancing markets. With electricity supply and mobility services becoming much more subject to consumer choice, one will soon find companies experimenting with different offerings. For instance, one could see electric utilities offering households electricity supply packages with e-mobility charging at home as well as charging at various guaranteed capacity levels and at particular times. Batteries could become partially owned by electric utilities (for grid services) and partially by households for improving self-consumption of domestic power generation. One could soon also find providers of micro-mobility giving users of a mobility platform a stipend or discount if they charge their vehicles, while varying the rewards depending on the electricity market conditions.

Digitalization, finally, will provide the technological foundation for the integration of the sectors on an operational level, as well as at the level of planning. As such, digitalization will be at the heart of unlocking the potential of widespread sector convergence and sector coupling. Indeed, many of the business models described above would not be possible without significant innovation in software and the ICTs, allowing for real-time monitoring and processing of information about both the mobility and electricity systems. The two systems, when physically interlinked, can be managed together by fully integrated digital platforms, something that was impossible before digitalization. For example, such platforms could provide (electricity) demand-side management offerings originating in the mobility system, such as modifying the power demand of electric car charging infrastructure or small changes in the speeds of moving electric vehicles to reduce (local) stress on the electricity grid. Furthermore, also the construction of both the electric and the mobility infrastructure could be optimized together, so as to create synergies and reduce system and legacy costs in the future. Digitalization already facilitates the use of public transport and MaaS. Public transport can already be operationally managed from a systemic and holistic point of view, unlike private transport in cars. Soon public transport and electricity provision could be managed jointly thanks to integrated digital "e-mobility services."

But, sector coupling is a highly complex matter: at its most basic, there is no coordination between transport and electricity so that electrified mobility simply represents an additional and highly problematic load for the electricity grid [39]. Similarly, renewables, especially photovoltaics, are producing during whenever the sun is shining or the wind is blowing, which does not necessarily match when electric vehicles are charging (for instance, at night or in the evening). This can cause extreme stress for both systems, especially around peak times when power demand is already high. In other words, if one wants the electrification of mobility to accelerate, this will necessitate building significant amounts of new grid and electricity supply infrastructure, most likely at extremely high costs. Will these costs be charged to electricity consumers or to the subsidized public transport sector? Will it be financed through regulated infrastructure fees that everyone pays, through taxation on the users of the new electrified services, or on the polluters, the users of the carbon-intensive services?

Inversely, there is also a very large potential for reaping the benefits of electric mobility that can take some of the stress away from the electricity grids [42]. The load that mobility represents on the electricity grid can be harnessed as a way to balance the grid with increasing levels of volatile supply from renewables, providing auxiliary services such as frequency regulation as well as demand response. The electric vehicles that are not permanently attached to a power grid but instead charged through batteries can similarly provide many system benefits through vehicle-to-grid schemes. This active management of the (stationary) vehicles can actually represent a (financial) benefit to the owner of the vehicles as well as to the operator of the grid, which can forego investments in new grid infrastructure or storage. But again, who makes sure that this actually happens?

12.6 Discussion

In the previous section, we have highlighted how sector convergence and sector coupling between transport and electricity might unfold. We have also identified three main drivers pushing for such convergence, namely deregulation, digitalization, and decarbonization. However, none of this is automatic and the power of these drivers might actually never be unleashed.

Let us refer back to our coevolution framework to support our argument: the framework posits that some coherence needs to exist between the state of technological advancement on the one hand and the prevailing institutions (i.e., rules) on the other, so that infrastructure markets and subsequent innovation can develop. As a matter of fact, some incoherence may also be beneficial for innovation. This can be illustrated by the developments in both the transport and the electricity sectors, respectively: deregulation—an institutional change—has led some market dynamics in each of these two sectors separately, which in turn has triggered technological developments, in particular dynamics in the ICTs (e.g., digitalization). Market dynamics combined with the ICTs have led to some innovative technologies both in mobility (e.g., electric vehicles and MaaS platforms) and electricity (e.g., energy communes and smart metering). However, for sector convergence and sector coupling to exist, such individual technologies will not suffice. The coevolution framework clearly states that technologies will have to be accompanied by new and appropriate policies and regulations for markets and innovation to evolve further. As a matter of fact, much of this innovation and much of these new market dynamics will precisely take place at the interface between sectors—notably the interface between the transport and the electricity sectors—as this is where the biggest potential for efficiency gains exists.

In other words, such efficiency gains lie precisely in the coupling of the transport and the electricity sectors or, rather, in the convergence of the mobility and the electricity systems. Or, still in other words, such convergence of infrastructure systems will not magically occur because of some gadget technologies, like smart meters or electrified self-driving vehicles. Rather, these systems have to be coupled by way of innovative and cross-sectoral regulation and accompanied by corresponding policies and support schemes. However, at the present moment, policies and regulations, including administrative structures and regulators are purely sector-specific. In Europe, sector-specific electricity-related policies and corresponding electricity regulators exist both at EU and national levels. Worse even, is the transport sector, which has sub-sectoral policies and regulators, again at EU and national levels, i.e., air policies and regulators, rail policies and regulators, road policies and regulators, and actually not much for urban mobility which is left to the cities to figure out. All this is heavily institutionalized already and we are about to institutionalize yet other policies, namely climate change policies and regulations and even more so digital policies and corresponding regulations. Technological developments have already hit sector-specific institutional limits. It is therefore highly likely that such a sector-specific institutional approach will slow down and, in the most pessimistic scenario, kill sector convergence and sector coupling. Even in the most optimistic scenario, such a sector institutional approach will optimize each sector for itself rather than together, while simultaneously disincentivizing cross-sectoral technological developments and behavior.

Yet the potential of sector coupling, be it only between transport and electricity, is huge: it is first of all a market potential for innovative firms capable of reaping the efficiency gains that result precisely from the fact that these sectors have evolved so far along different institutional and technological trajectories. It is also a huge potential for consumer benefit and public welfare, as coupled electricity and transport systems will simply be more efficient. It is finally an ecological and health potential, as coupled electricity and transport systems have the potential to be less polluting as well as contribute less to global warming.

So, how can we make sector coupling happen between electricity and transport? As a first step, we have to realize that market dynamics—while having led to some technological innovations in the aftermath of deregulation and thanks to the combination with digitalization—will, by itself, not suffice. As a matter, many of the technological developments, for example, photovoltaic electricity generation, have been subsidized and were not the result of market pull. Similarly, public transport in itself is heavily subsidized and would hardly exist without government, especially local government. At some point, all these subsidies will come to an end, but the market may well exhaust itself before. This is, on the one hand, because deregulation and parallel globalization are politically less and less acceptable. Digital markets, on the other hand, may also exhaust themselves, namely because of the trend toward monopolies (e.g., "winner-takes-it-all"). Without a holistic, comprehensive, and long-term-oriented regulatory framework, investing into sector coupling might simply be too risky for firms and investors. Policymakers and regulators need to fundamentally rethink the sector-specific institutional framework to effectively deal with the growing interdependencies between the sectors. The coupling of the electricity and the transport systems requires a system approach, which implies a comprehensive and long-term public policy view.

References

[1] Nelson, R.R. (1994). The co-evolution of technology, industrial structure, and supporting institutions. *Industrial and Corporate Change*, 3(1), 47–63. https://doi.org/10.1093/icc/3.1.47

[2] Saviotti, P.-P. (2005). On the coevolution of technologies and institutions. In M. Weber and J. Hemmelskamp (Eds.), *Towards Environmental Innovation Systems* (pp. 9–31). Berlin: Springer. https://doi.org/10.1007/3-540-27298-4_2

[3] von Tunzelmann, N. (2003). Historical coevolution of governance and technology in the industrial revolutions. *Structural Change and Economic Dynamics*, 14(4), 365–384. https://doi.org/10.1016/S0954-349X(03)00029-8

[4] Finger, M., Groenewegen, J., and Künneke, R. (2005). The quest for coherence between institutions and technologies in infrastructures. *Competition and Regulation in Network Industries*, 6(4), 227–259. https://doi.org/10.1177/178359170500600402

[5] Finger, M., Crettenand, N., and Lemstra, W. (2015). The alignment framework. *Competition and Regulation in Network Industries*, 16(2), 89–105. https://doi.org/10.1177/178359171501600202

[6] Crettenand, N. and Finger, M. (2013). The alignment between institutions and technology in network industries. *Competition and Regulation in Network Industries*, 14(2), 106–129. https://doi.org/10.1177/178359171301400202

[7] Urry, J. (2004). The "system" of automobility. *Theory, Culture & Society*, 21(4–5), 25–39. https://doi.org/10.1177/0263276404046059

[8] Plumer, B. (2016, September 26). Cars take up way too much space in cities. New technology could change that. *Vox*. https://www.vox.com/a/new-economy-future/cars-cities-technologies

[9] Allen, J.G. (2019). Reasons for commuter rail electrification: early 20th century and since 2000. *Transportation Research Record: Journal of the*

Transportation Research Board, 2673(7), 227–238. https://doi.org/10.1177/03611981198406 21

[10] Friedlander, G.D. (1968). Railroad electrification: Past, present, and future development of the great European systems. *IEEE Spectrum*, 5(9), 77–90. https://doi.org/10.1109/MSPEC.1968.5214393

[11] Feigon, S. and Murphy, C. (2016). *Shared mobility and the transformation of public transit*. https://trid.trb.org/view/1416012

[12] Wallop, H. (2020, February 24). E-scooters: Are they a menace to society? *The Times*. https://www.thetimes.co.uk/article/e-scooters-are-they-a-menace-to-society-ljwb9v9dl

[13] North, D.C. (1990). *Institutions, Institutional Change and Economic Performance by Douglass C. North*. Cambridge: Cambridge University Press. https://doi.org/10.1017/CBO9780511808678

[14] Kessides, I. and Willig, R. (1995). *Restructuring regulation of the rail industry for the public interest* (SSRN Scholarly Paper ID 614994). Social Science Research Network. https://papers.ssrn.com/abstract=614994

[15] van de Velde, D. (2008). A new regulation for the European public transport. *Research in Transportation Economics*, 22(1), 78–84. https://doi.org/10.1016/j.retrec.2008.05.011

[16] Audouin, M.U.J. (2019). *Towards mobility-as-a-service: A cross-case analysis of public authorities' roles in the development of ICT-supported integrated mobility schemes* [Ecole Polytechnique Federale de Lausanne (EPFL)]. https://doi.org/10.5075/epfl-thesis-9098

[17] Meyer, G. and Shaheen, S. (Eds.). (2017). *Disrupting Mobility—Impact of Sharing Economy and Innovative Transportation on Cities*. Berlin: Springer International Publishing.

[18] Hirschhorn, F., Paulsson, A., Sørensen, C.H., and Veeneman, W. (2019). Public transport regimes and mobility as a service: Governance approaches in Amsterdam, Birmingham, and Helsinki. *Transportation Research Part A: Policy and Practice*, 130, 178–191. https://doi.org/10.1016/j.tra.2019.09.016

[19] Aneke, M. and Wang, M. (2016). Energy storage technologies and real life applications—A state of the art review. *Applied Energy*, 179, 350–377. https://doi.org/10.1016/j.apenergy.2016.06.097

[20] Poullikkas, A. (2013). A comparative overview of large-scale battery systems for electricity storage. *Renewable and Sustainable Energy Reviews*, 27, 778–788. https://doi.org/10.1016/j.rser.2013.07.017

[21] Gür, T.M. (2018). Review of electrical energy storage technologies, materials and systems: Challenges and prospects for large-scale grid storage. *Energy & Environmental Science*, 11(10), 2696–2767. https://doi.org/10.1039/C8EE01419A

[22] Darby, S. (2010). Smart metering: What potential for householder engagement? *Building Research & Information*, 38(5), 442–457. https://doi.org/10.1080/09613218.2010.492660

[23] Asghar, M.R., Dán, G., Miorandi, D., and Chlamtac, I. (2017). Smart meter data privacy: A survey. *IEEE Communications Surveys Tutorials*, 19(4), 2820–2835. https://doi.org/10.1109/COMST.2017.2720195
[24] Yan, X., Ozturk, Y., Hu, Z., and Song, Y. (2018). A review on price-driven residential demand response. *Renewable and Sustainable Energy Reviews*, 96, 411–419. https://doi.org/10.1016/j.rser.2018.08.003
[25] Lavrijssen, S. and Carrillo Parra, A. (2017). Radical prosumer innovations in the electricity sector and the impact on prosumer regulation. *Sustainability*, 9(7), 1207. https://doi.org/10.3390/su9071207
[26] Vasconcelos, J. (2009). Energy regulation in Europe: Regulatory policies and politics of regulation. *European Review of Energy Markets*, 3(3), 12.
[27] van Baal, P.A. and Finger, M. (2019). The effect of European integration on Swiss energy policy and governance. *Politics and Governance*, 7(1), 6–16. https://doi.org/10.17645/pag.v7i1.1780
[28] Lam, L.H., Ilea, V., and Bovo, C. (2018). European day-ahead electricity market coupling: Discussion, modeling, and case study. *Electric Power Systems Research*, 155, 80–92. https://doi.org/10.1016/j.epsr.2017.10.003
[29] Bellekom, S., Arentsen, M., and van Gorkum, K. (2016). Prosumption and the distribution and supply of electricity. *Energy, Sustainability and Society*, 6(1), 22. https://doi.org/10.1186/s13705-016-0087-7
[30] Shaheen, S.A., Cohen, A.P., and Chung, M.S. (2009). North American carsharing: 10-Year retrospective. *Transportation Research Record*, 2110 (1), 35–44. https://doi.org/10.3141/2110-05
[31] Sutherland, B.R. (2019). Susceptibility of renewables to interest rates. *Joule*, 3(11), 2597–2599. https://doi.org/10.1016/j.joule.2019.11.003
[32] Kondziella, H. and Bruckner, T. (2016). Flexibility requirements of renewable energy based electricity systems—A review of research results and methodologies. *Renewable and Sustainable Energy Reviews*, 53, 10–22. https://doi.org/10.1016/j.rser.2015.07.199
[33] Armaroli, N. and Balzani, V. (2011). Towards an electricity-powered world. *Energy & Environmental Science*, 4(9), 3193–3222. https://doi.org/10.1039/C1EE01249E
[34] Coffman, M., Bernstein, P., and Wee, S. (2017). Electric vehicles revisited: A review of factors that affect adoption. *Transport Reviews*, 37(1), 79–93. https://doi.org/10.1080/01441647.2016.1217282
[35] Givoni, M. and Banister, D. (Eds.). (2010). *Integrated Transport: From Policy to Practice*. London: Routledge. https://www.taylorfrancis.com/books/e/9780203850886
[36] Gilchrist, A. (2016). *Industry 4.0: The Industrial Internet of Things*. London: Apress.
[37] Robinius, M., Otto, A., Heuser, P., *et al.* (2017). Linking the power and transport sectors—Part 1: The principle of sector coupling. *Energies*, 10(7). https://doi.org/10.3390/en10070956
[38] Schemme, S., Samsun, R.C., Peters, R., and Stolten, D. (2017). Power-to-fuel as a key to sustainable transport systems—An analysis of diesel fuels

produced from CO_2 and renewable electricity. *Fuel*, 205, 198–221. https://doi.org/10.1016/j.fuel.2017.05.061

[39] Glitman, K., Farnsworth, D., and Hildermeier, J. (2019). The role of electric vehicles in a decarbonized economy: Supporting a reliable, affordable and efficient electric system. *The Electricity Journal*, 32(7), 106623. https://doi.org/10.1016/j.tej.2019.106623

[40] Lopes, J.A.P., Soares, F.J., and Almeida, P.M.R. (2011). Integration of electric vehicles in the electric power system. *Proceedings of the IEEE*, 99(1), 168–183. https://doi.org/10.1109/JPROC.2010.2066250

[41] Saber, A.Y. and Venayagamoorthy, G.K. (2011). Plug-in vehicles and renewable energy sources for cost and emission reductions. *IEEE Transactions on Industrial Electronics*, 58(4), 1229–1238. https://doi.org/10.1109/TIE.2010.2047828

[42] Veldman, E. and Verzijlbergh, R.A. (2015). Distribution grid impacts of smart electric vehicle charging from different perspectives. *IEEE Transactions on Smart Grid*, 6(1), 333–342. https://doi.org/10.1109/TSG.2014.2355494

[43] Finger, M., Crettenand, N., and Lemstra, W. (2015). The alignment framework. Competition and regulation in network industries, 16(2), 89–105.

Chapter 13

Building renewable economies that maximize social welfare and innovation

Mary Collins[1,2]

Climate change is a global crisis that will pose irreversible damage to humankind and the earth that we inhabit. A 100% renewable[*] grid—coupled with other solutions in the transportation, agriculture, and other sectors—must be achieved as quickly as possible if we are to save our treasured ecosystems and preserve the planet for generations to come. However if people struggle in poverty, climate change won't be a priority. And if policymakers are not bought in due to short-term thinking and a laser focus on maintaining their position of power, climate change won't be priority. For those leaders that are willing to pursue renewables in their area to grow local climate wealth, a holistic industrial strategy is needed in order to lift all up in the economy and foster the next generation of innovation. This chapter outlines how renewables create social welfare, why industry clusters are important to grow wealth, and the elements of clean energy industry clusters that work together to create high-road jobs.

13.1 Introduction

13.1.1 The urgency of climate change

> *It's really quite simple. We've overloaded the atmosphere with heat-trapping gas and the rest are just details.*
>
> —Professor Jason Box, Geological Survey of Denmark and Greenland

According to the Intergovernmental Panel on Climate Change (IPCC), averting a climate crisis will require a "rapid and far-reaching" transformation in modern civilization [1]. The 2018 IPCC report concluded that current efforts to reduce greenhouse gas emissions will not stop global temperatures from rising to catastrophic levels, with a global climate crisis starting as early as 2040 [2,3]. Likewise,

[1]California Institute for Energy and Environment, University of California, Berkeley, CA, USA
[2]American Jobs Project, Berkeley, CA, USA
[*]This chapter uses the terms "clean energy," "renewable energy," and "advanced energy" interchangeably.

the Fourth National Climate Assessment released in November 2018 urges swift action on climate change "to avoid substantial damages to the US economy, environment, and human health and well-being over the coming decades" [4]. Such damages are predicted across the globe: with a 1.5°C rise in temperature by the end of this century, by 2022 there will be an estimated productivity loss of 80 million full-time jobs, which pencils out to US$2.4 billion in losses [5]. Not surprisingly, these losses will have the greatest impacts on the most vulnerable nations, with low-income countries losing up to 9% of GDP by 2100 [6]. Countries around the globe, in particular well-resourced ones, are called to critically evaluate their policies on issues such as energy generation, land use, building construction, and transportation to reduce global net human-caused CO_2 emissions by 45% from 2010 levels by 2030 [7]. While there are many solutions that must be deployed simultaneously, this chapter focuses on the generation of electricity, which accounts for 25% of global warming [8].

13.1.2 The growing clean energy market

The global clean energy industry earned over US$1.4 trillion in revenues in 2016, which is equal to that of the global apparel industry and nearly twice as much as the global airline industry [9]. This figure is projected to grow to US$2.2 trillion in 2025, and 2040 investments are expected to total US$7.4 trillion [10–12]. Furthermore, if global carbon dioxide emissions were decreased according to the Paris Agreement, GDP would grow 0.8% in 2050. Compared to the reference case, this would be an increase of US$19 trillion in GDP, which amounts to the market capitalization of all firms on the New York Stock Exchange [13].

13.1.3 Framing clean energy as an economic opportunity

For the massive shifts in our economy needed to save the planet, leadership from policymakers is key. However, for a myriad of reasons politicians may not have climate on their radar. In some instances, climate may not be a priority for politicians because they are dealing with urgent issues that are top of mind to voters: poverty, safety, and healthcare, among others. In other cases, policymakers deny climate change (e.g., Unites States President Trump) [14] or they may try to salvage waning industries like oil and gas in the Tabasco region of Mexico (e.g., President Lopez Obredor) [15] instead of prioritizing a climate-forward agenda. It is also possible that policymakers are pandering to voters that claim climate change or energy is not a top voting issue as is the case in the United States, where polls consistently reveal that energy issues are at the bottom of voter priorities while the economy consistently ranks at the top [16]. However, when clean energy is framed as an economic opportunity, 89% of voters believe that clean energy growth should be accelerated [17]. Therefore, framing clean energy as an economic opportunity cultivates buy-in from politicians to act on clean energy, especially those that are opposed to change or seek support of voters through a jobs-based message.

13.1.4 Renewables and social welfare

The growing clean energy industry provides not just climate benefits, but also increased social welfare for people around the globe. This increase in social welfare results from both the positive externalities of renewable development (e.g., job growth and higher wages) and the reduction of negative externalities from high-polluting fossil-fuel energy sources that overwhelmingly harm communities of color [18]. Globally, more than 10 million were people employed in clean energy in 2017 from all backgrounds, disciplines, and skill levels. That figure will continue to increase as the global market takes off, potentially supporting 24 million jobs by 2030 [19]. The increase in clean energy jobs will compensate for losses in legacy energy sectors—like oil and gas—with an expected net employment increase of 6 million workers in 2050 [13]. The implication of this net increase in clean energy jobs is that clean energy is better for the economy than traditional fossil-fuel energy sources.[†]

The sectoral shift away from fossil fuels and toward renewables results in another social welfare gain: higher wages than traditional industries. When compared to national wages in the United States, the Brookings Institution reports that average hourly wages for workers in the clean energy industry are 8%–19% higher than national average wages. When comparing jobs at the lower-paying end of the employment spectrum, clean energy workers earned US$5–10 more per hour than other jobs, indicating clean energy jobs have higher levels of equity. Despite earning higher wages, such jobs typically have lower education requirements, making them more accessible: half of workers in production and energy efficiency attaining no more than a high-school education [20, p. 16]. However, it is to be noted that clean energy jobs are currently dominated by White, older males. Less than 20% of clean energy jobs are filled by women and fewer than 10% are filled by Black workers in the United States [20, p. 25]. Thus, clean energy industrial strategies need to address the economic inequities that have played out at the onset of the industry. The United Kingdom is an example of a government that has incorporated equity goals into its industrial strategy (see the case study).

The rise of renewables also reduces the negative externality of pollution that results from the extraction and combustion of fossil fuels (see Figure 13.1). With renewables as the dominant energy source, ecosystems will be less pressured due to decreased water and air pollution. Coupled with this, people that live near fossil-fuel extraction and production will be less exposed to hazardous pollutants and will live healthier, more productive lives [21]. This is especially true for minorities and marginalized communities around the globe that have experienced environmental racism. In other words, either intentionally or out of neglect from policymakers, environmental racism results from the "actions and decisions that result in the

[†]While it is true that the global economy will be better off overall from the sectoral shift away from fossil fuel to renewables, the impacts on communities that relied heavily on fossil fuels, such as West Virginia in the United States (coal) or Tabasco in Mexico (oil and gas), must not be diminished. The implications of their hardship highlight the increased need for thoughtful economic development policy to help communities transition to right-fit industries.

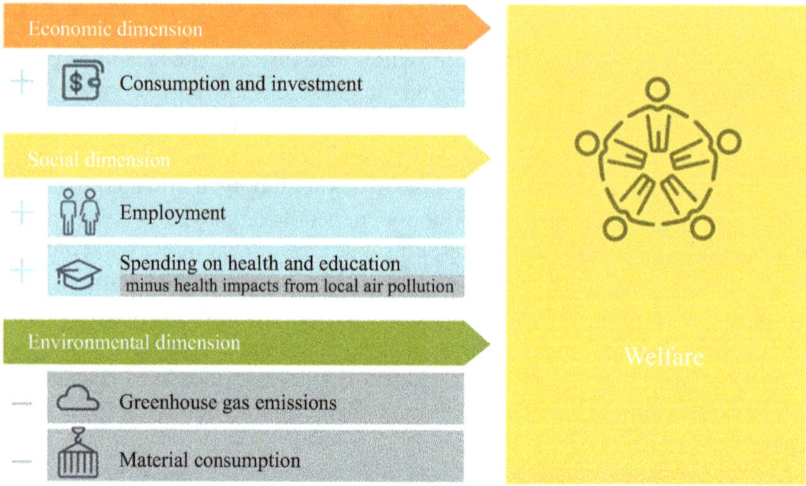

Figure 13.1 Reducing the negative externalities of fossil fuels, a net increase in jobs, and higher wages contribute to social welfare. Image from [21, p. 3]

disproportionate exposure of people of color to environmental hazards and environmental health burdens" [22]. In the United States, poor Black families are disproportionately impacted by rates of asthma as compared to poor White families [23]. Such disparities and their root cause can now be measured through a pollution inequity metric: the Proceedings of the National Academy of Sciences quantified the disparities between racial groups and pollution, finding that Black and Hispanic minorities suffer a disparate portion of pollution caused by Whites [24]. In Latin America, indigenous peoples have been subjected to mining, oil exploration, agriculture pesticides, and dam development, among other development activities from outside communities. Thus, indigenous communities in Latin America experience negative health impacts from air and water pollution and suffer the loss of traditional cultures [25]. The development of renewable energy economies is an opportunity to establish environmental justice, or "equal environmental protection under the law, and the right to live, work and play in communities that are safe, healthy and free of life-threatening conditions" [22].

The impacts of the energy industry on communities of color continue to evolve. Yet regardless if it is the fossil-fuel industry polluting communities of color or the renewable energy industry excluding them from the workforce, each persists as a market failure. This market failure—or moral failure on the part of society—can be addressed as new industries are built from the bottom-up. To correct such market failures, a holistic economic development plan can be created in partnership with government and industry.

Professor Robert Bullard: The founding father of environmental justice

In 1979, Robert Bullard was a professor at Texas A&M when he became intrigued about a case that his wife, an attorney, took on. In *Bean vs. Southwestern Waste Management Corp.*, a suburban neighborhood of mostly African American families were fighting a landfill in their community. After analyzing the data, Bullard found that there was a pattern of discriminatory dumping and pollution in Houston. For over 40 years, he has continued his studies, coining the term "environmental racism" and authoring 18 books. His keystone text *Dumping in Dixie: Race, Class and Environmental Quality* was the first environmental justice book published.

> *...whether by conscious design or institutional neglect, communities of color in urban ghettos, in rural "poverty pockets," or on economically impoverished Native-American reservations face some of the worst environmental devastation in the nation.*
>
> —Professor Robert Bullard
> Source: [26]

13.1.5 Economic clusters

Although the origin is debated, the concept of an economic cluster was popularized by Michael Porter's *Competitive Advantage of Nations* [27]. The idea of an economic cluster has been touted by scholars and policymakers alike for the past few decades [28]. Economic (or industry) clusters are regionally situated groups of interconnected companies and institutions that are engaged in a particular industry and supported by repeated exchanges of information and resources [29]. In today's competitive globalized economy, businesses are more likely to thrive in regions that cultivate the building blocks of cluster development: a rich innovation ecosystem, fertile ground for capital investment, a highly skilled workforce, clear policy signals, and a robust value chain. In fact, clusters enable businesses to leverage a trained workforce, a close network of suppliers, and a wealth of resources to overcome size limitations; improve productivity; and increase operational efficiency [30].

Typically, an economic cluster occurs on a subnational level and has an emphasis on a small number of industries and their interrelated firms [31]. However, a number of countries have pursued cluster-based strategies to maximize local content in the renewable energy value chain and thus to boost the domestic share of employment generation [32]. As is the case with the United Kingdom Offshore Wind Sector Deal (see the case study), national priorities and policies were established in partnership with industry to enable growth and eliminate market barriers [33].

Clusters inherently capitalize on a locality's comparative advantage, a principle that economists agree upon that states if each country (or region) were to focus on the industries in which it can be more productive, all are better off in the long run [34, pp. 33–34]. The aim of an economic cluster is to pair initiatives on both the demand and supply side of the industry in a strategic way that reinforces conditions for growth. Thus, it is not a particular policy that is responsible for fostering a cluster, but a series of comprehensive policies working together that result in success [35]. Such policies can be characterized in many ways,[‡] and take different forms depending on the region, scope, type of industry promoted, economic trends, and culture of policymaking [37].

According to Christian Ketels, a cluster-based approach also enables local industries to withstand shifting and often unpredictable economic forces due to value, geography, and business environment. Such economic forces include the cyclical changes, or ups and downs in the economy, and the secular shifts that are long-term, slower economic forces. Clusters create specific value to the region because firms collaborate to efficiently produce goods and services, often developing unique cultures and expertise. This applies not only to firms in the highlighted industry but for other firms in the region as well [30, p. 378]. The value inherent to clusters is linked to the localized geography, which facilitates "easy exchange of knowledge," a qualified workforce, access to markets and innovative ideas, and access to specialized services and workers [38]. The positive impacts of cluster-based strategies are particularly advantageous for small- and medium-sized enterprises (SMEs) that benefit from the positive business and policy environment [39]. A cluster may vary in size, from as large as a country to as small as a town, but proximity is nonetheless important [40]. The policy landscape, institutions, and universities make up the business environment, which is unique to the cluster, helping the industry cluster maintain a competitive edge despite shifts in the economy [30, p. 378].

The economic clusters that are particularly successful are those that have a high portion of manufacturing and innovation jobs, which is typically the case in renewable energy. The production of goods and services that are unique, innovative, and cannot be replicated enables industries to remain competitive in the economy [34, p. 55]. Moreover, the incidence of manufacturing and innovation will have huge ripple effects on other jobs in the region. For each one job in manufacturing, 1.6 local service jobs are created, and for each high-tech job, 5 additional long-run jobs are created outside of the tech field in a given region [34, pp. 60–61]. The key takeaway on employment impacts is that service jobs are the result of economic growth, not the driver of a region's prosperity [34, p. 12].

Studies have shown that wages are higher in industry clusters, with one concluding that workers typically receive wages that are 6% higher than those of workers employed in the same industry but outside a cluster [41, p. 42; 42].

[‡]In this chapter, I characterize economic cluster policies as those that create local demand, facilitate co-location in the value chain, train workers, foster connections in the innovation ecosystem, and provide access to capital and end-user finance for an industry. Others like Carley and Lawrence have categorized cluster policies as those that relate to business climate, information and coordination, export promotion and foreign direct investment, and R&D [36, p. 59].

Clusters also offer more stable employment and opportunities for workers due to opportunities to connect products and people to the larger market, and provide a framework to address regional deficiencies such as workforce gaps [41]. With respect to manufacturing clusters, census data show that doubling employment concentration within a particular industry is affiliated with a 2% increase in wages for employees [41]. Industry clusters have high numbers of SMEs or "independent enterprises that have fewer than 250 employees" and budgets at or below 27 million Euro [38, p. 55]. SMEs are responsible for 70% of jobs and 50%–60% of value in OECD economies [43]. In addition to contributing value to the local economy, SMEs also increase diversification and therefore economic resiliency in localities [43]. Clusters also typically have higher levels of innovation, which in turn brings prosperity to a region [40]. By one measure of innovation, patents per capita are highly correlated with economic performance and wages in a region [40].

Local governments can lay the foundation for economic clusters, but there is no silver bullet or one-size-fits-all solution. Typically, there are a number of policies working in harmony to achieve a number of interrelated objectives. Not all cluster growth strategies need to be driven by policy intervention. For instance, an offshore wind developer may invest directly in local workforce development efforts. Each region has its own unique competitive advantage, economy, and policy landscape, and local leaders should build on existing infrastructure and capabilities. Public funds must be used wisely: only to correct market failures and in instances where the industry will continue to flourish and create jobs after subsidies subside [34, p. 212].

> *When the government engages in industrial policies, it should do so mainly to correct some form of market failure.*
>
> —Enrico Moretti

Case study: Norway's Maritime CleanTech Cluster: Government support for growing clusters

Norway is an example of a country that takes an active role in fostering the development of economic clusters. Through government initiatives between Innovation Norway, the Industrial Development Corporation of Norway (Siva), and the Norwegian Research Council, there are three Norwegian cluster-based programs that aim to holistically support clusters through various levels of maturity. The Arena Programme supports emerging industry clusters [44], while the Norwegian Centres of Expertise (NCE) Programme supports mid-level clusters, and the established global clusters receive Global Centers of Expertise (GCE) classification [44].

Established in 2011, Norway's Maritime CleanTech Cluster initiated as a nine-member, Arena-classified member organization with participation from firms across the value chain. Companies in the cluster range from ship design to equipment suppliers to information and communications technology companies [45]. Goals of the Maritime CleanTech Cluster include increasing innovation, reducing market barriers, and growing capital and recruitment [46]. The cluster has grown to play a global leadership role in sustainable marine technology. A hybrid system for ferries was developed in 2012, and in 2014 its Urban Water Shuttle Project was deployed. The Maritime Cluster's status was upgraded to NCE in June 2014 and shortly after it initiated its Short Sea Pioneer project, aimed at developing and deploying new ways of transporting goods via sea [47]. In 2016 the cluster was awarded 11.7 million Euros to develop an all-electric high-speed passenger vessel [48].

There are now 75 firms participating in the Maritime CleanTech cluster across the value chain [48]. The cluster generates over 300 million Norwegian Krone in revenue (roughly US$(2019) 38 million), supports over 30,000 jobs in Norway, and exports products to 27 countries [46]. The Maritime CleanTech Cluster began as a small member-based organization, but due to Norway's cluster growth pathways and research and development (R&D) support, the cluster has grown to foster a new wave of maritime innovation and support local jobs.

Case study: United Kingdom offshore wind sector deal: A holistic industrial growth strategy

The United Kingdom's first offshore wind farm was installed in the year 2000, and since the UK has become the country with the largest installed capacity in the world. In March 2019, the UK government released its Offshore Wind Sector Deal. Referred to as "the Deal," it includes both sector and government goals and investments, with aims to build capacity in the supply chain, enhance UK's global competitiveness, and reach 30 GW of installed capacity by 2030. The Deal is a holistic industrial strategy and includes policies that support each the building blocks of an economic cluster. For instance, "Ideas" corresponds to our description of innovation ecosystem, "People" corresponds with our description of workforce development, and "Business Environment" and "Places" amounts to value chain.

To create clear local market signals, the UK government has established policies to create long-term market certainty to bolster investment from developers and investors. Every 2 years, the government will hold a lease auction for offshore wind, allowing the supply chain to prepare accordingly. Moreover, this will allow the government to plan for offshore and onshore transmission investments needed to integrate offshore wind in an efficient, cost-effective manner. Coupled with this is the Contract for Difference (CfD) support mechanism, a type of feed-in tariff (FIT) that will provide £557 million in support for the sale of offshore wind electricity. A contract for difference subsidizes the delta between the strike price and the wholesale market price (see Figure 13.2). For instance if a developer secures a 20-year contract at a strike price of £57/MWh, and the wholesale market price is £47/MWh, then the CfD will subsidize the gap (£10/MWh). If the wholesale market price is £69/MWh then the generator will pay the difference (£12/MWh) to the government.

The "Ideas" prong of the Deal aims to spur innovation and develop new technologies to support the growth of offshore wind. This includes working in partnership with universities to financially support research, development, and demonstration (RD&D) activities to meet the evolving needs and capabilities of the industry. For instance, turbines continue to grow in size and are five times larger than when they were first deployed. Technologies like artificial intelligence to help

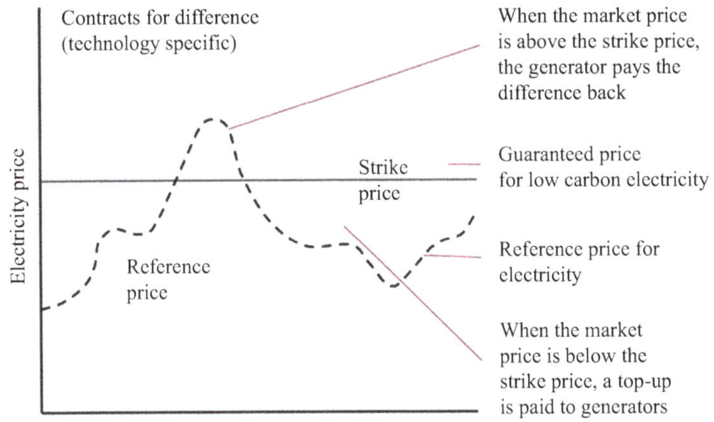

Figure 13.2 Contracts for differences in the United Kingdom drive the market for new technologies like offshore wind. Source: [49]

manage wind farm operations or autonomous vessels for subsea surveys are also in development in the UK. A System Management and Optimization Task Group will develop innovative grid integration solutions.

The Deal's "People" element aims to coordinate training efforts and increase job mobility by partnering with universities, industry programs, and training centers. An Investment in Talent Group has been created that will continue to identify skills gaps and ensure such gaps are addressed in early-stage training and apprenticeships. The Deal also has strategic goals for increasing participation of underrepresented communities. In 2018, only 16% of the offshore wind workforce was made up of women, and the Deal aims to have 33% women participating in the industry by 2030.

The Deal's "Bushiness Environment" outlines local content goals (60% for projects after 2030) and aim of increasing exports fivefold by 2030. Increasing business access for SMEs, sharing information in the supply chain and developing frameworks to support emerging technologies will come together to strengthen the UK supply chain and position itself in an increasingly global marketplace [50].

Economic cluster strategies encompass several elements working in harmony that encourage co-location of firms, bolster innovation, and train the workforce (see Figure 13.3). However, without a clear market signal for renewable technologies investments in research, workers, or a strong supply chain will be in vain. Whether driven by market forces or created by the government, a consistent market signal is key to drive growth and investment.

13.2 Local market

Creating a local market for products sends a market signal to businesses that encourages investment in new facilities and employees. High local demand can attract a local company base that could then expand to regional, national, and global markets. Clear utility and business regulatory environments coupled with

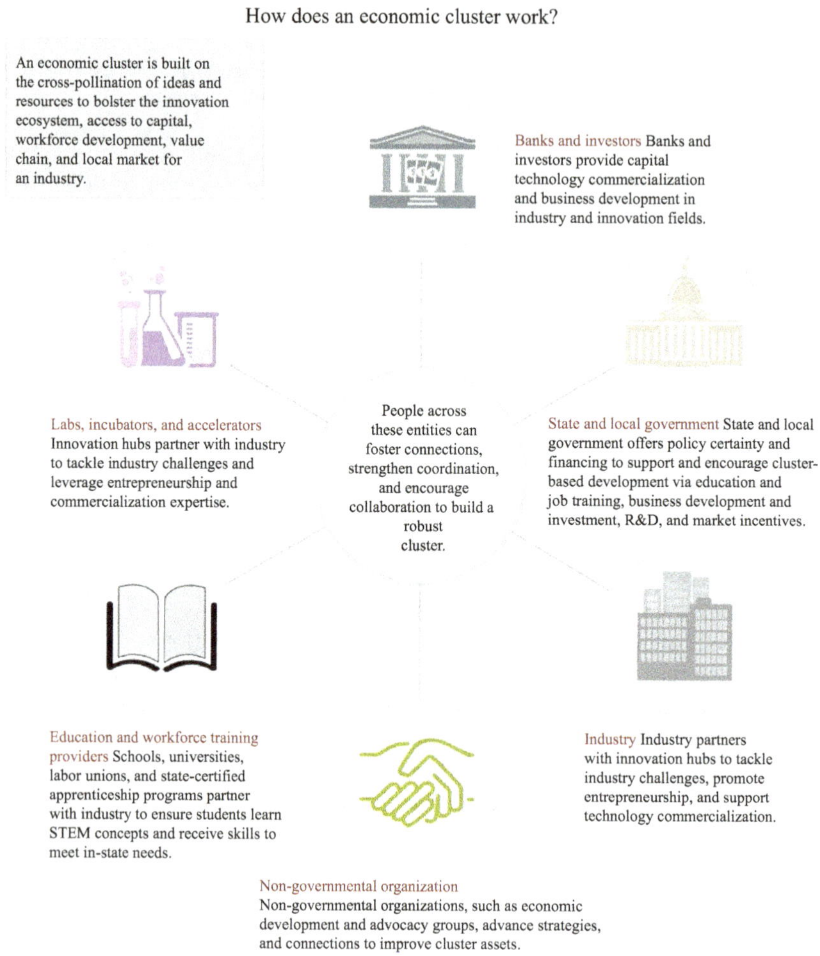

Figure 13.3 An economic cluster is built on the cross-pollination of ideas and resources to bolster the innovation ecosystem, access to capital, workforce development, and local market for industry. Source: [51, p. 20]

resources for project development and end-user adoption can create a strong local market. Next in this section are examples of policies used to enable renewable energy market certainty, which in turn allows for proper planning for infrastructure investments, supply chain capacity, and workforce training efforts.

13.2.1 Feed in tariffs: standard offer contracts to encourage technological growth

FITs are one of the most common policy instruments used by countries around the globe ranging from Germany to Uganda to Australia [36, p. 59]. FITs offer long-term support for the sale of clean energy, with contracts lasting 10–25 years. Long-term contracts provide businesses with the certainty they need to plan and invest. Payments per kWh may vary depending on the type of technology (e.g., offshore wind vs. biogas), project size (e.g., 500 kW project vs. 500 MW project) or project location (e.g., proximity to load centers). Such variables in payment can more accurately account for project costs. In some cases, there may be an FIT premium offered on top of strike prices or FIT may be calculated based on avoided costs [52].

Benefits of FITs include enhanced, stable market access for participants and investors, while encouraging technological growth along the maturity spectrum and supporting renewable energy targets. However, it is argued that FITs can distort wholesale market prices and, if not properly designed, can have unintended consequences. Poor planning or abrupt changes to tariff design can undermine investor confidence, as with the case of Spain in 2009.

Policy goals should drive the design of the FIT (see Table 13.1). The most successful policies ensure access to the grid, long-term contracts, and payments that reflect the cost of doing business in an area. If payments do not correlate with generation costs, deployments will not be optimal. If payments are too low, then the deployment levels will also be low. Conversely if payments are too high then the costs of clean energy development will be too high. Some FIT designs have floors and ceilings included to account for this, known as sliding premiums. Costs should be shared among all classes of ratepayers to transparently plan for activities such as grid balancing and distribution [52].

13.2.2 Net metering: increasing financial incentives for solar PV investment

Net metering is the primary mechanism for compensating residential and small-scale renewable energy projects[§]. Under net metering, customers with renewable electric generators can reduce their electric bill by generating some or all of their power and receive a credit from their electric provider for any excess generation. This improves the economics of investing in small-scale systems, therefore increasing their deployment and related grid benefits. While the term may vary, a number of other counties have some form of net metering that allows independent generators to connect to the grid and sell electricity back to utilities [53]. An increasing number of developing nations have put forth similar frameworks over the past few years [36, p. 59].

Benefits of net energy metering (NEM) includes the increase of distributed generation (DG) resources, which can offset the need for grid investments (e.g., additional power plants or transmission lines) and diversify a region's zero-carbon fuel mix. However, an increase of net energy metering may incur additional distribution-level investments for capabilities like two-way power flow [54].

One of the core policy design considerations is the rate of compensation. Customers are often compensated at the full retail rate; however, utilities take issue with this approach because they are not compensated for the costs incurred to manage

[§]See Chapter 5 on distributed generation for further discussions of the net metering scheme.

Table 13.1 Design elements of FIT reflecting policy goals

Goal	Design elements	Notes
Rapid renewable development to meet RE goals or targets	– Tariff prices based on RE project cost plus profit – Universal eligibility (including IPPs, IOUs, federal, state and local government agencies, non-profit agencies, and residential, commercial, and industrial customers) – Long contract terms (e.g., 15–25 years) – Tariff degression (higher payments this year than next) – No project size caps – Program caps based on long-term RE goals (i.e., not annual targets) – Inclusion of a broad diversity of RE technology types – Reduced administrative burden and streamlined approval processes	– In order to attract many investors and lots of capital to develop renewables rapidly, minimizing financial risk and maximizing options for generators is important – A diverse technology portfolio encourages a broader investor base – Stable and aggressive RE policies can help draw in-state manu-facturing, which can accelerate deployment
Jobs and economic development	– Long-term policy stability – Local content requirements – Preferential terms and/or tariffs for local ownership – Tariff differentiation, according to project size and technology type, ensures a broader diversity of projects developed in a wider variety of technology sectors, particularly DG	– A strong and stable FIT policy that uses bonus payments that can maximize local economic benefits. Incentives can also be offered to create a favorable environment for green manufacturing
Greenhouse gas reduction	– Aggressive tariff rates and terms – High tariffs for biomass co-firing with coal as well as for efficient CHP – Bonus payments for high-efficiency systems – *Note: Greenhouse gas design elements are likely to be more effective in regions that currently have high-carbon electricity production*	– Bioenergy, some hydro, and concentrated solar thermal power (with storage) have the potential to displace high-carbon coal electricity, because the technologies are used primarily for base-load generation

RE: Renewable energy, IPP: Independent power producer, IOU: Investor-owned utility, CHP: Combined heat and power.
Source: Couture *et. al.* [52]

the grid. Some utilities advocate for a per kilowatt or fixed monthly charge on NEM customers for infrastructure costs, or the purchase of additional equipment [55]. Additional NEM design considerations include caps on system size, total capacity cap, what technologies are eligible, and who owns any renewable energy credits [54].

As is the case with FITs, policy design will relate to goals. For instance if the goal is to increase local solar installation jobs, a strong net energy metering policy might be ideal. Additional NEM charges could however decrease the cost-competitiveness of photovoltaic (PV), limit the resiliency of customers to rate changes, and signal a higher-level risk to investors, all of which negatively impact the soft costs of solar deployment [56]. The state of Nevada in the US recently experienced a similar shock to the market when major solar installers, SolarCity and Sunrun, shut down operations in the state following the decision to cut rates and add fixed charges for existing net metering customers [57].

Other types of net metering include aggregate net metering for property owners with multiple meters on the same property or adjacent properties and community net metering for multiple users to benefit from a single net-metered system [58]. Aggregate and community net metering policies enable renters, multi-property owners, and customers in multi-unit residences, commercial spaces, and government-owned facilities to take advantage of net metering incentives typically only available to single-property owners.

13.2.3 Codes/standards to save energy in buildings and appliances

Efficiency standards for buildings and appliances enable continued growth while decreasing the need for energy, and also stimulate local job creation. In the United States alone, energy efficiency contributes to 2.35 million jobs [59]. Building codes establish minimum design and construction requirements for energy efficiency in buildings, like the International Energy Conservation Code (IECC), a residential building code created by the International Code Council. Codes established by the International Code Council are the most widely adopted codes for residential structures in many countries. Commercial building codes are largely based on the ASHRAE 90.1 standards developed by the American Society of Heating, Refrigerating, and Air Conditioning Engineers (ASHRAE) [60]. It is projected that building code policies are the most impactful energy efficiency tool in the built environment, given buildings will be around for decades to come [61].

Much like building codes, efficiency standards dictate a required efficiency level for eligible technologies. This could apply to fuel economies in vehicles, power plant efficiencies, or appliances. Appliance standards have been in place for about 30 years, with counties like South Korea and Mexico setting standards for lamps, refrigerators, and air conditioners [36, p.59]. As policymakers consider setting codes for buildings or appliances the establishment of baseline standards should be set as a minimum threshold, which should be updated frequently by a nominated policy committees [62].

13.3 Value chain

An industry value chain is composed of an array of companies engaged in the manufacturing, sale, marketing, and distribution of technologies. It also includes organizations that represent business interests across platforms. While a supply chain is composed of the entities and processes involved in production and distribution, a value chain is composed of all the activities used to create a competitive advantage. This base provides a solid foundation from which to attract more companies and customers.

13.3.1 Port innovation districts: clustering firms and research

Innovation districts are geographic areas where anchor institutions and companies cluster, connecting with start-ups, incubators, and accelerators. Proximity between these entities has been shown to be a large driver in innovation, especially between networks of manufacturers, logistics firms, customers, and researchers—all key players in the offshore wind value chain. Innovation districts perform well when associated with unique infrastructure such as ports. Other factors that promote successful innovation districts include on-site university research programs, incubators that provide facilities and support for start-ups, anchor company facilities, common areas, and flexible office space. Examples of successful innovation districts include the newly established Rotterdam Smart Manufacturing Innovation District, which is a burgeoning entrepreneurial hub located at the Port of Rotterdam. The Rotterdam Innovation District (RID) was established by the port and city of Rotterdam in 2015 as a response to a Dutch policy assessment that identified the need for more spending on innovation. Industrial buildings of former shipyards and abandoned waterfronts were redeveloped to form RID. RID includes an experimental living and working space for innovators in cleantech and medical clusters, a start-up accelerator, prototyping facilities, and an investment fund for port-related innovations. Universities and research centers such as Albeda College, Rotterdam University of Applied Sciences, Delft University of Technology, and Yes!Delft have co-located at RID. If mid-sized manufacturers want to locate at RID, they must engage with researchers and students and use the space actively for production [51].

Brief case study: Wisconsin Supply Chain Marketplace

The Supply Chain Marketplace is a dynamic online platform for Wisconsin suppliers to engage new customers and facilitate buyer connections, supporting local business growth and access to new market opportunities. It is open to all businesses free of charge so that suppliers can showcase their business capabilities; be readily searchable to potential buyers by targeted industries, certifications, and ownership; and access requests for proposals and calls for innovation. The Supply Chain Marketplace was initially launched by New North, the northeast Wisconsin economic development group, as a regional economic diversification asset, with support from the US Department of Defense's Office of Economic Adjustment. In 2017, the state awarded US$99,000 to New North to expand the platform statewide and include other industries, and it now boasts over 1,000 local businesses. With the planned arrival of Foxconn Technology Group, a Taiwanese electronics manufacturer, the platform also hosted a page where 500 companies pitched their skills to support Foxconn's future construction and supply chain needs (Source: [63]).

13.4 Workforce development

Trained and skilled workers are fundamental to industry success, and strategic workforce development can support talent recruitment and retention. Workforce development requires collaboration across schools, businesses, labor unions, and government offices to integrate STEM education, foster industry-ready skills via apprenticeships and career-integrated curriculum, enable stackable credentials that offer multiple entries and exits, and provide resources that match skills to available jobs. Adequate funding from public and private sources is essential to ensure workers are trained for the jobs of the future. The transition to the clean energy economy will require over 320 occupations across skill levels, technologies, products, and services [20, p. 4]. Workers in industries such as semiconductors, shipbuilding, and oil and gas can transfer knowledge and skills to other clean energy industries like offshore wind [64]. By leveraging existing industries, more than half the jobs created in renewable energy can be localized [65].

Green-collar job: Blue-collar employment that has been upgraded to better respect the environment; family-supporting, career-track, vocational, or trade-level employment in environmentally friendly fields; examples include electricians who install solar panels, farmers engaged in organic agriculture, etc. [66].

13.4.1 Fostering apprenticeships through support and incentives

Apprenticeships offer significant benefits to local workers and businesses. They empower students to develop valuable on-the-job skills while still bringing home a paycheck, and they help employers and industry clusters meet the demand for trained workers. Government investments in apprenticeships have significant social returns: Apprenticeship programs can produce more than US$27 in tax returns for every US$1 invested by the state over the career of an apprentice [67]. Apprenticeship programs and uptake vary by country and subnational region, with countries such as Germany known to have strong programs.

Improved engagement of the private sector in hands-on training can help build a skilled workforce and retain talent in homegrown businesses. In some regions with low private-sector apprenticeship engagement, tax incentives could spur more businesses to hire apprentices. Incentives could even target certain demographics or industries, as US states Montana and Rhode Island have done [67].

South Carolina has demonstrated how a small investment in apprenticeships can have significant payoffs for workers and the state. The state's apprenticeship system offers a modest US$1,000 tax credit per apprentice per year to help offset planning and administration costs [68]. In order to count toward the credit, an apprentice must be employed for at least 7 months each year of their program, for up to 4 years [68]. In addition to the tax credit, administrative assistance from the state's Apprenticeship Consultants helps to increase program uptake by providing dedicated experts who work with businesses interested in establishing apprenticeships in the state [68]. The system has served over 26,000 apprentices [69].

13.4.2 Stackable credentials: a career ladder for middle-skill clean energy jobs

Middle-skill jobs require more than a high-school education, but less than a bachelor's degree. In the advanced energy economy, middle-skill workers such as solar panel manufacturers and installers are needed, and training must keep up with evolving technology. However challenges arise in places such as the United States, where middle-skill workers account for 53% of the US labor market, but only 43% of American workers are trained to that level [70].

To bring skill levels up to the needs of the evolving cleantech industry, stackable credentials can be deployed. These comprise of an organized sequence of certificates that can be accumulated over time to strengthen individual qualifications, and to advance a career pathway or a career ladder to different and higher-paying jobs [71]. This sequence includes shorter-term skill development blocks which allow students to exit and enter while still having gained marketable skills. It reduces educational and employment barriers for nontraditional and disadvantaged students, which can be a tool to increase diversity in the clean energy sector [72]. Stackable certificates should be portable, or independently verified and accredited, so student credentials are recognized by employers and educational institutions across regions. This system can be utilized to support more clean energy technology manufacturing by establishing specific career pathways with portable certificates.

The North Carolina Community College System has a green jobs curriculum focused on critical growth areas such as buildings, energy, engineering, environment, and transportation. This project has streamlined over 80 curriculums and affected thousands of students, and students can now earn multiple credentials endorsed by the National Association of Manufacturers while working toward an associate degree. The project has conducted over 40 professional development events for more than 680 faculty participants to provide experience with new technologies and employer-desired skill sets [73].

13.5 Access to capital and end-user finance

For clean energy to thrive, capital is essential so an active investment environment can attract more entrepreneurs and investors to a region. Regardless if it is an entrepreneur commercializing a new technology, or a first adopter on the block ready to install solar panels, access to financing is key to success. For entrepreneurs and innovators, access to investors and competitively priced non-dilutive capital (i.e., that does not impact ownership of the company) can be the difference between success and failure for a new or expanding business. It is also important for consistent access to capital across the development cycle, from the seed and early/growth stages to the late stage. Innovations in end-user finance can open up new markets for distributed technologies because technologies are more readily accessible for consumers. Governments can consider actively facilitating access to capital to stoke the entrepreneurial clean technology environment. Enabling end users with access new distributed technologies, such as solar panels, can be achieved through policies like tax liens and government loans.

To address the negative externality affiliated with the R&D phase of clean energy, governments can support new technologies through the access to capital "valley of death" (see Figure 13.4). Encouraging potentially high-risk investment through policies like financing and tax credits and can support firms throughout various phases of the commercialization phase and attract talent to a region. A number of countries incentivize start-up investment through taxes. In the US, almost 50% of state governments have an angel tax credit scheme [36]. In some instances, states invest public funds or pooled funds in a fund of funds. A fund of funds is a fund that invests in other investment funds as opposed to investing directly in stocks, bonds, and other securities [74]. Because a fund of funds has a diverse portfolio, investors are better protected from high-risk investments and exposed to more ventures [74]. A fund of funds not only fosters the state's investment community but also increases access to capital for in-state businesses.

These companies tend to lack physical assets that can be leveraged for bank loans, which is an issue that is particularly acute for new technologies that may be costly to develop and exhibit longer commercialization periods [75]. Policymakers could establish a fund of funds and encourage top investors to devote their capital to

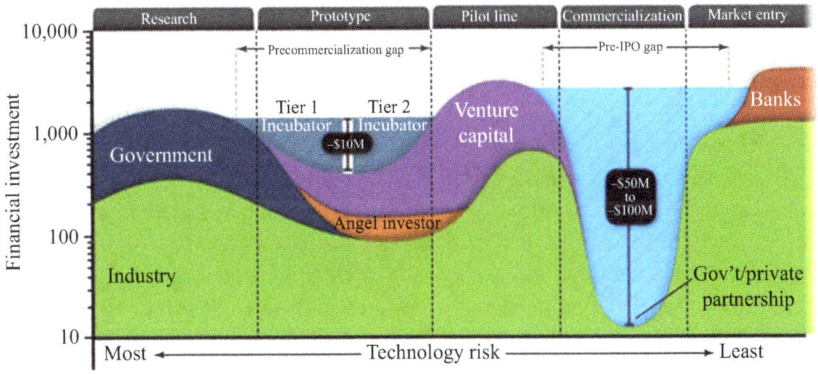

Figure 13.4 The commercialization phase of the RD&D process is subject to gaps in funding due to high capital costs and long time horizons. Government programs such as loan guarantees help fill the gap. Source: [87]

companies in the region, thereby increasing access to capital for in-state businesses. Fund-of-funds managers could make patient capital investments. Patient capital refers to funds invested in a company for the long term to allow technologies with longer development horizons to grow into valuable investments. The fund-of-funds managers could also look to evaluate fund performance by looking at comprehensive returns to the economy and state revenues, as private funds in other states have done [76,77]. Comprehensive returns include both direct financial returns from an investment and the indirect economic development returns from that investment. These returns are important to value and track since the goal of a fund of funds would be to produce both financial and economic development returns.

Case study: Illinois' fund of funds

Illinois first began investing in venture funds in 2002, using funds from the US$13 billion invested by the state treasurer. The fund invests no more than 15% in any given venture or private-equity fund, which then, in turn, invests in high-potential start-ups. The state receives a financial return for its investments, with past rates of financial return around 6%. Since 2005, Illinois has channeled US$66 million of state funds into 18 private investment funds. These investments have supported almost 6,300 new Illinois jobs. In 2016, Illinois refinanced its fund of funds with US$222 million, which is expected to create over 18,000 jobs. An advisory board oversees the investments, which are targeted toward local technology and bioscience companies (Source: [78]).

13.5.1 Loan guarantees: bridging the valley of death for renewable energy technologies

While grants may be appropriate for researchers in early-stage research, later-stage development requires more capital and longer-time horizons to get firms past the valley of death. Longer timelines may deter venture capitalists and traditional funders from investing in commercialization activities. Government-granted loan guarantees, like those granted by ARPA-e in the United States, can provide the patient capital for high-risk, high-reward technologies. Since establishment in 2009, ARPA-e has supported the formation of 76 companies, 346 patents. The US$2 billion spent on R&D has yielded US$2.9 billion in follow-on funds from the private sector [79]. The federal loan guarantee program resulted in some large successes, such as Tesla. The firm paid off its loan 10 years early with interest, and has since become an icon of American innovation and manufacturing [80]. However, the case of Solyndra, the now defunct and bankrupt solar manufacturing company, remains in the psyche of US policymakers. Solyndra received a US$535 million loan guarantee in 2009 for its Silicon-free, thin-film PV technology. Manufacturing in a cost-competitive manner became impossible for Solyndra in light of the collapse in silicon prices, and a wave of subsidized Chinese solar manufacturers. Several years later, the firm filed for Chapter 11 bankruptcy.

13.5.2 Brief case study: addressing challenges with market entry: Maryland's Offshore Wind Business Development Grant Program

The Offshore Wind Business Development Grant Program provides financial assistance to Maryland companies, especially minority-owned businesses, seeking to enter the offshore wind industry [81]. The program was established by the Maryland Offshore Wind Energy Act of 2013 and is managed by the Maryland Energy Administration (MEA) [81,82]. The program targets two key areas of business development: For market entry, funds can offset bidding costs and associated administrative expenses, and for retooling, funds can cover equipment and facility upgrades needed for offshore wind-related exports [83]. In 2018, MEA awarded US$700,000 in grant funding to businesses specializing in welding and fabrication as well as development and construction [84–86]. This infusion of capital is helping catalyze offshore wind development among the 1,300 businesses located around the Port of Baltimore [84].

13.5.3 Opening end-user markets to distributed technology

Advances in business models, like energy-as-a-service or property-assessed clean energy (PACE) financing, allow home owners to finance renewable energy and energy efficiency projects with a loan that is repaid through a property tax bill. However, this method may require enabling legislation that outlines guidelines such as eligible technologies, as is the case in the United States. The state of California's PACE enabling legislation includes a variety of eligible clean energy technologies, as well as protocols for homes defaulting on their mortgage [88–91]. Another manner to encourage adoption of distributed technologies, governments can also provide loans to purchase solar panels, which can then be rented to homeowners like in the Dominican Republic. Businesses that provide accessible credit lines and financing may also be eligible for loans by governments, as is the case with Grameen Shakti in Bangladesh [36, p. 59].

13.6 Innovation ecosystem

Innovation is essential for business and industry competitiveness, and a strong knowledge hub can be a beacon for talent and investment. Innovative companies and the availability of talent will continue to drive global competition [34, p. 215]. The innovation ecosystem supports fundamental research across universities and labs, fosters an entrepreneurial culture that seeks to advance and disrupt industries, and brings ideas to market.

Clean energy technology requires large amounts of capital-intensive R&D; thus, it makes sense for the government to subsidize basic and applied research [34]. In fact, the social rate of return on R&D is twofold that of private returns, indicating that R&D investments are worthwhile investments of public funds [92]. Many countries directly fund R&D activities through government-funded research institutions including the German Fraunhofer Institutes, United States Department of Energy National Lab Network, and the Indian Council on Scientific and Industrial Research (CSIR). CSIR was established in 1942 and is the largest R&D facility in India. Over the years, CSIR adapted to the demands of a competitive global market, and established a market-driven R&D approach by placing more emphasis on outputs and performance in demand-heavy technologies [93].

In addition to directly funding research efforts, governments and universities can provide protections and streamline red tape with technology transfer. Policies like points toward tenure track for faculty at universities can encourage academics to pursue entrepreneurial actives. Fostering connections between the businesses in the cluster and research institutes via vouchers (see the case study) can help small- to medium-sized businesses in the value chain develop next-generation products and processes. Requiring businesses to engage with research institutes to meet technical standards is another way to create connections between firms and researchers.

Case studies: New Mexico and Tennessee innovation vouchers

The New Mexico Small Business Assistance Program is a successful model that has helped 2,341 businesses gain access to technology at the Sandia or Los Alamos National Labs through a competitive application process that matches qualified businesses to lab scientists. The state government provides funding for the program, incentivizing the labs to engage with local small- and medium-sized businesses. In Tennessee, the state created a US$2.5 million innovation voucher program whereby local manufacturers of varying sizes can "purchase" services from Oak Ridge National Lab to develop new products and processes [94].

Case study: Denmark's knowledge-based wind cluster

After the oil crisis of the 1970s, Denmark began investing in renewable energy technologies to facilitate energy independence. The country has put forth a number of policies to support wind development such as financial support through capital grants and government-funded R&D, pricing strategies such as renewable portfolio standards (RPS) and FITs, priority access to the grid, long-term targets for wind electricity consumption, and subsidies provided to companies that use renewable energy and increase their energy efficiency [95]. During the 1980s, Denmark paired market incentives with research-industry partnerships through legislation provided a 30% investment grant to purchase wind turbines that were certified by a national laboratory [96]. The grant expired in 1989, but the "Danish Certification Scheme," in which the government ensures product quality, remains today. The impacts on the industry cluster are twofold: not only do wind manufacturers have "access to universities, research and test centers with cutting-edge know-how and extensive experience in collaborating with the industry," but the certification scheme also ensures product differentiation based on local knowledge [97]. This not only keeps standards high, but it makes Danish products unique and place-based, which continually keep the industry competitive in a changing global market. To ensure that smaller enterprises have access to the Danish knowledge cluster, the government subsidizes SMEs to develop and test their products by facilitating access to universities and research institutes. Given the large size of wind turbines, infrastructure is essential for wind manufacturers. Companies can opt in to support the development of physical and electrical infrastructure [98,99]. As of 2015, the country has a clear competitive advantage in the sector, with the world's top wind energy manufacturer being Danish [100].

13.7 Conclusion

Much like baking a cake, growing clean energy industries requires right mix of ingredients in proportion to each other. Policies to support R&D, increase access to capital, facilitate alignment in the supply chain, train workers for the jobs of the future, and create steady local demand for clean energy must work together in order to achieve success. Proper planning and timing is also key: investing in training programs for workers in a technology that has no clear market trajectory is an example of a poor investment. Whenever public funds are used, sound economic rationale and policy analysis are the required precursor. Questions for planners to consider include the following:

- What comparative advantages does my region have, and how does that align with global cleantech market trends?
- Does this investment facilitate a first-mover advantage in a growing global market?
- What market failure is being addressed by the use of government funds?
- When should subsidies be removed?
- Who has traditionally benefitted from an industry's growth and who is being left behind?

A regions' social, technological, and economic goals must drive policy, and there is no one-sized-fits-all approach to renewable energy clusters. Clusters can be tailored to reach the social welfare agenda of a locale, which may include environmental justice or increased jobs in a struggling neighborhood. With the continued economic growth, leaders can continue to garner buy-in for the new climate agenda: one that creates distributed wealth and lifts all boats.

Chapter exercises

1. This chapter discusses social welfare, or the co-benefits other than emissions resulting from renewable energy development. Three examples of renewable energy social welfare are provided: (1) increased employment; (2) higher wages, and (3) greater levels of environmental justice. What else could be included in the definition of social welfare?
2. Should governments "pick winners" through policies such as loan guarantee? Why or why not?
3. How do support mechanisms for industry growth vary between large-scale technologies (e.g., offshore wind) and distributed technologies (e.g., solar PV)?

4. The shift away from fossil fuels to renewable energy reduces or eliminates a negative externality: air pollution. Could the shift to renewables cause unintended consequences or create other negative externalities?

Definitions

Industry Cluster: Clusters are regionally situated groups of interconnected companies and institutions that are engaged in a particular industry and supported by repeated exchanges of information and resources.

Environmental Justice: All people and communities have the right to equal environmental protection under the law, and the right to live, work, and play in communities that are safe, healthy, and free of life-threatening conditions.

Environmental Racism: Whether, by conscious design or institutional neglect, actions and decisions that result in the disproportionate exposure of people of color to environmental hazards and environmental health burdens [22].

Green-collar Job: Blue-collar employment that has been upgraded to better respect the environment, family-supporting, career-track, vocational, or trade-level employment in environmentally friendly fields; examples include electricians who install solar panels, farmers engaged in organic agriculture, etc. [66].

Web resources

American Jobs Project: http://Americanjobsproject.us
CNEE clean energy legislative guide: http://cnee.colostate.edu/cleanenergypolicyguide/
Cluster mapping: http://www.clustermapping.us/
The Drawdown: https://www.drawdown.org/
Third Way: https://www.thirdway.org/issue/clean-energy
Advanced Energy Economy: https://www.aee.net/
Green for All: https://www.greenforall.org/
Brookings Institution: https://www.brookings.edu/
IRENA: https://www.irena.org/
World Resources Institute: https://www.wri.org/our-work/topics/energy
Environmental Justice Principles per Columbia University: http://www.columbia.edu/cu/EJ/Reports_Linked_Pages/EJ_principles.pdf
BlueGreen Alliance: https://www.bluegreenalliance.org/
Database of State Incentives for Renewables & Efficiency: https://www.dsireusa.org/
E2: https://www.e2.org/
American Council on Renewable Energy: https://acore.org/
Regulatory Assistance Project: https://www.raponline.org/
American Council for Energy Efficient Economy: https://aceee.org/

Energy Futures Initiative: https://energyfuturesinitiative.org/
Workforce Development Institute: https://wdiny.org/

References

[1] "Global Warming of 1.5°C: Summary for Policymakers," *Intergovernmental Panel on Climate Change,* p. 17, 2018, accessed January 17, 2019, https://report.ipcc.ch/sr15/pdf/sr15_spm_final.pdf

[2] C. Joyce, "Climate Report Warns of Extreme Weather, Displacement of Millions Without Action," NPR, October 8, 2018, accessed October 25, 2018, https://www.npr.org/2018/10/08/655360909/grim-forecast-from-u-n-on-global-climate-change

[3] C. Davenport, "Major Climate Report Describes a Strong Risk of Crisis as Early as 2040," *The New York Times,* October 7, 2018, accessed November 26, 2018, https://www.nytimes.com/2018/10/07/climate/ipcc-climate-report-2040.html

[4] "Fourth National Climate Assessment, Volume II: Impacts, Risks, and Adaptation in the United States," *U.S. Global Change Research Program,* p. 34, 2018, accessed November 26, 2018, https://nca2018.globalchange.gov/downloads/NCA4_2018_FullReport.pdf

[5] "Working on a Warmer Planet: The Impact of Heat Stress on Labour Productivity and Decent Work," International Labour Office – Geneva, ILO, July 1, 2019, p. 13, accessed July 8, 2019, https://www.ilo.org/wcmsp5/groups/public/—dgreports/—dcomm/—publ/documents/publication/wcms_711919.pdf

[6] IMF (International Monetary Fund). 2017. Seeking Sustainable Growth: Short-Term Recovery, Long-Term Challenges, Washington, DC, Oct.

[7] "IPCC Press Release," *Intergovernmental Panel on Climate Change,* p. 2, October 8, 2018, accessed October 25, 2018, https://www.ipcc.ch/site/assets/uploads/2018/11/pr_181008_P48_spm_en.pdf

[8] M. Bloomberg and C. Pope, Climate of Hope: How Cities, Businesses and Citizens Can Save the Planet, p. 54, April 2017, St. Martin's Press.

[9] "Advanced Energy Now 2017 Market Report," *Advanced Energy Economy,* p. 4, 2017, accessed July 30, 2018, http://info.aee.net/aen-2017-market-report

[10] J. Doerr, "The Green Road to Prosperity," Scientific American, March 1, 2009, accessed November 26, 2018, https://scientificamerican.com/article/the-green-road-to-prosperity/

[11] A. Roy and P. Khuje, "Global Renewable Energy Market," *Allied Market Research,* November 2017, accessed November 19, 2018, https://www.alliedmarketresearch.com/press-release/renewable-energy-market.html

[12] "New Energy Outlook 2017," *Bloomberg New Energy Finance,* p. 2, June 2017, accessed July 30, 2018, https://data.bloomberglp.com/bnef/sites/14/2017/06/BNEF_NEO2017_ExecutiveSummary.pdf

[13] IRENA (2017a), "Chapter 3: Global Energy Transition Prospects and the Role of Renewables", In International Energy Agency and IRENA (Eds.),

Perspectives for the Energy Transition: Investment Needs for a Low-carbon Energy System, pp. 121–189, OECD/IEA and IRENA, Paris and Abu Dhabi.

[14] A. de la Garza, "Trump Mocks Climate Change," *Time*, January 20, 2019, https://time.com/5508259/trump-climate-change-defense-department/

[15] S. Eschenbacher and D.A. Garcia, "Mexico's Rising Oil Nationalism Faces Debt-rating Crisis," *Reuters*, May 14, 2019, Accessed July 14, 2019, https://www.reuters.com/article/us-mexico-pemex-insight/mexicos-rising-oil-nationalism-faces-debt-rating-crisis-idUSKCN1SK1TE.

[16] Morning Consult, "Political Intelligence, Midterms 2018," https://morningconsult.com/wp-content/uploads/2018/05/Morning-Consult-Political-Intelligence-May-Release.pdf, slide 16.

[17] K.S. Anderson, W. Ayres, and G. Bolger, "Conservative Republicans Open to Clean Energy," September 2015, https://assets.clearpath.org/2016/09/clearpath_pollster_memo.pdf

[18] B.S. Ihab Mikati, A.F. Benson, T.J. Luben, J.D. Sacks, and J. Richmond-Bryant, "Disparities in Distribution of Particulate Matter Emission Sources by Race and Poverty Status," American Journal of Public Health, April 2018, https://ajph.aphapublications.org/doi/abs/10.2105/AJPH.2017.304297

[19] "Renewable Energy and Jobs: Annual Review 2018," International Renewable Energy Agency, p. 24, 2018, accessed November 19, 2018, http://irena.org/-/media/Files/IRENA/Agency/Publication/2018/May/IRENA_RE_Jobs_Annual_Review_2018.pdf

[20] M. Muro *et. al.*, "Advancing Inclusion through Clean Energy Jobs," April 2019, Brookings Institution, https://www.brookings.edu/wp-content/uploads/2019/04/2019.04_metro_Clean-Energy-Jobs_Report_Muro-Tomer-Shivaran-Kane.pdf

[21] IRENA, "Renewable Energy Benefits: Understanding the Socio-Economics," November 2017, https://www.irena.org/-/media/Files/IRENA/Agency/Publication/2017/Nov/IRENA_Understanding_Socio_Economics_2017.pdf?la=en&hash=C430B7EF772BA0E631190A75F7243B992211F102.

[22] R. Bullard "Environmental justice: Grassroots activism and its impact on public policy decision making," *Journal of Social Issues*, 56, no. 3 (2000).

[23] S. Assari and M.M. Lankarani, "Poverty status and childhood asthma in White and Black families: National survey of children's health," *Healthcare (Basel)*, 6, no. 2 (2018), p. 62. Published online 2018 Jun 12. doi: 10.3390/healthcare6020062.

[24] C.W. Tessum *et. al.*, "Inequity in consumption of goods and services adds to racial–ethnic disparities in air pollution exposure," *Proceedings of the National Academy of Sciences*, 116 no. 13 (2019), pp. 1–6; DOI:10.1073/pnas.1818859116.

[25] E. Castro Diaz, "Environment Issues in the Latin American Region," United Nations International Expert Group Meeting On Indigenous Peoples and Protection of the Environment, 2007, https://www.un.org/esa/socdev/unpfii/documents/workshop_IPPE_castro.pdf

[26] T. Lohan, "Dr. Robert Bullard: Lessons from 40 Years of Documenting Environmental Racism," April 2019, https://therevelator.org/bullard-environmental-justice/; see also https://drrobertbullard.com/

[27] M. Porter, "Location, competition, and economic development: local clusters in a global economy," *Econ Dev Q*, 14 (2000), pp. 15–34.

[28] G.M. Spencer, T. Vinodrai, M.S. Gertler, and D.A. Wolfe, "Do clusters make a difference? Defining and assessing their economic performance," *Reg Stud*, 44 (2010), p. 698

[29] M. Porter, 1998. "Clusters and the New Economics of Competition," *Harvard Business Review* (November–December 1998).

[30] C.H.M. Ketels and O. Memedovic, "From clusters to cluster-based economic development," *International Journal of Technological Learning, Innovation, and Development*, 1, no. 3 (2008), pp. 378–380, accessed November 19, 2018, https://www.clustermapping.us/sites/default/files/files/resource/From_clusters_to_cluster-based_economic_development.pdf

[31] Rodrik D, *One Economics, Many Recipes*, 2007, Princeton University Press, Princeton.

[32] J.C. Kuntze and T. Moerenhout, "Local Content Requirements and the Renewable Energy Industry – A Good Match?" International Centre for Trade and Sustainable Development, 2013, Geneva.

[33] T.J.A. Roelandt, V.A. Gilsing, and J van Sinderen, "New Policies for the New Economy: Cluster Based Innovation Policy: International Experiences," *Paper Presented at the 4th Annual EUNIP Conference*, Tilburg, The Netherlands, December 7–9, 2000.

[34] E. Moretti, *The New Geography of Jobs*, 2013, First Mariner Books, Houghton Mifflin Harcourt Publishing Company, New York.

[35] E.J. Feser, "Old and New Theories of Industry Clusters. Selected Works," accessed 5 June 2013, works.bepress.com/edwardfeser/3/

[36] S. Carley and S. Lawrence, *Energy-Based Economic Development*, 2014, Springer-Verlag London, DOI: 10.1007/978-1-4471-6341-1_4.

[37] H. Pack and K. Saggi, "The Case for Industrial Policy: A Critical Survey," World Bank Policy Research Working Paper 3839, World Bank, Development Research Group, Washington, DC.

[38] S. Nousala, "The sustainable development of industry clusters: emergent knowledge networks and socio complex adaptive systems," *Systematics, Cybernetics, and Informatics*, 7 no. 5 (2009), p. 2.

[39] A. Karev, S.C.L. Koh, and L.T. Szamosi "The cluster approach and SME competitiveness: A review," Journal of Manufacturing Technology Management, September 2007, https://www.emerald.com/insight/content/doi/10.1108/17410380710817273/full/html.

[40] M. Porter, *Cluster of Innovation: Regional Foundations of US Competitiveness*, 2001, Washington, DC: Council on Competitiveness, p. 14.

[41] J. Cortright, "Making Sense of Clusters: Regional Competitiveness and Economic Development," The Brookings Institution, March 2006, accessed November 19, 2018, https://www.brookings.edu/wp-content/uploads/2016/06/20060313_Clusters.pdf

[42] "Employment in New York State," New York Department of Labor, p. 2, August 2014, accessed November 27, 2018, https://labor.ny.gov/stats/PDFs/enys0814.pdf

[43] Organisation for Economic Co-operation and Development (OECD), 2017, "Enhancing the Contributions of SMEs in a Global and Digitalized Economy," Meeting of the OECD Council at the Ministerial Level, Paris, 6.

[44] Eurofound, n.d., Norwegian Innovation Clusters, accessed March 1, 2018, https://www.eurofound.europa.eu/observatories/emcc/erm/supportinstrument/norwegian-innovation-clusters

[45] H. Okland, 2016, "Developing Clean Maritime Solutions by Cluster Cooperation," NCE Maritime CleanTech, April 13. Accessed April 4, 2018. http://www.mlit.go.jp/common/001127984.pdf

[46] About NCE Maritime CleanTech Cluster. Accessed March 19, 2019. https://maritimecleantech.no/about-us/

[47] NCE Maritime CleanTech. n.d., A Brief History of Our Cluster, accessed March 1, 2018, https://maritimecleantech.no/about-us/our-history/

[48] NCE Maritime CleanTech. Awarded 11.7 Million Euros to Develop All-Electric High-Speed Passenger Vessel, January 29, accessed March 1, 2018, https://maritimecleantech.no/2018/01/29/awardednok-110-million-develop-fully-electric-high-speed-passenger-vessel/

[49] "UK Offshore Wind: Opportunities for trade and investment," https://www.gov.uk/government/publications/uk-offshore-wind-opportunities-for-trade-and-investment/uk-offshore-wind-opportunities-for-trade-and-investment

[50] Industrial Strategy: UK Offshore Wind Sector Deal, HM Government, March 2019, https://assets.publishing.service.gov.uk/government/uploads/system/uploads/attachment_data/file/790950/BEIS_Offshore_Wind_Single_Pages_web_optimised.pdf

[51] American Jobs Project, The California Offshore Wind Project: A Vision for Industry Growth, February 2019, http://americanjobsproject.us/wp/wp-content/uploads/2019/02/The-California-Offshore-Wind-Project.pdf

[52] T. Couture *et al.*, "A Policymaker's Guide to Feed-in Tariff Policy Design," National Renewable Energy Laboratory, July 2010, https://www.nrel.gov/docs/fy10osti/44849.pdf

[53] E. Martinot, A. Chaurey, D. Lew, J. Moriera, and N. Wamukonya, "Renewable energy markets in developing countries," *Annu Rev Energy Environ*, 27(2002), pp. 309–348.

[54] The State Clean Energy Cookbook: A Dozen Recipes for State Action on Energy Efficiency and Renewable Energy, Stanford's Steyer-Taylor Center for Energy Policy and Finance, 2013.

[55] "Best Practices in State Net Metering Policies and Interconnection Procedures," *Interstate Renewable Energy Council and the Vote Solar Initiative,* November 2013, p. 21, http://freeingthegrid.org/wp-content/uploads/2013/11/FTG_2013.pdf

[56] J. Kennerly, K. Wright, C. Laurent, W. Rickerson, and A. Proudlove, "Rethinking Standby & Fixed Cost Charges: Regulatory & Rate Design Pathways to Deeper Solar PV Cost Reductions," NC Clean Energy Technology Center and Meister Consultants Group, August 2014, pp. 4–6,

https://nccleantech.ncsu.edu/wp-content/uploads/Rethinking-Standby-and-Fixed-Cost-Charges_V2.pdf

[57] J.S. John, "Nevada's Solar Job Exodus Continues, Driven by Retroactive Net Metering Cuts," Greentech Media, January 8, 2016, accessed February 14, 2016, http://www.greentechmedia.com/articles/read/nevadas-solar-exodus-continues-driven-by-retroactive-net-metering-cuts

[58] J. Durkay, "Net Metering: Policy Overview and State Legislative Updates," National Conference of State Legislatures, last modified December 18, 2014, accessed February 14, 2016, http://www.ncsl.org/research/energy/net-metering-policy-overview-and-state-legislative-updates.aspx

[59] The 2019 US Energy and Employment Report, NASEO and Energy Futures Initiative, https://static1.squarespace.com/static/5a98cf80ec4eb7c5cd928c61/t/5c7f3708fa0d6036d7120d8f/1551849054549/USEER+2019+US+Energy+Employment+Report.pdf

[60] American Jobs Project, The Georgia Jobs Project, 2016, http://american-jobsproject.us/wp-content/uploads/2016/04/GA-Jobs-Project-Full-Update.pdf

[61] R.A. Athalye, B. Liu, D. Sivaraman, R. Bartlett, D.B. Elliott, "Impacts of Model Building Energy Codes Pacific Northwest National Laboratory," October 2016, https://www.energycodes.gov/sites/default/files/documents/Impacts_Of_Model_Energy_Codes.pdf

[62] "Clean Energy Policy Guide for State Legislatures," Center for the New Energy Economy, Colorado State University, 2017.

[63] American Jobs Project, The New Mexico Jobs Project, April 2018, http://americanjobsproject.us/wp/wp-content/uploads/2018/04/New-Mexico-Jobs-Project-4.3.18-WEB.pdf

[64] D. Fornahl, R. Hassink, C. Klaerding, I. Mossig. and H. Schröder, "From the old path of shipbuilding onto the new path of offshore wind energy? The case of Northern Germany", *European Planning Studies*, 20, no. 5 (2012), pp. 835–855, http://dx.doi.org/10.1080/09654313.2012.667928

[65] IRENA and CEM (Clean Energy Ministerial) (2014), The Socio-economic Benefits of Solar and Wind Energy, http://www.irena.org/menu/index.aspx?mnu=Subcat&PriMenuID=36&CatID=141&SubcatID=418

[66] V. Jones, *The Green Collar Economy: How One Solution Can Fix Our Two Biggest Problems*, 2008, Harper Collins, New York.

[67] "An Effectiveness Assessment and Cost-Benefit Analysis of Registered Apprenticeship in 10 States," *Mathematica Policy Research,* p. 40, July 25, 2012, accessed August 28, 2017, http://wdr.doleta.gov/research/FullText_Documents/ETAOP_2012_10.pdf

[68] "Frequently Asked Questions," *Apprenticeship Carolina,* accessed August 28, 2017, http://www.apprenticeshipcarolina.com/faqs.html

[69] "By the Numbers," *Apprenticeship Carolina,* accessed November 7, 2017, http://www.apprenticeshipcarolina.com/by-the-numbers.html

[70] "United States' Forgotten Middle," National Skills Coalition, 2017, https://www.nationalskillscoalition.org/resources/publications/2017-middle-skills-fact-sheets/file/United-States-MiddleSkills.pdf

[71] McCarthy, "Stackable Credentials: What Are They?" *United States Department of Labor and Training Administration*, http://www.doleta.gov/taaccct/pdf/presenters/McCarthy.pdf

[72] E. Ganzglass, "Scaling 'Stackable Credentials,'" Center for Postsecondary and Economic Success, March 2014, http://www.clasp.org/resources-and-publications/files/2014-03-21-Stackable-Credentials-Paper-FINAL.pdf

[73] "Code Green Super CIP (Curriculum Improvement Project)," *SuccessNC,* accessed December 4, 2015, http://www.successnc.org/initiatives/code-green-super-cip-curriculum-improvement-project

[74] "Fund of Funds," *Investopedia*, accessed October 19, 2016, http://www.investopedia.com/terms/f/fundsoffunds.asp

[75] J. Freed, "Making a Case for Innovation Investment in an Era of Austerity," *Third Way,* May 19, 2011, accessed February 22, 2018, http://www.thirdway.org/memo/making-a-case-for-innovation-investment-in-an-era-of-austerity

[76] "Empowering Oregon's Trailblazing Entrepreneurs," Oregon Venture Fund, accessed July 10, 2018, http://www.oregonangelfund.com/

[77] "Ben Franklin Technology Partners Releases Independent Report on its Economic Impact on Pennsylvania," *PR Newswire,* June 3, 2013, accessed July 10, 2018, https://www.prnewswire.com/news-releases/ben-franklin-technology-partners-releases-independent-report-on-its-economic-impact-on-pennsylvania-209957351.html

[78] American Jobs Project, The Wyoming Jobs Project, August 2018, http://americanjobsproject.us/wp/wp-content/uploads/2018/08/Wyoming-Jobs-Project-8.13.18-WEB.pdf.

[79] ARPA-e, ARPA-e Impact, accessed July 17, 2019, https://arpa-e.energy.gov/?q=site-page/arpa-e-impact

[80] C. Simpson, "Tesla Is First Green Tech Company to Pay Back Its Department of Energy Loan," *The Atlantic*, May 22, 2013, https://www.theatlantic.com/business/archive/2013/05/tesla-first-green-tech-company-pay-back-its-department-energy-loan/314981/

[81] "Offshore Wind Business Development Grant Program," Maryland Energy Administration, accessed April 24, 2018, http://energy.maryland.gov/Pages/Info/renewable/offshorewindbusinessdevelopment.aspx

[82] "§ 9-20C-03 Maryland Offshore Wind Business Development Fund," Justia, accessed April 24, 2018, https://law.justia.com/codes/maryland/2013/article-gsg/section-9-20c-03/

[83] "Offshore Wind Business Development Grant Program (FY 2018)," Maryland Energy Administration, p. 2, March 2017, accessed April 24, 2018, http://energy.maryland.gov/Documents/OSW%20Business%20Program%20FOA%20-%20FINAL.pdf

[84] "2018 Award Winners for Maryland's Offshore Wind Business & Workforce Development Program Announced," Maryland Energy Administration, January 10, 2018, accessed April 24, 2018, http://news.maryland.gov/mea/2018/01/10/2018-award-winners-for-marylands-offshore-wind-business-workforce-development-program-announced/

[85] "Strum Contracting Company, Inc.," accessed April 24, 2018, http://strumcontracting.com/
[86] "DEVRECO," accessed April 24, 2018, http://www.devreco.com/
[87] *US Department of Energy*, http://energy.sandia.gov/sandia-will-host-pv-bankability-workshop-at-solar-power-international-spi-2013/
[88] "CA Pub Res Code § 26054 (2016)," accessed December 17, 2018, https://law.justia.com/codes/california/2016/code-prc/division-16/chapter-4/article-1/section-26054/
[89] "Commercial PACE Financing," Dividend, accessed December 17, 2018, https://www.dividendfinance.com/commercial-pace-financing
[90] "Eligible Product List," CaliforniaFIRST, 2018, accessed December 17, 2018, https://renewfinancial.com/file/3866/download?token=HUXCYlQK
[91] "Solar PV & Energy Storage Contractors and Developers," K2 Clean Energy Capital, accessed December 17, 2018, http://k2cleancap.com/solar-pv-energy-storage-contractors-and-developers/
[92] N. Bloom, M. Schankerman, and J. Van Reenan. "Identifying Technology Spillovers and Product Market Rivalry," NBER Working Paper 13060, National Bureau of Economic Research, April 2007.
[93] M. Dutz, "Unleashing India's Innovation," *The World Bank, 2007,* p. 64, http://siteresources.worldbank.org/SOUTHASIAEXT/Resources/223546-1181699473021/3876782-1191373775504/indiainnovationfull.pdf
[94] American Jobs Project, Wisconsin Jobs Project, January 2018, http://americanjobsproject.us/wp/wp-content/uploads/2018/03/Wisconsin-Jobs-Project-1.17.18-WEB.pdf
[95] S. Shukla and S. Sawyer, "30 Years of Policies for Wind Energy," Global Wind Energy Council and International Renewable Energy Agency, 2012, p. 62, accessed March 24, 2016, https://www.irena.org/DocumentDownloads/Publications/IRENA_GWEC_WindReport_Full.pdf
[96] B.T. Madsen, "Public Initiatives and Industrial Development after 1979," in WIND POWER – the Danish Way: From Poul la Cour to Modern Wind Turbines, The Poul la Cour Foundation, Askov 2009, pp. 52–59, https://www.amazon.com/Wind-Power-Danish-Way-Turbines/dp/8799318806/ref=sr_1_1?dchild=1&keywords=9788799318803&linkCode=qs&qid=1592761893&s=books&sr=1-1&tag=x_gr_w_bb_sout-20
[97] "A Global Wind Power Hot Spot," Ministry of Foreign Affairs of Denmark, accessed June 28, 2016, http://www.investindk.com/Clusters/Cleantech/Wind-power
[98] "A World-Leader in Wind Energy," Ministry of Foreign Affairs of Denmark, 2015, accessed June 28, 2016, http://denmark.dk/en/green-living/wind-energy/
[99] Interview with Birger T. Madsen, Former Managing Director of the Danish Wind Industry Association, May 3, 2016.
[100] "World Wind Energy Market Update 2015," Navigant Research, 2015, p. 213, accessed June 28, 2016, http://www.provedor.nuca.ie.ufrj.br/estudos/navigant1.pdf

Chapter 14
Challenges ahead for a clean energy transition worldwide

Sergio Castellanos[1,2,3] *and Daniel M. Kammen*[2,4,5]

Developing a road map for future generations to transition to fully clean energy systems is not an easy task. There are no "one-size-fits-all" specific solutions, and the approaches that different world regions will undertake will be commensurate to their current infrastructures, population needs, energy demands, financial resources, cultural perception, and immediate risk to climate crisis effects. There are, however, broader commonalities that, if correctly appropriated, could enable multiple regions to chart a decarbonized transition. The goal of this chapter is to highlight transition elements required to take place in different regions of the world, as well as a selection of technical, economic, and societal hurdles in our quest toward a clean energy transition worldwide.

14.1 Introduction

The Intergovernmental Panel on Climate Change's (IPCC's) Special Report on Global Warming of 1.5 °C lays out the detrimental impacts on the planet that are expected to occur should we exceed our carbon budget and reach global mean surface temperature (GMST) above 1.5 °C. The consequences are yet more dire under a prospect of reaching higher temperature increases, as some models also describe in detail [1].

The ultimate bottom line is a clear call for climate action. This urgent call is also echoed by the United Nations through their Sustainable Development Goals (SDGs), specifically via Goals 7 and 13 [2]. The dimension of the challenge becomes clearer upon observing the rate of CO_2 emissions required to decline, starting in 2020 and through the next 20–35 years, if we are to achieve a net-zero global economy by mid-century, as seen in Figure 14.1. In fact, different versions

[1]California Institute for Energy and Environment, University of California, Berkeley, CA, USA
[2]Energy and Resources Group, University of California, Berkeley, CA, USA
[3]Civil, Architectural and Environmental Engineering, The University of Texas at Austin, Austin, TX, USA
[4]Goldman School of Public Policy, University of California, Berkeley, CA, USA
[5]Department of Nuclear Engineering, University of California, Berkeley, CA, USA

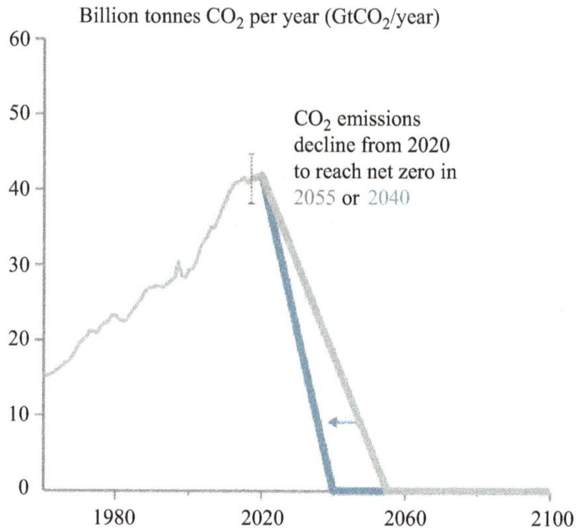

Figure 14.1 *CO_2 emissions decline required to reach net zero by mid-century. Image from [1]*

of this graph point to the drastic reality that the longer the action is delayed, the steeper the curve required to decarbonize. Delaying immediate action can set us in a path where we will have to go from a 3% to a 5% declining decarbonization rate, maintained continuously over decades [3]. This task is unprecedented, and despite some economies achieving that rate for short periods of time, no one place has sustained such decarbonization rate for full decades.

The most well-known accord aimed at deep emission cuts is The Paris Agreement [4]. More than 1,200 local governments spread across 25 countries have declared a climate emergency and are simultaneously developing road maps that can collectively put us onto a path toward achieving zero greenhouse gas (GHG) emissions [5].

The drastic behavioral, economical, and technological changes that must take place in order for such global clean-energy transition to occur are many, and herein we attempt to summarize some of the most pressing ones.

14.2 Decarbonization pathways

The monumental task of achieving net-zero emissions worldwide will require major sustained action in both adapting and developing new institutional bodies and regulatory frames, transforming societal behavior, and increasing technological efforts in the research, development, commercialization, and large-scale deployment of clean energy systems.

Planning tools, such as integrated assessment models (IAMs), provide a consistent, holistic modeling framework to analyze complex problems—such as climate change—by integrating economic, social, and institutional dimensions with atmospheric and environmental characteristics [6–8]. In the context of our discussions, IAMs elucidate paths that describe future trajectories, and aim to educate decision-makers on the options, benefits, costs, time scales, and trade-offs of creating different mitigation pathways without carbon-intensive resources, or decarbonized. Ideally, IAMs would perfectly weave in the entirety of the SDGs; however, despite ongoing efforts to do so [9], this modeling-integration effort does not come without its challenges [10].

Sustainable Development Goals

The Sustainable Development Goals (SDGs) and the 2030 Agenda for Sustainable Development are a set of 17 universal goals (Figure 14.2) and 169 targets built by thousands of participants and stakeholders to develop an agenda broader than only ending poverty and hunger, promoting protection on the environment, and foster an inclusive economic growth [2]. SDGs refer to the "development that meets the needs of the present while safeguarding Earth's life-support system, on which the welfare of current and future generations depend" [9].

The SDGs include the "need for a peaceful and inclusive societies, access to justice for all, and effective, accountable, and inclusive institutions at all levels" [11]. The SDGs are interconnected, and accomplishing any one of them requires addressing issues associated with others.

In light of the IPCC's Special Report on 1.5 °C, an increasing body of literature is focusing on modeling mitigation pathways consistent with the aspirational goal of the Paris Agreement.

Figure 14.2 Map of 17 SDGs from the United Nations

For example, the work by Grubler *et al.* [12] establishes an efficiency-focused global scenario called Low Energy Demand (LED), which describes rapid changes from institutions and society as to how energy is transformed and consumed but relaxes the dependency on stringent climate policies. In fact, consumer goods are modeled as increasing in devices per capita, but with lower average electricity intensity. Similar increases are envisioned for mobility (in terms of modes) with decreased energy intensity weighted by modal share, and in food demand, without modeled energy intensity. In the case of commercial buildings and public buildings, an increase in m^2 $capita^{-1}$ is envisioned with overall lower energy per m^2; industry demand in terms of gigatons of steel, aluminum, cement, paper, petrochemicals and feedstocks is expected to fall, just as the energy intensity for each activity and process, too. Lastly, the freight transport sector is assumed to increase in ton-kilometers, but with a reduced average intensity ($MJtkm^{-1}$).

This efficiency-oriented model renders the lowest energy demand scenario (with still similar climate targets) compared to other works in the field, and equally important, the results elucidate pathways with increased productivity in agriculture, accompanied by lower demand on biofuels and higher reforestation levels. The reliance on fossil fuels is evident by the increase in electrification trends via renewables.

The work by McCollum *et al.* [13] highlights the demanding actions required to meet a 1.5 °C GMST increase target. As a key insight derived from comparing six different IAM frameworks, it is observed that investment reallocations required within the next lustrum are at consistent levels with other studies but distributed differently. Of importance is to note that the total energy investments under a 1.5 °C target scenario are expected to be close to 3% of global GDP, or somewhere between US$2.3 trillion and US$4.7 trillion yearly (for a 2050 horizon), with significant changes in investment schedules if the GMST increase target is 0.5 °C higher and placed at 2 °C.

Yet another work, which contrasts other models, is that of Wilson *et al.* [14] which emphasizes "novel energy-using goods and services which may diffuse rapidly among consumers... [constituting] potential means of pushing the speed and magnitude of emission reductions beyond... 1.5 °C mitigation." In essence, IAMs can also be heavily weighted toward disruptive low-carbon innovation schemes and paint a more creative—yet still unknown—path of what the future might look like.

In the end, more than looking for a forecast of the future, the equally (if not more) important aspect of these road maps is their role of being an element for discussion on how to best address the issues and forge a path with concrete actions taking place with due time.

To complement these IAMs, other existing frameworks provide different optics to help us understand the challenges ahead, and the possible routes we, as society, can take.

For example, Davis *et al.* [15] detail how net-zero emissions energy systems require immediate attention to champion technologies and conduct efforts to research, develop, demonstrate, and deploy them as soon as possible. The detailed energy sectors and their difficult-to-eliminate emissions analyzed in [15] constitute a fundamental piece—and a needed granular layer—that bridges very specific technological needs, the opportunities, and their impacts, into the bigger energy systems discussions.

At a broader level, Geels *et al.* [16] introduce a sociotechnical framework to accelerate transitions for deep decarbonization, for which they detail three processes: (a) increase momentum of niche innovations; (b) weaken existing systems; and (c) strengthen exogenous pressure. The premise is that a window of opportunity can come about when all three levels are combined in a mutually reinforcing process. New technologies and business models grow via a strong supporting ecosystem but require widespread social acceptance. Furthermore, aggressive policies that avoid locking in emissions-intensive technologies and speed their phasing out can accelerate clean energy transitions. This framework, exemplified in [16] with case examples for the UK, France, and the US, allows one to better understand the interplay of social, technical, and political effects in systems decarbonization.

The decarbonization problem is complex along many dimensions: from the appropriate technologies, to the right formulation of policies, and making sure these are well suited for each of the regions where they are being implemented. There are arguments to be made for a strong focus on developing nations first. For one, they possess more than 1.4 billion inhabitants with no access to electricity, and by consequence, any intervention—even if marginal—would create significant impacts in terms of quality of life, health, and economic well-being [17]. Moreover, doing so would support reducing the energy access gap for the underserved.

On the other hand, some of these nations have already demonstrated great success and leadership in decarbonizing their systems [18,19]. Drivers that might be adequate for some locations might not have the socioeconomic lever in other places. Whether due to geopolitical, environmental, or cultural differences, these examples highlight the complexity of addressing a decarbonization transition in a linear fashion, and without country-specific context.

In adapting and creating pathways, the frequent assumption of an ever-increasing energy demand might continue to be challenged [12,20]. For one, we have observed that significant energy intensities can be developed for many sectors and end uses. Furthermore, demand growth by no means has to be the norm, especially under a resource-constrained planet. There is a sense that many countries might be forecasting exceedingly aggressive growth projections, which contrast with more moderate growth scenarios once analyzed in retrospect. Nevertheless, given the multiple factors impacting energy demand, this variable will continue to be significantly be affected by both our use of it, and the climate change effects.

In the current context of our social dynamics, a place for regulatory certainty, order, and structure is required to fund and provide the tools required to transition to low-carbon energy systems. Part of the challenge will be to reconcile that order and structure, with the unfolding chaos that rising GMST, ocean acidification, and other climate change-induced natural disasters will create on our social, institutional, and economic systems.

The energy choices both industrialized and nonindustrialized countries make in the years to come will determine whether a true path toward a 1.5–2 °C world is achievable. But this shouldn't necessarily be a task that compromises the well-being of citizens. In fact, implementing actions to reach the Paris Agreement can unlock more than US$13 trillion before 2050 [21], leading to dignified jobs and other benefits in security, health, equity, and development, as we simultaneously reinvent our approach to sustainably transform and consume energy.

14.3 The role of financing in transitions for decarbonization

As IAM's results converge toward higher deployment of clean fuels, clean energy technologies, and sustainable afforestation and reforestation approaches, all of this will require a strong financing infrastructure that can support their development and aggressive deployment through large volumes of capital influx. A barrier for clean energy financing in some cases is its lower rate of return compared to other fossil-fuel projects [22], or even other non-energy industries, such as software, and in countries like the US, the healthcare industry [23].

A few key priorities on the financial approaches that must take place in the context of a decarbonized energy system, inspired by Mazzucato and Semieniuk [24] and summarized by Brown and Granoff [25], are the following:

- Shifting from lowest-cost marginal abatement opportunities to those that maximize near-term emissions reductions and open opportunities for long-term decarbonization pathways.
- Reassess how we measure progress in climate finance, beyond the typical volumetric financial flows, emphasizing the type, source, and target of finance needed for all sectors involved in the decarbonization strategy.
- Focus on the effectiveness, or the transformative nature, of finance reform, as opposed to the quantity of it, and its most effective use.

The range of possibilities for investment in a decarbonized future are quite broad and will necessitate some degree of consensus to better target and amplify each dollar invested. Following again the work from Brown and Granoff [25], the topics they identify as important to engage and work on are the following:

1. *Prioritization of investments along the technology development spectrum, and how should different financial actors be targeted.*

   This line of priority addresses the different risk appetite of investors, and the type of incentives that should be created toward increasing investments along different segments of the technology development curve.
2. *Tools to target and prioritize investment for rapid diffusion of low-carbon innovations, across technology development spectrum and geographies.*

   A continued approach to develop metrics and cost analyses to better understand where the areas for opportunities are, and funnel resources accordingly for appropriate interventions.
3. *Accounting for investment in low-carbon technologies in developed and developing countries.*

   There is a complex web of interactions documented when countries from different income groups invest in both research and development (R&D) and tech deployment. The argument therefore calls for replicating R&D and deployment efforts regardless of the income group, but with better tracking to understand the best approaches in supporting initiatives in developed and developing countries.

4. *Need to focus on high-cost technologies, beyond marginal abatement cost curves.*

   This topic aims to highlight the fact that decarbonization takes time, and if the end goal is known (i.e., decarbonization), then investments should be made early on to minimize adjustment costs.

5. *Transactional or financial structuring innovations to support investments in new technologies.*

   The uncertain landscape of what technologies might look like in the future, their revenue streams and potential risks will require financial innovations to support them as they develop.

Just as decarbonization discussions center around urgent radical changes along infrastructure and social dimensions in the next decades, an enabling pillar will undoubtedly be the capacity to finance and sustain the investment levels that will be required to develop, pilot, and deploy clean technologies across different geographies and political landscapes.

Financing programs: Examples from California and Connecticut

Increasing financing options to the broader community entails approaches that could be linked to commonly owned assets (e.g., housing). An example of a financing option initiative that originated in California and has been successful in the United States is the property assessed clean energy (PACE). PACE financing enables upfront capital for residential property owners to install energy-efficient systems, including electric and thermal solar energy systems, as well as energy efficiency retrofits. Building owners utilize a voluntary property tax assessment to finance energy-saving projects (akin to finance mechanisms for road paving and sewer projects), allowing them to retain the assessment with the property upon sale and pass the payments through to tenants. Programs like PACE validate energy efficiency technologies and contribute to their widespread dissemination and measurement.

The Connecticut Green Bank (CGB), a quasi-public corporation, was established in 2011 to develop and implement financing strategies to increase clean energy accessibility and affordability to residential, commercial, and industrial consumers [26,27]. CGB offers incentives, co-investments, credit support, and other market-based tools.

In terms of investment, CGB's capital is typically deployed through (a) their residential solar incentive programs (upfront residential solar system payment in exchange for a stream of renewable energy credits), (b) loan loss reserve credit enhancements (risk reduction mechanism to support the direct lending of a private entity), and (c) direct loans and leases [28,29].

In an interesting coevolution of policies and mechanisms, CGB has developed a commercial-based PACE program, called C-PACE, which "allows building managers to pay for green energy improvements over time through a voluntary benefit assessment on their property tax bill and programs are structured so that monthly energy savings are more than offset the monthly loan repayment." [30]

14.4 The electricity grid

Within the energy space, much has been written in this book about the electric and transportation sectors. In fact, this is not surprising when one looks at the world inventory for GHG emissions, shown in Figure 14.3. This figure [31] details the sectors, the end uses of such sectors, and the emitted GHG associated with them, all summing up to 100% at each category. From this GHG emissions visualization, it is easy to notice that "electricity and heat" account for 30.4% of the emissions, followed by transportation with 15.9%. These two sectors represent the two main drivers in GHG emissions worldwide.

So much has advanced to date in terms of technological approaches that one could be optimistic about the potential tools available to achieve a decarbonized economy within the next decades [32]. That sense of optimism should not be confused with complacency, though, as there is much to be done yet to fully decarbonize the electric grid.

A decarbonized electric grid can be understood for practical purposes as a grid that is beyond 80% clean in terms of CO_2 emissions associated. A proposed way of achieving deep decarbonization is through electric systems running 100% on renewable energy, as described in Chapter 2 of this book, and by many other authors [33–35].

Some authors argue a decarbonization pathway via 100% renewables might be more costly compared to a balanced system containing a mix of renewables with other low-carbon resources [36]. The identified challenges in 100% renewable energy scenarios, as summarized by Jenkins *et al.* [37], consist of the following:

- Significant deployment of high-voltage direct-current transmission lines, connecting intermittent resources spread geographically across large distances to smooth resources variations.
- Uncertainty regarding the ability and willingness of entities to provide demand flexibility that goes beyond hourly and daily intermittency variations, and into longer time spans (e.g., weeks, or even seasons).
- Potential overcapacity deployment to compensate for periods of time with below-average renewable energy resource availability. The effects can be multiple in terms of curtailment, or if stored, in high total system costs.
- Reliance on long-durational seasonal storage to bridge periods of low wind and solar resource availability. Again, requiring costly technologies that are yet to be scalable.

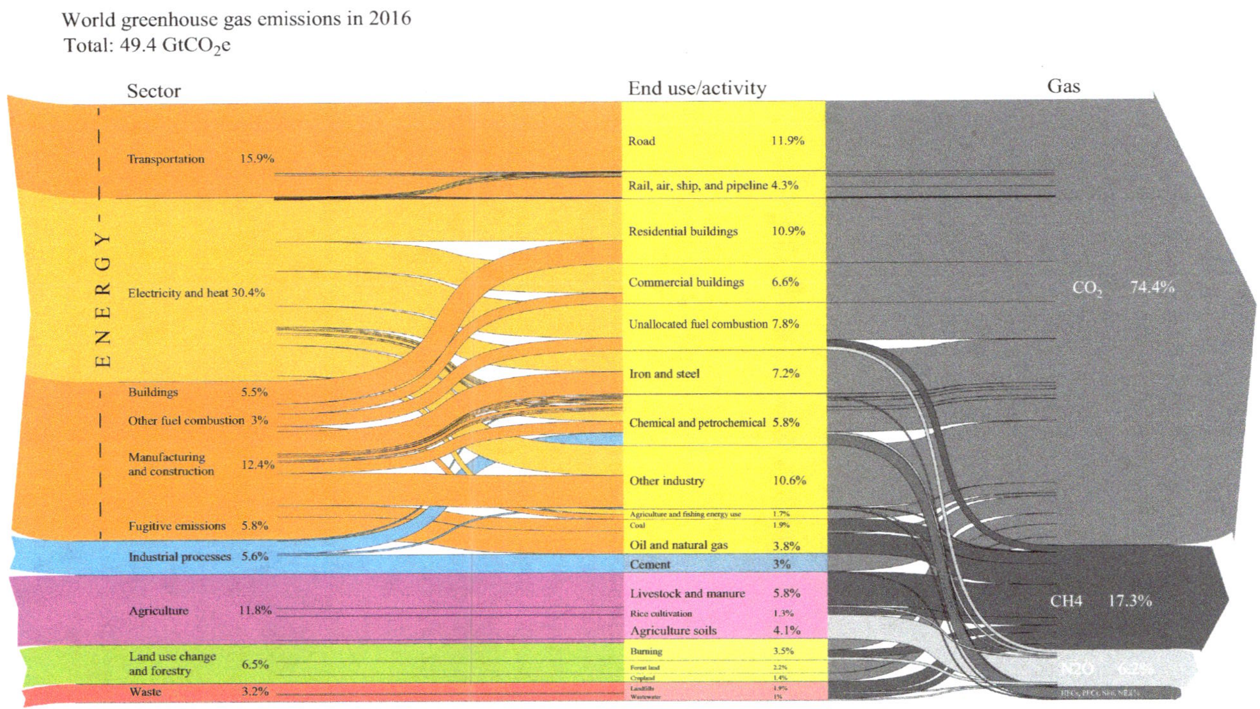

Figure 14.3 *Visual summary of the world's association between sectors, end use and activities, and emitted GHG type. Image Source: Climate Watch, based on raw data from IEA 2018, CO_2 Emissions from Fuel Combustion, www.iea.org/statistics; modified by World Resource Institute [31]*

If financing mechanisms strengthen and increase accordingly, R&D efforts are amplified, and bold policies are enacted, there are no engraved natural limitations that prohibit a 100% renewable energy grid. The challenge, however, is that we ought to be making large and aggressive investments now to achieve the Paris Agreement goals. It seems that there is room for reconciling different perspectives: increase R&D to develop novel, affordable technologies, paired with market-oriented policies that can allow for current technologies to continue their cost decrease, while implementing reinforcement loops (e.g., mandates and bans) that catalyze the transition. This, in fact, is an interpretation of the framework previously described and proposed by Geels *et al.* [16]. How the financing challenges presented in Section 14.3 are addressed will be crucial to ensure affordable and equitable pathways moving forward.

As means of complementing and touching on some topics briefly mentioned in previous chapters of this book, we highlight a few we believe are worth continuing having discussions about.

14.4.1 Nuclear energy

Ongoing arguments on technology costs, commissioning times, concerns about nuclear waste and the risk of nuclear accidents, and the consequences on public perception continue to fuel discussions among the scientific community, producing arguments for and against considering nuclear as a viable technology for decarbonizing the power sector.

Nuclear energy projects have been reported in many cases to have increasing costs over time. This cost increase has been attributed to the complexity and experience in operating nuclear facilities, land-related and fuel-accessibility constraints, among other drivers; and although some authors have found this effect to be mildly present and not generalized [38], many others argue that in fact the "forgetting curves" (an upward trend in costs over time) are symptomatic of the industry, and there are clear impediments limiting the cost-declining trajectory required to deploy at faster rates [39–41].

Some proponents have put forward that light water small modular reactors are among the few options that could be scaled up in faster way compared to conventional reactors [42,43]. At the same time, there is a degree of acceptance that even this type of technology (and others) might not be able to be deployed in a rapid manner, especially under the structure of the current R&D paradigm in some countries [43]. This all goes to show that the expectations for nuclear to grow aggressively enough and play a pivotal role in a decarbonized future are rather limited.

In fact, there are ongoing cases where nuclear is being phased off, mainly due to security concerns. Germany is one example. In this country, nuclear is scheduled to be phased out by 2022, and in the process of decommissioning nuclear plants, studies have suggested a replacement in electricity generation coming from coal-based generators which have increased mortality risks due to degraded air quality from burnt coal [44,45]. The coal plants are also now being scheduled to be retired, significantly reducing these externalities upon their closure, but it does serve as a cautionary planning tale for regions attempting similar transitions.

The radioactive danger, however, cannot be ignored and should be heavily weighted in these discussions [46]. Furthermore, when it comes to waste disposal, nuclear facilities, "whether for disposal or interim storage, take decades to plan, license and build" [47], which puts them at further disadvantaged when contextualized in the climate emergency we are facing.

The challenges for nuclear are obvious, yet the evolution of this debate should be based on facts and figures and be open to be molded as technologies evolve and costs decline, risks are significantly mitigated (or not), while avoiding a reduction in R&D resources for this technology. In other words, we are in a crisis and excluding potential low-carbon technologies is a luxury we don't have, especially if viable paths exist to deploy those technologies in a timely and cost-effective way. Only progress along the social, technical, and economic lines will be able to answer that question for us in the years to come.

14.4.2 Energy efficiency

As important it is to focus on the supply side by planning a clean turnover of the existing grid infrastructure and replacing aging and polluting electricity generators, an element on the demand side often missed in the discussion of electricity is energy efficiency. Its relevance is best described by Lovins *et al.* [32], when referring to saved energy as "the world's largest decarbonizer and energy 'source' (bigger than oil)."

The International Energy Agency, as part of its Sustainable Development Scenario [48], places efficiency measures as one of the most significant drivers to achieve drastic CO_2 emissions reductions, along with renewable energy technologies. The expected investments in end uses also scale accordingly, with efficiency measures requiring the larger investment share by 2050, and placed into smart buildings, manufacturing processes, transportation, etc. Other models show that between energy efficiency and renewable energy, these can achieve almost the entirety of a reduction down to 9.7 GT/year in 2050.

Saving energy (or implementing energy efficiency measures) can be a concept difficult to relate to as you cannot see saved energy like, say, a solar module or a wind turbine that can generate electricity. Energy efficiency's potential, and returns are, however, comparable to renewables and can be implemented via cost-effective approaches that are currently available. Furthermore, these measures can be applied in buildings, and beyond into industry and vehicles.

Among two of the most important factors in energy efficiency's role in a decarbonized future, as described in [20], are the following:

- Depending on the share of renewable energy, degree of electrification in the economy, and time of the day, there will be significant variations in the carbon footprint of electricity use.
- Implementing energy efficiency measures focused on appliances and equipment optimized to occur at times when the power sector is the most polluting in terms of CO_2 emissions. These measures fit well with the features and opportunities described in Chapter 7 ("Demand response technologies in buildings for curbing and shifting electric loads").

Energy efficiency will also gain further relevance upon the unfolding of climate change. In a world with increasing GMST, a similar increase in load demands is expected over the coming decades to achieve thermal comfort, especially in subtropical developing nations [49,50]. Efficiency measures, such as solar passive design, low absorptance, and thermal insulation are some strategies aimed at counteracting the increase in cooling load.

Inversely, for studies with a focus on the heat demand, the projections point toward a general decrease in demand over the long run given ever-increasing temperatures [51]. This trend, however positive, might seem in terms of lower energy consumption, increases when contextualized in a fully electrified scenario. As shown in a study for a decarbonized US scenario, a significant build-out of electricity capacity would be required to accommodate peak electricity due to heating demand [52]. This fully electrified scenario is not that far-fetched. Cities in California have recently led the charge by banning natural gas in new building constructions. While there is support by city residents in general, this regulation has not been very welcomed by restaurant owners who believe it is too restrictive and expensive.

In short, a strong energy efficiency implementation can not only achieve drastic reductions, but also buy enough time for other clean technologies to be developed, tested, and deployed [53].

14.5 Transportation

The transportation sector is quite broad, encompassing road, rail, aviation, and shipping, with significant challenges in each of these modes.

The potential to reduce energy use in transportation depends on the end use and form of journey. Some mode shifts in shorter-duration trips can be implemented faster (e.g., ridesharing and bicycle use) than shifts in longer-duration ones (e.g., flying).

Aviation and long-distance transportation, such as shipping, are known for their high reliance on hydrocarbons, and have a challenge of developing fuel substitutes with high volumetric and gravimetric density [15]. This substitution process with low-carbon alternatives cannot be directly solved through electrification alone and requires large-scale development to match the current and forecasted transportation needs in the years to come.

As described by Davis *et al.* [15], the fuel substitutes for these transportation sub-sectors are hydrogen, ammonia, and alcohols and ethers, among others. Hydrogen can be more efficient than internal combustion, although it has a low volumetric energy density compared to other liquid fuels, requiring higher storage capacity (e.g., liquid hydrogen would require ~8× larger volume compared to diesel). Ammonia, despite its carbon-free nature, has a low gravimetric energy density (MJ/kg) and would weight ~2× to hold similar energy per unit weight, compared to diesel or jet fuel, and with the associated externality of releasing NO_x when processed through thermolysis. Biofuels (e.g., biomass conversion, ethanol, and renewable diesel) have been developed at smaller scales, but have reported

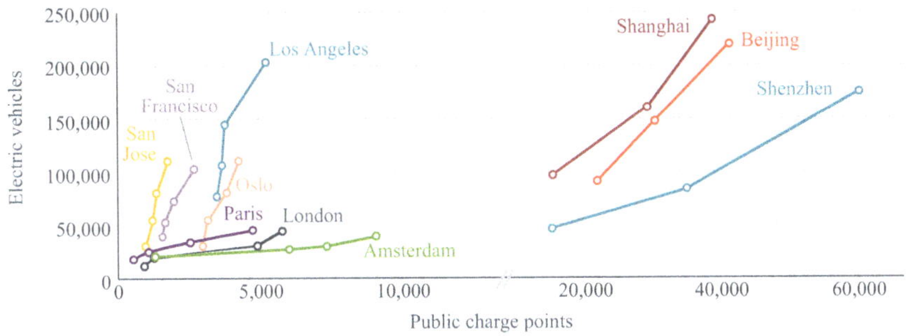

Figure 14.4 Electric vehicles vs. Public charge points for cities in the US, Europe, and China. Image reproduced from [54]

issues with costs, scalability, food security, land availability, impact on ecosystems, and life-cycle carbon emissions, so the challenges ahead are multiple.

In terms of CO_2 reduction levels, the emissions for shipping and aviation in 2018 were 878, and 982 Mt CO_2, respectively, with IEA Sustainable Development scenarios expecting a reduction down to 435 Mt CO_2 for shipping, and 625 Mt CO_2 for aviation by 2050 [20].

In road transportation, specifically in urban settings, an accelerated transition to electrify the transportation sector is currently taking place mostly in developed countries. An increased push for the electrification of light duty vehicles (LDVs) and buses has been motivated both by GHG, and by air quality concerns of the existing global internal combustion-based fleet. The environmentally benign impact of electrifying this portion of the transportation sector must be paired with the deployment of clean electricity generation sources so that the emissions are avoided all the way from the vehicle's exhaust tailpipe, to the generation plants that produce electricity to charge those vehicles.

Cities speeding ahead

In the race to electrify the transportation sector, marked differences can be observed across cities in Europe, the United states, and China, as shown in Figure 14.4, from [54]. While urban form, population, cultural norms, vehicle mix, commuting patterns, and driving patterns all vary by region, the infrastructure rollout between cities has been quite different even for similar amount of EVs on the road. Some of the most important cities in China had between 40,000 and 60,000 charging stations by 2018, compared to a range between 1,700 and 5,200 in the US for that same period. For cities, the volumes and type of charging stations, coupled with the city population characteristics and behaviors, will dictate the appropriate infrastructure rollout strategy and policies needed to support and sustain EVs growth.

China and the countries across Europe are leading the electric vehicle (EV) market growth through the design and implementation of policies at national, regional, and local levels, coupled with the rollout of supporting urban and grid-related infrastructure [55]. These policies contain affordability, convenience, and awareness at the core [55]. These elements take the form of actionable drivers for EV's uptake, such as financial incentives (e.g., tax credits, rebates, and fee exemptions); non-financial incentives (e.g., free parking, access to high-occupancy vehicle lanes and low-emission zones, tunnels and ferry fees exemptions); charging infrastructure (e.g., public charging stations); campaigns and research (e.g., designation of EV regions, study of consumer behaviors, smart charging studies, promote awareness and education about EV developments); transit and fleets (e.g., electrifying notorious fleets for increased visibility such as taxis, buses, car-sharing programs), as reported by [56].

The convergence of the electricity sector with the transportation sector seems to be inevitable and will bring undoubtedly more benefits to both sectors: higher integration with electricity markets, and wider adoption and cost-reduction of EVs (and batteries). The clearest example is the vehicle-to-grid (V2G) interaction, in which EVs can exchange electricity with the grid in two directions, by optimizing distinct objective functions, such as charge capacity vs. revenue from selling electricity and/or providing ancillary services. While this integration is in its early stages, the uptake is expected to occur rapidly.

The potential in electrifying, however, is just one aspect. LDVs could also use hydrogen as fuel source, as previously described for the aviation and shipping sectors. Furthermore, measures such as regenerative braking, logistical improvements, lighter-weight vehicles, switching to other modes (e.g., rail), and better vehicle aerodynamics and tires are some of the reductions possible with today's available technologies that do not strictly necessitate an electrified fleet [57].

14.6 Other energy-intensive sectors and end uses

Some concerns have been voiced that industry has resisted drastic changes and has, to a large extent, avoided policy interventions to minimize a decrease in their competitiveness [58]. The job prospects of a clean global energy transition are obvious benefits of allowing industries to grow, as presented in Chapter 13, representing a means to achieve sustainable economic development and increased social equity. At the same time, decarbonizing the industry sector should not be mutually exclusive with its capacity to be competitive and provide the positive benefits described in the previous chapter.

Heavy industry emits its largest portion of emissions from the processes themselves rather than from combustion [58]. In terms of industrial processes in general, around one-third of the total energy is devoted to high-temperature processes. Given the composition mix of the construction of infrastructure, a growth in demand for these products in the following decades is expected.

Narrowing done the energy-intensive sectors, we find iron and steel, cement, aluminum, chemical and petrochemical, pulp and paper, lead the list [48]. In fact, just the production of steel, cement, and aluminum accounts for 10% of global energy consumption [20], drawing attention for these industries to be prioritized. A brief summary of the challenges for steel and cement is presented next.

14.6.1 Industry: steel and cement

To produce steel there is a direct step in the process (reduction of iron oxide) that emits CO_2 directly as a side product, which needs to be accounted on top of the CO_2 associated to burning fossil fuels for high-temperature process [15]. Wesseling *et al.* [59] provide an overview of low-carbon innovations in the iron and steel that need to take place to meet The Paris Accord by 2050, as shown in Table 14.1.

The cement industry is responsible for approximately 6% of the world's anthropogenic CO_2 emissions [60]. More than two-thirds of the cement manufacturing emissions come from a chemical process that calcinates limestone, and hence switching to alternative fuels would not do much in mitigating those emissions [20]. In terms of the energy required, more than 70% can come from fossil fuels like coal and petroleum coke, as is the case for California [60].

Innovations in this category could be the replacement of feedstock with geopolymers which hold the promise to significantly reduce CO_2 emissions. This approach, however, still has unproven long-term performance and has a long road ahead to increase its (market) readiness level [59]. Similarly, there are other ways to reduce CO_2 emissions and energy usage in cement manufacturing, such as improving ovens, adding preheaters, improving the calcination process, and improving milling and grinding equipment [61]. These measures significantly reduce fuel usage, CO_2 by-products, and electricity usage.

For both cases, a reduction in GHG emissions can be achieved starting at the planning stage by designing for lower material usage in mind, where less steel and cement are employed and substituted with improved alternatives. Furthermore, the opportunities for industry are not limited to the ones shown in Table 14.1, and could also be in terms of them taking a more active role and in an integrated-way with the electric system, providing peak power and load management [58].

While these approaches can have a positive impact, most likely pairing with other technologies will be required to achieve the ambitious CO_2 target goals by 2050. An example is the use of carbon capture and storage (CCS) technologies.

CCS has been identified as a crucial component in the pathways scenarios highlighted in the IPCC's Special Report 1.5 °C [1]. CCS along with GHG removal (e.g., bioenergy with CCS, direct air capture, and afforestation) is needed to reach the (now) ambitious goal of 1.5 °C increase [62]. The IEA states that CCS "is one of the most cost-effective solutions available to reduce emissions for some industrial and fuel transformation processes, especially those that inherently produce a relatively pure stream of CO_2, such as natural gas and coal-to-liquids processing, hydrogen production from fossil fuels, and ammonia production" [20].

The IPCC report outlines a set of CCS technologies with their CO_2 removal and deployment potential (GT CO_2 $year^{-1}$) and their costs (US\$ tCO_2^{-1}), which range between 0–10 and 5–250, respectively [63]. Their main attractiveness will

Table 14.1 Low-carbon innovations necessary to meet the 2050 GHG emission target for the iron and steel industry. Adapted from Wesseling et al. *[59]*

Technology	Level of innovation	Description	Benefits	Bottlenecks to diffusion of innovation
Recirculating blast furnace & CCS	+	Under R&D. Requires high integration into existing plants	Less CO_2 (++)	Higher energy demand, costs, infrastructure, acceptance
Smelt reduction & CCS	++	Makes obsolete coke ovens, BF[1], and BOF[2] of conventional steel factories	Less CO_2 (++/+++)	Costs, infrastructure, acceptance
Direct reduction with H_2	++	Makes obsolete coke ovens, BF, and BOF of conventional steel factories, but is combined with electric arc furnace; needs H_2 supply infrastructure	Less CO_2 (+++, potentially excess electricity converted to H_2)	Costs, infrastructure & technology
Electrowinning	+++	Makes obsolete coke ovens, BF, and BOF of conventional steel factories, needs large electricity supply; technology only on lab scale available	Less CO_2 (+++ with renewable energy sources electricity) smaller, probably lower capital expenditure	Only available in lab; low coal/CO_2-prices and high electricity prices

[1]BF = blast furnace
[2]BOF = basic oxygen furnace

come from coupling these technologies in industrial processes at large scales, but this has yet to happen.

The support for CCS has not lived up to the expectations required to combat climate change. Among the many challenges identified for the scalability of these technologies, thus far, are the following:

- Concerns of elongating the viability and profitability of fossil-fuel-based projects [64].
- Safety and environmental concerns, especially in populated onshore regions [64].
- Inadequate assumptions on societal support and governance structures for modeled pathways [64].
- Questions as to what paths could help transition from small commercial projects to industrial levels that could have a meaningful role—as expected—in most decarbonization pathways [65].
- Lack of support for technologies due to lack of general ambition and pressure, and CCS technologies' disconnect with industrial policy rationales [65].
- The "commercial-scale" nature of CCS projects, which contrast with other more accessible technologies that can be found in the kW ranges, hence contrasting a much higher initial capital expenditure [65].
- Continuous over-optimistic cost estimates, a focus on energy projects as opposed to industry-related projects, and underestimating issues in the transport and storage segments, among others [65,66].

In terms of CCS's role in the power sector, some studies contemplate them in their long-term planning procedure [67,68], while other ones completely exclude them given their low probability of being available at the scales required in the next decades. The assumption in the former is that base load technologies will not be available, only intermittent renewable technologies [67]. For those that exclude CCS as having a role in the power sector, they still contemplate these technologies but coupled directly to energy-intensive sectors and industries (e.g., cement and steel) [69].

Achieving existing goals of the Paris agreement requires more than deep emission cuts from the power, transportation, and industry sectors, as has been described and also shown in Figure 14.2. Additional efforts and resources need to be funneled toward the agriculture sector and managing natural and working lands that constitute net sinks of carbon to increase carbon storage capacity from the existing carbon atmospheric emissions.

Recent reports estimate that global tree cover loss reached record highs in recent years, due to various factors, including forestry, shifting agriculture, wildfire events, and the largest contributor being commodity-driven deforestation, where forest is converted to non-forest land use such as agriculture, mining, or energy infrastructure [70,71]. Despite these effects might seem to be disconnected from the energy sector at first glance, much of the recent funding efforts in several countries have been directed toward research at the intersection of food, energy, and water [72,73].

14.6.2 Data and digitalization

The future of energy systems is said to have 3D's: *Decarbonized*, *Decentralized*, and *Digitalized*. The first two have been covered in this and previous chapters of this book. Progress to date on the digitalization characteristic leaves no doubt that the future of the energy systems will be digital indeed. The ways in which this is happening—and will continue in the future—are multiple; from blockchain technologies, to pay-as-you-go business models built on the high worldwide penetration of cell phones, there are many opportunities and changes that will transform our approach to energy services.

In the case of distributed ledgers, also known as blockchain, these can automatically execute smart contracts in peer-to-peer networks, while chronologically linking all transactions in a decentralized way via cryptography [74]. The applicability in the transportation and energy sectors ranges: either for EV charging schemes in smart communities [75,76], to smart grids where "consumers and prosumers can trade energy within their community, offer (near) real-time pricing, and facilitate a local balance of supply and demand" [77]. In an increased scale, conventional utilities are facing "demands for increasing transparency by the regulatory authorities. As a result, any possibility of cost savings and efficiency improvement in the operation of energy systems and markets... is worth investigating" [74]. The increase in transparency, competition, and decentralized accounting systems is therefore very promising for the future of energy systems, and one that will likely be supported by blockchain technologies.

On a more traditional way, business models have also leveraged on the use of digitalized payments and evolved into the pay-as-you-go scheme. Examples abound in developing nations and specifically in energy for solar photovoltaics use. These models use mobile money and are "a combination of hardware and software systems that typically rely on mobile phones as a platform for making payments (or verifying the transfer of money), and most include a remotely activated cut-off switch in the system hardware that prevents use when fees or loan payments have not been completed" [17].

In recent times, a considerable amount of buzz with "big data" has been generated. This growth in the generation, manipulation, use, and interpretation of large volumes of data is not particular to a specific sector, although it does have its particular role to play in the sectors discussed in this chapter.

One can logically assume that improving many of the foreseen challenges, say in the energy systems, will require substantial amounts of data with novel algorithms to discover new materials, products, and processes, while optimizing current ones.

Artificial intelligence holds the promise to shed light on many of these unknown—and improved—ways of conducting our business as usual. Already studies are pushing the frontiers of what's been done in the past: from materials discovery for energy purposes [78,79], to evaluating millions of satellite images to act on climate change [80], among many other examples [81].

But as this growing trend of data-intensive studies continues, a common concern with the increasing computationally intensive algorithms and use cases is that training computational models tend to require relatively large computing power, and therefore significantly drive energy demand.

This concern might not be that well-founded, as a 2020 study calculated that despite a rise in service demands, data centers are responsible for only 1% of the electricity consumption worldwide [82]. The number is the more impressive once stated in terms of energy per compute instance given that "the energy intensity of global data centers has decreased by 20% annually since 2010" while "global data center compute instances have increased by 550% over the same time period." The reasons for this energy intensity decrease, as detailed by Masanet *et al.* [82], are that both the electricity per computation of a volume server and the watts per storage unit installed have decreased 4× and 9×, respectively. Simultaneously, an increase in the number of compute instances hosted per server and a lower data center power usage effectiveness have contributed to this efficiency gain.

Challenges ahead remain in the form of potential energy demand increase if the currently available efficient resources run out, as well as difficulty in predicting the energy use of novel technologies, such as quantum computing. More research in this topic is required, but without a doubt, the power of data can help us move ahead in acting on climate change, as long as the costs (e.g., energy use) do not outweigh the benefits (e.g., energy savings).

14.7 Articulating a human-centered approach

One key aspect that should remain as part of these deeper technological and regulatory conversations, however, is that of equity. Addressing climate change with a technological mindset only, while ignoring the social justice side of the story, will only lead to social unrest as the climate gets more uninhabitable, potentially widening existing income, racial, and ethnic disparities along the process. These social, technical, and economical issues beg for approaches inspire to create grassroot movements, leading the charge in bottom-up efforts. As summarized in [83], the bottom line on having a social justice approach to sustainable transitions is to protect vulnerable people from the expected climate change impacts, from disruptions of transformation, and to create an equitable society in a decarbonized world.

An embodiment of this human-centered and equitable approach is the push for energy and environmental justice. As per California's state law definition, environmental justice is the principle of "fair treatment of people of all races, cultures, and incomes with respect to the development, adoption, implementation, and enforcement of environmental laws, regulations, and policies" [84]. Yet another definition from Robert D. Bullard, known as the "father of environmental justice" in the United States defines environmental justice as the principle that "all people and communities are entitled to equal protection of environmental and public health laws and regulations" [85].

A great body of literature is found on the topic of energy and environmental justice. From communities of color and low income living in close proximity to toxic sites and landfill siting [86], to air pollution exposure disparity [87,88], to increase in urban noise levels [89,90], to disparities in the adoption of solar rooftop photovoltaics

[91], among many other examples. However, these approaches need to be properly planned, too; in fact, some studies have shown that energy injustices can also be associated with low-carbon transitions and can be related to injustices in distribution, equity, and fairness [92]. Such was the case reported by Sovacool *et al.* [92] for four cases and technologies: two in generation (nuclear and solar power) and two in demand/end-use case studies (EVs, smart meters).

These cases tend to insinuate that corrective action should be taken chiefly by governments, or industry. But the reality is that much of the changes have been spurred by civil society taking matter into their hands.

Nowadays, an increasing number of social movements are gaining momentum, fueled by a sense of despair from younger generations in an unfolding crisis where the speed of change required has not been met with the required urgency. These movements have taken the form of protests; one of them being initially spearheaded by Swedish young climate activist, Greta Thunberg, with her strike movement "Fridays for Future," which started as a protest outside the Swedish parliament against inaction on the climate crisis. This protest very quickly made waves around the globe, becoming now an international recurrent event that since its inception in August 2018 and by February 2020 has already drawn more than 13 million people—mainly younger generations—across 228 countries, to protest and march in a call for climate action [93,94]. In the academic realm, an intensified movement of sit-outs across universities [95,96] (at times successful, [97–100]) has created a wave of pressure for universities to divest their endowments from fossil-fuel-related investment vehicles and assets. Similarly, employees in the private sector have expressed their concerns in the role their companies are playing in regards to climate change, sparking activism movements [101].

More recently, financial institutions have started to follow suit with announcements to halt their loans on infrastructure projects in the fossil-fuel space (e.g., arctic drilling [102] and coal companies [103]). Despite these announcements, however, these commitments to stop arctic drilling have been dwarfed by their appetite for profit-maximization and recurrent investment on other emissions-intensive projects despite their acknowledging of the Earth's unsustainable trajectory.

It is our hope that the momentum gained thus far will continue to grow, leading to deep and meaningful change to address climate change with the boldness and urgency required, but human-centered with an equitable mindset that allows for an inclusive involvement of all people and communities.

14.8 Closing remarks

The complexity of transitioning to a fully decarbonized economy requires multiple coordinated efforts from industry, governments, academia, and society to ensure a rapid and just transition, commensurate with our current climate crisis.

As we make our way toward cleaner economies, we will also have to learn and teach the younger generations the set of skills, values, knowledge, and behaviors that will allow us, as society, to collectively work toward creating a more sustainable world. From a return-on-investment point of view, educating about climate change has been demonstrated to reduce carbon emissions, in some cases in a similar capacity as other strategies such as the adoption of EVs and rooftop solar photovoltaic [104].

The challenges ahead for a clean energy transition worldwide are many and quite complex. The time for arguing is long gone now, and what seems to be the greater challenge is overcoming our still tepid global response to the greatest threat we are facing as a species.

References

[1] Masson-Delmotte TWV, Zhai P, Pörtner HO, *et al.* IPCC, 2018: Summary for policymakers. In: Global Warming of 1.5 °C. An IPCC Special Report on the impacts of global warming of 1.5 °C above pre-industrial levels and related global greenhouse gas emission pathways, in the context of strengthening the global. Geneva, Switzerland. 2018.

[2] Desa UN. Transforming our world: The 2030 agenda for sustainable development. 2016. Available from: https://sustainabledevelopment.un.org/post2015/transformingourworld/

[3] Figueres C, Schellnhuber HJ, Whiteman G, Rockström J, Hobley A, and Rahmstorf S. Three years to safeguard our climate. Nature. 2017;546(7660):593–5.

[4] Climate Change [Internet]. 2016 [cited 2020 Mar 15]. Available from: https://www.un.org/en/sections/issues-depth/climate-change/

[5] Urban Governance for a Sustainable Future. One Earth. 2020;2(2):117–9.

[6] Cantore N and Padilla E. Equality and CO_2 emissions distribution in climate change integrated assessment modelling. Energy. 2010;35(1):298–313.

[7] Van Asselt MB and Rotmans J. Uncertainty in integrated assessment modelling. Climatic Change. 2002;54(1–2):75–105.

[8] Easterling WE. Why regional studies are needed in the development of full-scale integrated assessment modelling of global change processes. Global Environmental Change. 1997;7(4):337–356.

[9] Integrated Sustainable Development Goals (iSDG) Model—United Nations Partnerships for SDGs platform [Internet]. [cited 2020 Mar 15]. Available from: https://sustainabledevelopment.un.org/partnership/?p=11520

[10] van Soest HL, van Vuuren DP, Hilaire J, *et al.* Analysing interactions among Sustainable Development Goals with integrated assessment models. Global Transitions. 2019;1:210–25.

[11] Kamau M, Chasek P, and O'Connor D. Transforming multilateral diplomacy: The inside story of the Sustainable Development Goals. Routledge; 2018.

[12] Grubler A, Wilson C, Bento N, *et al.* A Low Energy Demand scenario for meeting the 1.5 °C target and Sustainable Development Goals without negative emission technologies. Nature Energy. 2018;3(6):515–27.

[13] McCollum DL, Zhou W, Bertram C, *et al.* Energy investment needs for fulfilling the Paris Agreement and achieving the Sustainable Development Goals. Nature Energy. 2018;3(7):589–599.

[14] Wilson C, Pettifor H, Cassar E, Kerr L, and Wilson M. The potential contribution of disruptive low-carbon innovations to 1.5 °C climate mitigation. Energy Efficiency. 2019;12(2):423–40.

[15] Davis SJ, Lewis NS, Shaner M, *et al.* Net-zero emissions energy systems. Science. 2018;360(6396):*eaas9793*.

[16] Geels FW, Sovacool BK, Schwanen T, and Sorrell S. Sociotechnical transitions for deep decarbonization. Science. 2017;357(6357):1242–4.

[17] Alstone P, Gershenson D, and Kammen DM. Decentralized energy systems for clean electricity access. Nature Climate Change. 2015 5(4):305–14.

[18] Barido DP de L, Avila N, and Kammen DM. Exploring the enabling environments, inherent characteristics and intrinsic motivations fostering global electricity decarbonization. Energy Research & Social Science. 2020;61: 101343.

[19] Sutherland BR. Supporting low-carbon transitions. Joule. 2019;3(12):2894–6.

[20] IEA. World Energy Outlook 2019 [Internet]. Paris; 2019. Report No.: IEA (2019). Available from: https://www.iea.org/reports/world-energy-outlook-2019

[21] Perspectives for the Energy Transition: Investment Needs for a Low-carbon Energy System [Internet]. [cited 2020 Mar 15]. Available from: https://www.irena.org/publications/2017/Mar/Perspectives-for-the-energy-transition-Investment-needs-for-a-low-carbon-energy-system

[22] Yoshino N, Taghizadeh–Hesary F, and Nakahigashi M. Modelling the social funding and spill-over tax for addressing the green energy financing gap. Economic Modelling. 2019;77:34–41.

[23] Gaddy BE, Sivaram V, Jones TB, and Wayman L. Venture capital and cleantech: The wrong model for energy innovation. Energy Policy. 2017;102:385–395.

[24] Mazzucato M and Semieniuk G. Financing renewable energy: Who is financing what and why it matters. Technological Forecasting and Social Change. 2018;127:8–22.

[25] Brown J and Granoff I. Deep decarbonization by 2050: Rethinking the role of climate finance. San Francisco: Climate Policy Initiative. 2018.

[26] Ameli N and Kammen DM. Clean energy deployment: Addressing financing cost. Environmental Research Letters 2012;7(3):034008.

[27] Managan K and Klimovich K. Setting the PACE: Financing Commercial Retrofits. 28.

[28] Accelerating Green Energy Adoption in CT | CT Green Bank [Internet]. [cited 2020 Mar 15]. Available from: https://ctgreenbank.com/

[29] Connecticut Green Bank—Green Bank Network [Internet]. [cited 2020 Mar 15]. Available from: https://greenbanknetwork.org/connecticut-green-bank/

[30] Connecticut Green Bank Fact Sheet [Internet]. Coalition for Green Capital. 2017 [cited 2020 Mar 15]. Available from: http://coalitionforgreencapital.com/2017/04/05/connecticut-green-bank-fact-sheet/

[31] World Greenhouse Gas Emissions: 2016 [Internet]. World Resources Institute. 2020 [cited 2020 Mar 15]. Available from: https://www.wri.org/resources/data-visualizations/world-greenhouse-gas-emissions-2016

[32] Lovins AB, Ürge-Vorsatz D, Mundaca L, Kammen DM, and Glassman JW. Recalibrating climate prospects. Environmental Research Letters. 2019;14(12):120201.

[33] Jacobson MZ, Delucchi MA, Cameron MA, and Frew BA. Low-cost solution to the grid reliability problem with 100% penetration of intermittent wind, water, and solar for all purposes. Proceedings of the National Academy of Sciences. 2015;112(49):15060–5.

[34] Frew BA, Becker S, Dvorak MJ, Andresen GB, and Jacobson MZ. Flexibility mechanisms and pathways to a highly renewable US electricity future. Energy. 2016;101:65–78.

[35] Jacobson MZ, Delucchi MA, Bazouin G, *et al.* 100% clean and renewable wind, water, and sunlight (WWS) all-sector energy roadmaps for the 50 United States. Energy & Environmental Science. 2015;8(7):2093–2117.

[36] Jenkins JD and Thernstrom S. Deep decarbonization of the electric power sector: Insights from recent literature. Energy Innovation Reform Project. 2017.

[37] Jenkins JD, Luke M, and Thernstrom S. Getting to zero carbon emissions in the electric power sector. Joule. 2018;2(12):2498–510.

[38] Lovering JR, Yip A, and Nordhaus T. Historical construction costs of global nuclear power reactors. Energy Policy. 2016;91:371–82.

[39] Lang PA. Nuclear power learning and deployment rates; disruption and global benefits forgone. Energies. 2017;10(12):2169.

[40] Koomey J, Hultman NE, and Grubler A. A reply to "Historical construction costs of global nuclear power reactors." Energy Policy. 2017;102:640–3.

[41] Portugal-Pereira J, Ferreira P, Cunha J, Szklo A, Schaeffer R, and Araújo M. Better late than never, but never late is better: Risk assessment of nuclear power construction projects. Energy Policy. 2018;120:158–66.

[42] Morgan MG, Abdulla A, Ford MJ, and Rath M. US nuclear power: The vanishing low-carbon wedge. Proceedings of the National Academy of Sciences. 2018;115(28):7184–89.

[43] Ford MJ and Schrag DP. The case for a tortoise approach to US nuclear research and development. Energy Policy. 2019;135:111013.

[44] Jarvis S, Deschenes O, and Jha A. The Private and external costs of Germany's nuclear phase-out. National Bureau of Economic Research; 2019.

[45] Kharecha PA and Sato M. Implications of energy and CO_2 emission changes in Japan and Germany after the Fukushima accident. Energy Policy. 2019;132:647–53.

[46] Prăvălie R and Bandoc G. Nuclear energy: Between global electricity demand, worldwide decarbonisation imperativeness, and planetary environmental implications. Journal of Environmental Management. 2018;209:81–92.

[47] Macfarlane A and Ewing R. Déjà vu for U.S. nuclear waste. Science. 2017;356(6345):1313.

[48] IEA (2019). World Energy Model [Internet]. Paris; Available from: https://www.iea.org/reports/world-energy-model

[49] Wong SL, Wan KKW, Li DHW, and Lam JC. Impact of climate change on residential building envelope cooling loads in subtropical climates. Energy and Buildings. 2010;42(11):2098–103.

[50] Invidiata A and Ghisi E. Impact of climate change on heating and cooling energy demand in houses in Brazil. Energy and Buildings. 2016;130: 20–32.

[51] Andrić I, Gomes N, Pina A, *et al.* Modeling the long-term effect of climate change on building heat demand: Case study on a district level. Energy and Buildings. 2016;126:77–93.

[52] Waite M and Modi V. Electricity load implications of space heating decarbonization pathways. Joule. 2020;4(2):376–94.

[53] Förster H, Schumacher K, De Cian E, *et al.* European energy efficiency and decarbonization strategies beyond 2030—A sectoral multi-model decomposition. Climate Change Economics. 2013;4(supp01):1340004.

[54] Electric vehicle charging guide for cities | International Council on Clean Transportation [Internet]. [cited 2020 Mar 15]. Available from: https://theicct.org/publications/city-EV-charging-guide

[55] Analyzing policies to grow the electric vehicle market in European cities | International Council on Clean Transportation [Internet]. [cited 2020 Mar 15]. Available from: https://theicct.org/publications/electric-vehicle-policies-eu-cities

[56] Hall D, Moultak M, and Lutsey N. Electric Vehicle Capitals of the World. Washington, DC: International Council on Clean Transportation; 2017.

[57] Allwood JM, Dunant CF, Lupton RC, *et al.* Absolute zero: Delivering the UK's climate change commitment with incremental changes to today's technologies. University of Cambridge; 2019. Report No.: DOI: 10.17863/CAM.46075.

[58] Åhman M, Nikoleris A, and Wyns T. Decarbonizing industry: Emerging roadmaps point to major need for financing radical innovation. Carbon Management. 2013;4(1):5–7.

[59] Wesseling JH, Lechtenböhmer S, Åhman M, Nilsson LJ, Worrell E, and Coenen L. The transition of energy intensive processing industries towards deep decarbonization: Characteristics and implications for future research. Renewable and Sustainable Energy Reviews. 2017;79:1303–13.

[60] Hasanbeigi A and Springer C. Deep Decarbonization Roadmap for the Cement and Concrete Industries in California. San Francisco, CA: Global Efficiency Intelligence; 2020. Available from: https://www.globalefficiencyintel.com/decarbonization-roadmap-california-cement-concrete

[61] Napp TA, Gambhir A, Hills TP, Florin N, and Fennell PS. A review of the technologies, economics and policy instruments for decarbonising energy-intensive manufacturing industries. Renewable and Sustainable Energy Reviews. 2014;30:616–40.

[62] Bui M and Dowell NM. Chapter 1: Introduction—Carbon Capture and Storage. 2019; pp. 1–7.

[63] Rogelj J, Shindell D, Jiang K, *et al.* Chapter 2: Mitigation pathways compatible with 1.5 °C in the context of sustainable development. In: Global Warming of 15 °C an IPCC special report on the impacts of global warming of 15 °C above pre-industrial levels and related global greenhouse gas emission pathways, in the context of strengthening the global response to the threat of climate change [Internet]. Intergovernmental Panel on Climate Change; 2018 [cited 2020 Mar 15]. Available from: https://www.ipcc.ch/report/sr15/

[64] de Coninck H, Revi A, Babiker M, *et al.* Strengthening and implementing the global response. In: *Global Warming of 1.5 °C an IPCC special report on the impacts of global warming of 1.5 °C above pre-industrial levels and related global greenhouse gas emission pathways, in the context of strengthening the global response to the threat of climate change.* Intergovernmental Panel on Climate Change; 2018.

[65] Ozawa M, Chaplin J, Pollitt M, Reiner D, and Warde P. In Search of Good Energy Policy. Cambridge University Press; 2019. 389 pp.

[66] Von Hirschhausen C, Herold J, and Oei P-Y. How a "low carbon" innovation can fail—Tales from a "lost decade" for carbon capture, transport, and sequestration (CCTS). Economics of Energy & Environmental Policy. 2012;1(2):115–24.

[67] Mileva A, Johnston J, Nelson JH, and Kammen DM. Power system balancing for deep decarbonization of the electricity sector. Applied Energy. 2016;162:1001–9.

[68] Heuberger CF and Mac Dowell N. Chapter 12. CCS in electricity systems. In: Bui M and Mac Dowell N, editors. Energy and Environment Series [Internet]. Cambridge: Royal Society of Chemistry; 2019 [cited 2020 Mar 15]. p. 392–425. Available from: http://ebook.rsc.org/?DOI=10.1039/9781788012744-00392

[69] Gielen D, Boshell F, Saygin D, Bazilian MD, Wagner N, and Gorini R. The role of renewable energy in the global energy transformation. Energy Strategy Reviews. 2019;24:38–50.

[70] Curtis PG, Slay CM, Harris NL, Tyukavina A, and Hansen MC. Classifying drivers of global forest loss. Science. 2018;361(6407):1108–1111.

[71] Agriculture Drove Recent Record-Breaking Tree Cover Loss [Internet]. World Resources Institute. 2020 [cited 2020 Mar 15]. Available from: https://www.wri.org/blog/2020/02/agriculture-drove-recent-record-breaking-tree-cover-loss

[72] Home | Energy | Food and Agriculture Organization of the United Nations [Internet]. [cited 2020 Mar 15]. Available from: http://www.fao.org/energy/home/en/

[73] Innovations at the Nexus of Food, Energy and Water Systems | NSF - National Science Foundation [Internet]. [cited 2020 Mar 15]. Available from: https://www.nsf.gov/funding/pgm_summ.jsp?pims_id=505241

[74] Andoni M, Robu V, Flynn D, *et al.* Blockchain technology in the energy sector: A systematic review of challenges and opportunities. Renewable and Sustainable Energy Reviews. 2019;100:143–74.

[75] Liu C, Chai KK, Zhang X, Lau ET, and Chen Y. Adaptive blockchain-based electric vehicle participation scheme in smart grid platform. IEEE Access. 2018;6:25657–65.
[76] Su Z, Wang Y, Xu Q, Fei M, Tian Y-C, and Zhang N. A secure charging scheme for electric vehicles with smart communities in energy blockchain. IEEE Internet of Things Journal. 2018;6(3):4601–13.
[77] Mengelkamp E, Notheisen B, Beer C, Dauer D, and Weinhardt C. A blockchain-based smart grid: Towards sustainable local energy markets. Computer Science-Research and Development. 2018;33(1–2):207–14.
[78] Pilania G, Wang C, Jiang X, Rajasekaran S, and Ramprasad R. Accelerating materials property predictions using machine learning. Scientific reports. 2013;3(1):1–6.
[79] Butler KT, Davies DW, Cartwright H, Isayev O, and Walsh A. Machine learning for molecular and materials science. Nature. 2018;559(7715):547–55.
[80] Demir I, Koperski K, Lindenbaum D, *et al.* Deepglobe 2018: A challenge to parse the earth through satellite images. In: 2018 IEEE/CVF Conference on Computer Vision and Pattern Recognition Workshops (CVPRW). IEEE; 2018. pp. 172–17209.
[81] Rolnick D, Donti PL, Kaack LH, *et al.* Tackling climate change with machine learning. arXiv:190605433 [cs, stat] [Internet]. 2019 Nov 5 [cited 2020 Mar 15]; Available from: http://arxiv.org/abs/1906.05433
[82] Masanet E, Shehabi A, Lei N, Smith S, and Koomey J. Recalibrating global data center energy-use estimates. Science. 2020;367(6481):984–6.
[83] Patterson JJ, Thaler T, Hoffmann M, *et al.* Political feasibility of 1.5 °C societal transformations: the role of social justice. Current Opinion in Environmental Sustainability. 2018;31:1–9.
[84] Bill Text - SB-115 Environmental justice. [Internet]. [cited 2020 Mar 15]. Available from: https://leginfo.legislature.ca.gov/faces/billNavClient.xhtml?bill_id=199920000SB115
[85] Bullard RD. The legacy of American apartheid and environmental racism. John's J Legal Comment. 1993;9:445.
[86] Yandle T and Burton D. Reexamining environmental justice: A statistical analysis of historical hazardous waste landfill siting patterns in metropolitan Texas. Social Science Quarterly. 1996;77(3):477–92.
[87] Su JG, Apte JS, Lipsitt J, *et al.* Populations potentially exposed to traffic-related air pollution in seven world cities. Environment International. 2015;78:82–9.
[88] Jerrett M, Burnett RT, Kanaroglou P, *et al.* A GIS—Environmental justice analysis of particulate air pollution in Hamilton, Canada. Environment and Planning A. 2001;33(6):955–73.
[89] Casey JA, Morello-Frosch R, Mennitt DJ, Fristrup K, Ogburn EL, and James P. Race/ethnicity, socioeconomic status, residential segregation, and spatial variation in noise exposure in the contiguous United States. Environmental Health Perspectives. 125(7):077017.
[90] Seltenrich Nate. Inequality of noise exposures: A portrait of the United States. Environmental Health Perspectives. 125(9):094003.

[91] Sunter DA, Castellanos S, and Kammen DM. Disparities in rooftop photovoltaics deployment in the United States by race and ethnicity. Nature Sustainability. 2019;2(1):71–6.

[92] Sovacool BK, Martiskainen M, Hook A, and Baker L. Decarbonization and its discontents: A critical energy justice perspective on four low-carbon transitions. Climatic Change. 2019;155(4):581–619.

[93] FridaysForFuture [Internet]. [cited 2020 Mar 15]. Available from: https://www.fridaysforfuture.org/

[94] Thunberg G. Our house is on fire. The Guardian. 2019;20.

[95] MIT sit-in for climate action & fossil fuel divestment [Internet]. MIT Sit In. [cited 2020 Mar 15]. Available from: http://www.mitsitin.org/

[96] Irfan U. Why climate activists disrupted the Harvard-Yale football game [Internet]. Vox. 2019 [cited 2020 Mar 15]. Available from: https://www.vox.com/2019/5/13/18282438/harvard-yale-football-game-climate-change-fossil-fuel-divestment-finance

[97] Fossil Fuels Divestment Continues Georgetown's Commitment to Sustainability [Internet]. Georgetown University. 2020 [cited 2020 Mar 15]. Available from: https://www.georgetown.edu/news/fossil-fuels-divestment-continues-georgetown-commitment-to-sustainability/

[98] "We live on a finite planet": Gonzaga students hold sit-in to push university to divest from fossil fuel industry | The Spokesman-Review [Internet]. [cited 2020 Mar 15]. Available from: https://www.spokesman.com/stories/2020/feb/13/we-live-on-a-finite-planet-gonzaga-students-hold-s/

[99] Moss J. Live blog: Fossil free Pitt coalition's sit-in in the Cathedral of Learning [Internet]. The Pitt News. 2020 [cited 2020 Mar 15]. Available from: https://pittnews.com/article/155561/news/live-blog-fossil-free-pitt-coalitions-sit-in-in-the-cathedral-of-learning/

[100] University Announces Fossil Fuel Divestment Plans after Years of Student Advocacy [Internet]. 2020 [cited 2020 Mar 15]. Available from: https://thehoya.com/university-announces-fossil-fuel-divestment-plans-after-years-of-student-advocacy/

[101] Employee Activism Works—Even When It Doesn't. Bloomberg.com [Internet]. 2020 Feb 26 [cited 2020 Mar 15]; Available from: https://www.bloomberg.com/news/articles/2020-02-26/employee-activism-works-even-when-it-doesn-t

[102] Davies R. Norway's $1tn wealth fund to divest from oil and gas exploration. The Guardian [Internet]. 2019 Mar 8 [cited 2020 Mar 15]; Available from: https://www.theguardian.com/world/2019/mar/08/norways-1tn-wealth-fund-to-divest-from-oil-and-gas-exploration

[103] BlackRock Client Letter | Sustainability [Internet]. BlackRock. [cited 2020 Mar 15]. Available from: https://www.blackrock.com/corporate/investor-relations/blackrock-client-letter

[104] Cordero EC, Centeno D, and Todd AM. The role of climate change education on individual lifetime carbon emissions. PLOS One. 2020;15(2):e0206266.

Index

www.ingramcontent.com/pod-product-compliance
Lightning Source LLC
LaVergne TN
LVHW011829200926
848484LV00002B/341
9781839530210